CW00819552

Conserving Africa's Mega-Diversity in the Anthropocene

The Hluhluwe-iMfolozi Park Story

Centring on South Africa's Hluhluwe-iMfolozi Park, this book synthesizes a century of insights from the ecology and conservation management of one of Africa's oldest protected wildlife areas. The park provides important lessons for conservation management as it has maintained conservation values rivalling those of much larger parks, sometimes through and sometimes despite strong management interventions, including the rescue of the white rhino from extinction. In addition, the book highlights the ecological science produced in the park, much of which has become widely influential, including the megaherbivore concept, new functional approaches to understanding biomes, and new understandings about the role of consumers in shaping ecosystems. The volume is ideal for researchers and policy makers interested in the conservation of relatively small, isolated, protected areas.

JORIS P. G. M. CROMSIGT is an Associate Professor in wildlife ecology at the Swedish University of Agricultural Sciences. His research spans the broad field of the ecology of large mammals and their role in functioning of ecosystems. He has over 16 years of experience in working in South African savanna systems, much of this based in Hluhluwe-iMfolozi Park.

SALLY ARCHIBALD works on understanding the dynamics of savanna ecosystems in the context of global change. Her work integrates field ecological data, remote sensing, modelling, and biogeochemistry. She is involved in collaborative research projects on fire–grazer interactions, inter-continental savanna comparisons, the importance of land–atmosphere feedbacks, and pursuing a global theory of fire. Prof. Archibald was a finalist in the NSTF emerging researcher awards in 2016 and is on the steering committee of several scientific programmes including iLEAPS, the Miombo Network, and SASSCAL. She has authored and co-authored more than 30 publications achieving an H-index of 18.

NORMAN OWEN-SMITH received his PhD degree from the University of Wisconsin for his study on the behavioural ecology of the white rhinoceros. His research has covered the ecology of large mammalian herbivores and their interactions with vegetation. His awards include Gold Medals from the Zoological Society of South Africa and the Southern African Association for the Advancement of Science, life membership in the Ecological Society of America, the Bill Venter/Altron Literary Award and the Harry Oppenheimer Fellowship, and he is a Fellow of the Royal Society of South Africa. He has written or edited five books.

ECOLOGY, BIODIVERSITY AND CONSERVATION

The world's biological diversity faces unprecedented threats. The urgent challenge facing the concerned biologist is to understand ecological processes well enough to maintain their functioning in the face of the pressures resulting from human population growth. Those concerned with the conservation of biodiversity and with restoration also need to be acquainted with the political, social, historical, economic and legal frameworks within which ecological and conservation practice must be developed. The new Ecology, Biodiversity, and Conservation series will present balanced, comprehensive, up-to-date, and critical reviews of selected topics within the sciences of ecology and conservation biology, both botanical and zoological, and both 'pure' and 'applied'. It is aimed at advanced final-year undergraduates, graduate students, researchers, and university teachers, as well as ecologists and conservationists in industry, government, and the voluntary sectors. The series encompasses a wide range of approaches and scales (spatial, temporal, and taxonomic), including quantitative, theoretical, population, community, ecosystem, landscape, historical, experimental, behavioural, and evolutionary studies. The emphasis is on science related to the real world of plants and animals rather than on purely theoretical abstractions and mathematical models. Books in this series will, wherever possible, consider issues from a broad perspective. Some books will challenge existing paradigms and present new ecological concepts, empirical or theoretical models, and testable hypotheses. Other books will explore new approaches and present syntheses on topics of ecological importance.

Ecology and Control of Introduced Plants
Judith H. Myers and Dawn Bazely

Invertebrate Conservation and Agricultural Ecosystems
T. R. New

Risks and Decisions for Conservation and Environmental Management
Mark Burgman

Ecology of Populations
Esa Ranta, Per Lundberg, and Veijo Kaitala

Nonequilibrium Ecology
Klaus Rohde

The Ecology of Phytoplankton
C. S. Reynolds

Systematic Conservation Planning
Chris Margules and Sahotra Sarkar

Large-Scale Landscape Experiments: Lessons from Tumut
David B. Lindenmayer

Assessing the Conservation Value of Freshwaters: An international perspective
Philip J. Boon and Catherine M. Pringle

Insect Species Conservation
T. R. New

Bird Conservation and Agriculture
Jeremy D. Wilson, Andrew D. Evans, and Philip V. Grice

Cave Biology: Life in darkness
Aldemaro Romero

Biodiversity in Environmental Assessment: Enhancing ecosystem services for human well-being
Roel Slootweg, Asha Rajvanshi, Vinod B. Mathur, and Arend Kolhoff

Mapping Species Distributions: Spatial inference and prediction
Janet Franklin

Decline and Recovery of the Island Fox: A case study for population recovery
Timothy J. Coonan, Catherin A. Schwemm, and David K. Garcelon

Ecosystem Functioning
Kurt Jax

Spatio-Temporal Heterogeneity: Concepts and analyses
Pierre R. L. Dutilleul

Parasites in Ecological Communities: From interactions to ecosystems
Melanie J. Hatcher and Alison M. Dunn

Zoo Conservation Biology
John E. Fa, Stephan M. Funk, and Donnamarie O'Connell

Marine Protected Areas: A multidisciplinary approach
Joachim Claudet

Biodiversity in Dead Wood
Jogeir N. Stokland, Juha Siitonen, and Bengt Gunnar Jonsson

Landslide Ecology
Lawrence R. Walker and Aaron B. Shiels

Nature's Wealth: The economics of ecosystem services and poverty
Pieter J. H. van Beukering, Elissaios Papyrakis, Jetske Bouma, and Roy Brouwer

Birds and Climate Change: Impacts and conservation responses
James W. Pearce-Higgins and Rhys E. Green

Marine Ecosystems: Human Impacts on Biodiversity, Functioning and Services
Tasman P. Crowe and Christopher L. J. Frid

Wood Ant Ecology and Conservation
Jenni A. Stockan and Elva J. H. Robinson

Detecting and Responding to Alien Plant Incursions
John R. Wilson, F. Dane Panetta and Cory Lindgren

Conserving Africa's Mega-Diversity in the Anthropocene

The Hluhluwe-iMfolozi Park Story

Edited by

JORIS P. G. M. CROMSIGT

Department of Wildlife, Fish, and Environmental Studies, Swedish University of Agricultural Sciences, Umeå, Sweden;
Centre for African Conservation Ecology, Department of Zoology, Nelson Mandela Metropolitan University, Port Elizabeth, South Africa

SALLY ARCHIBALD

School of Animal, Plant and Environmental Sciences, University of the Witwatersrand, Johannesburg, South Africa

NORMAN OWEN-SMITH

School of Animal, Plant and Environmental Sciences, University of the Witwatersrand, Johannesburg, South Africa

CAMBRIDGE
UNIVERSITY PRESS

University Printing House, Cambridge CB2 8BS, United Kingdom

One Liberty Plaza, 20th Floor, New York, NY 10006, USA

477 Williamstown Road, Port Melbourne, VIC 3207, Australia

4843/24, 2nd Floor, Ansari Road, Daryaganj, Delhi - 110002, India

79 Anson Road, #06-04/06, Singapore 079906

Cambridge University Press is part of the University of Cambridge.

It furthers the University's mission by disseminating knowledge in the pursuit of education, learning and research at the highest international levels of excellence.

www.cambridge.org
Information on this title: www.cambridge.org/9781107031760

First published 2017

Printed in the United Kingdom by TJ International Ltd, Padstow, Cornwall

A catalogue record for this publication is available from the British Library

ISBN 978-1-107-03176-0 Hardback
ISBN 978-1-107-62799-4 Paperback

Contents

Contributors

SALLY ARCHIBALD
Centre for African Ecology, School of Animal, Plant and Environmental Sciences, University of the Witwatersrand, South Africa.

RANDAL ARSENAULT
Department of Biological Sciences, University of Alberta, Edmonton, Canada.

DAVE A. BALFOUR
Independent, South Africa.

PENNY A. BECKER
Washington Department of Fish and Wildlife, Olympia, WA, USA.

HEATH BECKETT
Department of Biological Sciences, University of Cape Town, Rondebosch, South Africa.

MARISKA TE BEEST
Department of Ecology and Environmental Science, Umeå University, Umeå, Sweden.

WILLIAM J. BOND
South African Environmental Observation Network, Cape Town, South Africa.

MARCUS J. BYRNE
Centre for African Ecology, School of Animal, Plant and Environmental Sciences, University of the Witwatersrand, South Africa.

TRISTAN CHARLES-DOMINIQUE
Department of Biological Sciences, University of Cape Town, Rondebosch, South Africa.

GEOFF CLINNING
Ezemvelo KZN Wildlife, Hluhluwe-iMfolozi Park, Hluhluwe, South Africa.

CORLI COETSEE
Scientific Services, Kruger National Park, Skukuza, South Africa.

DAVID COOPER
Ezemvelo KwaZulu-Natal Wildlife, St Lucia, South Africa.

MICHAEL D. CRAMER
Department of Biological Sciences, University of Cape Town, Rondebosch, South Africa.

JORIS P.G.M. CROMSIGT
Department of Wildlife, Fish, and Environmental Studies, Swedish University of Agricultural Sciences, Umeå, Sweden.

L. ALEXANDER DEW
Department of Ecology and Environmental Science, Umeå University, Umeå, Sweden.

DAVE J. DRUCE
Ezemvelo KZN Wildlife, Hluhluwe-iMfolozi Park, Hluhluwe, South Africa.

HELEEN DRUCE
School of Life Sciences, University of KwaZulu-Natal, Westville, Durban, South Africa.

LIHLE DUMALISILE
Gauteng Nature Conservation, Johannesburg, South Africa.

JIM M. FEELY
Centre for African Conservation Ecology, Department of Zoology, Nelson Mandela Metropolitan University, Port Elizabeth, South Africa.

GABRIELLA FLACKE
University of Western Australia, School of Animal Biology, Crawley, WA, Australia.

CLAIRE GEOGHEGAN
Mammal Research Institute, Department of Zoology and Entomology, University of Pretoria, Pretoria, South Africa.

CLEO M. GOSLING
Conservation Ecology, Groningen Institute for Evolutionary Life Sciences, University of Groningen, Groningen, The Netherlands.

JAN A. GRAF
Association for Water and Rural Development, Hoedspruit, South Africa.

MICAELA SZYKMAN GUNTHER
Department of Wildlife, Humboldt State University, Arcata, California, USA.

NICOLE HAGENAH
South African Environmental Observation Network, Grasslands–Forests–Wetlands Node, Pietermaritzburg, South Africa.

MATT W. HAYWARD
Schools of Biological Sciences; and School of Environment, Natural Resources and Geography, College of Natural Sciences, Bangor University, Bangor, UK.

OWEN HOWISON
Conservation Ecology, Groningen Institute for Evolutionary Life Sciences, University of Groningen, Groningen, The Netherlands.

RUTH A. HOWISON
Conservation Ecology, Groningen Institute for Evolutionary Life Sciences, University of Groningen, Groningen, The Netherlands.

ANNA E. JOLLES
College of Veterinary Medicine and Department of Zoology, Oregon State University, Corvallis, OR, USA.

LAURENCE M. KRUGER
Organization for Tropical Studies, Skukuza, South Africa.

WAYNE L. LINKLATER
Centre for Biodiversity and Restoration Ecology, School of Biological Sciences, Victoria University, Wellington, New Zealand.

DAVID G. MARNEWECK
Centre for Wildlife Management, University of Pretoria, Pretoria, South Africa.

MANDISA MGOBOZI POSWA
Faculty of Science and Agriculture, University of Zululand, Richardsbay, South Africa.

JEREMY J. MIDGLEY
Department of Biological Sciences, University of Cape Town, Rondebosch, South Africa.

ABEDNIG MKHWANAZI
Ezemvelo KZN Wildlife, Hluhluwe-iMfolozi Park, Hluhluwe, South Africa.

MARCOS MOLEÓN
Department of Conservation Biology, Doñana Biological Station (CSIC), Seville, Spain.

HAN OLFF
Conservation Ecology, Groningen Institute for Evolutionary Life Sciences, University of Groningen, Groningen, The Netherlands.

NORMAN OWEN-SMITH
Centre for African Ecology, School of Animal, Plant and Environmental Sciences, University of the Witwatersrand, Johannesburg, South Africa.

CATHERINE L. PARR
School of Environmental Sciences, University of Liverpool, Liverpool, UK.

ROGER PORTER
Ex Natal Parks Board and Ezemvelo KZN Wildlife, Pietermaritzburg, South Africa.

SUSAN JANSE VAN RENSBURG
South African Environmental Observation Network, Grasslands–Forests–Wetlands Node, Pietermaritzburg, South Africa.

NICKI LE ROEX
Division of Molecular Biology and Human Genetics, Faculty of Health Sciences, Stellenbosch University, Cape Town, South Africa.

ELIZABETH LE ROUX
Centre for African Conservation Ecology, Department of Zoology, Nelson Mandela Metropolitan University, Port Elizabeth, South Africa.

ADRIAN M. SHRADER
School of Life Sciences, University of KwaZulu-Natal, Scottsville, South Africa.

MICHAEL J. SOMERS
Centre for Wildlife Management, Centre for Invasion Biology, University of Pretoria, Pretoria, South Africa.

A. CARLA STAVER
Department of Ecology and Evolutionary Biology, Yale University, New Haven, USA.

WILLIAM D. STOCK
Centre for Ecosystem Management, School of Natural Sciences, Edith Cowan University, Joondalup, WA, Australia.

COLETTE TERBLANCHE
Independent, South Africa.

MARTINA TRINKEL
School of Life Sciences, University of KwaZulu-Natal, Westville, Durban, South Africa.

MICHIEL P. VELDHUIS
Conservation Ecology, Groningen Institute for Evolutionary Life Sciences, University of Groningen, Groningen, The Netherlands.

JULIA L. WAKELING
Silverstreet Capital, Cape Town, South Africa.

Foreword

Ezemvelo KZN Wildlife is a biodiversity conservation organization with the challenging but rewarding responsibility for nature conservation and development and promotion of ecotourism activities within the province of KwaZulu-Natal (KZN), South Africa. Its core aims are biodiversity conservation, wise and sustainable use of natural resources, the creation and management of partnerships with stakeholders and communities, and the provision of affordable ecotourism destinations within KZN. Hluhluwe-iMfolozi Park (HiP) is a figurehead for our entire organization – being the genetic home to the southern white rhino and a world-famous ecotourism venue. We are very proud of the conservation story of HiP, and of the generations of park managers and scientists who have worked to ensure that this park conserves our natural resources, provides educational opportunities, and creates wealth for the people living in the region. This book is the culmination of many years of work and is fully endorsed by our organization. No biodiversity agency can operate without scientific input, and Ezemvelo is no exception. We are proud of the scientific advances that have been enabled by the work in HiP, and we are especially pleased to see evidence in this book of how many of these advances have led to tangible improvements in management operations.

We believe that our experiences in HiP have much to offer the world – especially other small protected areas in Africa. For this reason we hope that this book will be widely read.

Dr M. D. Mabunda, CEO of Ezemvelo KZN Wildlife

Preface

The area that was to become the Hluhluwe-iMfolozi Park (HiP) was among the first in Africa to be formally protected: the Hluhluwe and Umfolozi Game Reserves were proclaimed in 1895, a few years ahead of the game reserve that became the Kruger National Park. They were separated by a large stretch of land that functioned as a corridor for animals, which has since been amalgamated to form a 950-km^2 conservation area spanning a diversity of land forms, climates, and vegetation types. The game reserves were established because of concerns about the disappearance of wildlife as a result of hunting in the region then known as Zululand, especially triggered by how few white rhinos remained. HiP is considerably smaller than Africa's flagship national parks, the Kruger National Park (19,500 km^2) and Serengeti National Park (14,763 km^2), and unlike most protected areas in eastern Africa, its boundaries are completely fenced. However, despite its small size, the park hosts a diversity of vegetation types and animal species that can rival much larger protected areas. Its steep rainfall gradient (550–950 mm) means the park's vegetation ranges from semi-arid to mesic savanna, and the park supports a full suite of the megaherbivores (animals weighing more than 1000 kg when adult) and large mammalian carnivores typical of African savanna ecosystems. HiP's rolling mix of grassland and forest in the north-east and more gently undulating thorn savanna in the south conserves key habitats, including several threatened and endemic plant species.

HiP shares with Kruger and Serengeti national parks a long history of ecological monitoring and scientific research spanning close to a century. This experience has been well documented for both Kruger (du Toit *et al*., 2003) and Serengeti (Sinclair and Norton-Griffiths, 1979 and subsequent volumes) and we now contribute a similar synthesis for HiP. Because of the small size and turbulent history of the two game reserves following their proclamation, a laissez-faire management policy has never been adopted. Nevertheless, the park managers attempted to retain or restore the ecological processes that had formerly operated on a much larger scale. Much has been learnt from the success and failures of conservation management and by the pioneering ecological research that has been undertaken to gain better understanding of the intrinsic dynamics of this microcosm of Africa. These lessons are particularly relevant for attempts to conserve and restore savanna systems elsewhere in protected areas that represent small relicts of vaster ecosystems.

We have divided this book into three parts. *Part I* sets the scene by covering historical and prehistorical human influences (Chapter 1), the heterogeneous biophysical template (Chapter 2), and documentation of long-term vegetation (Chapter 3) and large herbivore dynamics (Chapter 4). *Part II* records how research conducted within HiP has contributed to advancing ecological understanding. Much of this research has had significant impacts on our understanding of the structure and function of savannas globally, as well as their response to anthropogenic and other drivers of change. A study on white rhinos led to the concept of megaherbivores, their substantial impacts on the vegetation, and consequences for coexistence of other herbivore species (Chapter 5). Moreover, HiP has been the testing bed for seminal research on the roles of climate and consumers – both fire and herbivory – in impacting savanna vegetation dynamics (Chapters 6, 7, and 10). This has led to new perspectives on functional trait syndromes of woody plants and alternative biome states (Chapter 8). Functional contributions by smaller organisms, particularly termites, dung beetles, and rodents, are covered in Chapter 9. Many of these scientific advances have contributed towards the management of the reserve. *Part III* shifts attention to these management interventions more broadly, highlighting several examples of effective collaboration between science and management. Contrasting management strategies for black and white rhinoceros are described in Chapter 11. Problems encountered in the restoration of the large carnivore community and their resolutions are covered in Chapter 12, while interventions to contain the impacts of both indigenous and alien wildlife diseases are presented in Chapter 13. Chapter 14 describes the re-introduction of elephants in HiP and potential responses to their burgeoning population, while Chapter 15 addresses measures used to control alien invasive plants. Finally, Chapter 16 synthesizes findings from these various studies and management actions, evaluates the 'success story' of the HiP, and looks ahead to future challenges in coping with the pervasive human influences typifying the 'Anthropocene' epoch.

A magical transformation is experienced once you cross through the gate into HiP and encounter elephants, rhinos, buffalos, and large predators in place of the domestic livestock and human settlements pervasive outside. This small African park therefore captures the mind and soul of all who visit it, and has driven generations of managers, scientists, and their students to devote their time and energy to understanding and protecting it. It is our hope that some of their passion will reach you, the reader of this book, and that this compendium of science and conservation management will contribute towards ensuring that the next generation will still have this experience both within HiP and elsewhere in Africa.

Explanation of Some of the Names Used in this Book

Some explanation of the names and naming conventions adopted in this book is needed. The area that the park encloses is rich in local Zulu names, indicating the long history of human presence in the landscape (see Appendix). At the time of the first proclamation of the two reserves, the names derived from the local Zulu language were rendered as 'Hluhluwe' and 'Umfolozi'. However, it became recognized that the latter spelling was incorrect according to Zulu orthography, because of a distinction between the prefix and the word that follows. Different classes of nouns are associated with distinct prefixes, and in this case the correct prefix should be 'i', not 'u'. This means that the name of the game reserve should be rendered as iMfolozi, with the second letter capitalized. Hence the acronym 'HiP' became adopted for the combined Hluhluwe-iMfolozi Park. This did not fully resolve the naming issue. If one wants to be consistent, Hluhluwe should be rendered with its prefix as 'umHluhluwe'. To add to the confusion, the official proclamation of the combined park in 2012 spelt the name as Hluhluwe-Imfolozi Park, incorrectly capitalizing the prefix. In this book, we follow the widely adopted convention of referring to the combined protected area as the Hluhluwe-iMfolozi Park (acronym HiP). In contexts prior to the consolidation of the park, we use the original names applied to the Hluhluwe and Umfolozi game reserves. Furthermore, we apply the spelling 'Mfolozi' (omitting the prefix) to the White and the Black Mfolozi rivers as well as for the region of the park south of the Black Mfolozi river. The 'Corridor' refers to the region between the two original game reserves (see Chapter 1).

Another important naming issue that we had to deal with in the book is the still controversial splitting of the genus *Acacia* into multiple genera. African species have been assigned to new genera *Vachellia* and *Senegalia* (Kyalangalilwa *et al.*, 2013). We will continue to use *Acacia* as a genus name throughout the book to avoid confusion with the preceding ecological literature concerning this group of species.

References

du Toit, J. T., Rogers, K. H., & Biggs, H. C. (2003) *The Kruger experience: ecology and management of savanna heterogeneity*. Island Press, Washington, DC.

Kyalangalilwa, B., Boatwright, J. S., Daru, B. H., Maurin, O., & Bank, M. (2013) Phylogenetic position and revised classification of *Acacia* sl (Fabaceae: Mimosoideae) in Africa, including new combinations in *Vachellia* and *Senegalia*. *Botanical Journal of the Linnean Society* **172**: 500–523.

Sinclair, A. R. E., & Norton-Griffiths, M. (1979) *Serengeti, dynamics of an ecosystem.* The University of Chicago Press, Chicago.

Further Details on Zulu Place Names in the Hluhluwe-iMfolozi Park

JIM M. FEELY

Zulu Place Names in Hluhluwe-iMfolozi Park: Some Recurring Features

This appendix presents a list of Zulu place names and their location in the Hluhluwe-iMfolozi Park that probably originated during the pre-Colonial era and thus are relics of the Late Iron Age (Tables 0.1–0.3; see also a map with these Zulu names in the online Supplementary Material). It excludes names from the twentieth century. There are recurring features in the Zulu place names in the Park which refer to topographic features or to wild animals (see below). Among the old names are those of the Park itself: Hluhluwe probably referring to the climbing plant *Dalbergia armata* (thorny rope), and Mfolozi to the zigzag pattern on baskets and pottery, among other meanings. The game reserves were named for the main rivers traversing the park, the Hluhluwe and Mfolozi rivers. Both river names are probably very old, so that now there can only be speculation concerning any connection between them and their meaning. The climbing plant, for example, is distributed along many rivers in KwaZulu-Natal, so why is this one so-named?

Topographic Features

(S) from Black Mfolozi river southward, (C) between Black Mfolozi river and Hlabisa-Mtubatuba road (R618), (N) from Hlabisa road northward.

Cairn (stones) *isivivane*, accumulated by travellers along a footpath, usually over a hill, to avoid bad luck on a journey: (S, C, N) *eSivivaneni* (3).

Table 0.1 *Places southward of Black Mfolozi river*[a] *(Magqubu Ntombela)*

Place name	Feature	Place name	Feature
uBhocozi	stream	**iMbulunga**	hill
uBizo	thicket/stream	**uMduba**	hill
eCekaneni	ridge	**uMeva**	area
iChibi elibomvu	pan	**iMfolozi eMhlophe**	river
iChibilembube	pan	**iMfolozi eMnyama**	river
iChibilentungunono	pan	**uMfulamkhulu**	stream
iChibilenyathi (2)	pans	**eMgqizweni**	area/pan
iChibilethangwe	pan	**uMhlanganobhedu**	stream/area
iChibilokumbiwa	pan	**eMhlanganweni**	confluence
uCiyane	hill	**uMhlolokazana**	hill
uDadethu	stream/pan/area	**uMhlolwana**	hill
oDakaneni	stream	**uMhluzi**	stream
uDengezi	hill	**eMndindini**	area
iDlaba	stream	**uMomfu**	hill/stream/cliff
eDuduseni	area/ridge	**uMpekwa**	area
eFuyeni	stream	**uMphafa**	stream
uGidiyoni	stream	**uMphanjana**	hill/area
uGome (2)	streams	**iMpila**	ridge
eGqolweni (= eMapulankweni)	ridge	**iMpila encane**	hill
oGqoyini	stream	**iMpila enkulu**	hill
oGunqweni	area	**eMsasaneni**	hills
kwaHlathikhulu	bush	**uMthombokandleke**	spring
iKhandaledube	stream	**eMthonjenikakhaya**	stream
oKhetheni	stream	**iMunywane**	stream
uKhukho	hill	**eMzaneni**	area
uLubisana	hill/stream	**uNcoki**	hill
uLuthelezi	hill	**uNdleke**	hill
eMachwetshaneni	hill	**iNdlovuma**	stream
eMachitshaneni	area	**uNdlovusiyashikana**	stream/area
uMadlozi	stream	**uNdomba**	stream
eMadwaleni	ridge	**eNgonyamaneni**	hill
eMahobosheni	ridge/area	**iNgwenyama**	stream
uMagunda	area/bush	**iNgwenyemnqini**	pan
kwaMakhamisa	place (R. H. T. P. Harris' camp)	**uNkawu**	stream
uMakhamisa (= uBulunga)	stream (Harris worked with donkeys)	**uNkobenkulu**	area/stream/ thicket
uMakhokhelweni	ridge/area	**uNobiya**	stream
iMantiyana (2)	hills	**uNoma**	hill
uManya	hill/stream	**uNozibunjana**	bush
uMasango	stream	**eNqabaneni**	hill
aMatshemhlophe	hill	**uNqokotshane**	stream
aMatshemnyama	hill	**uNqolothi**	hill
eMawuzi	area	**eNqutshini**	area
iMbhuzane	hill/stream	**eNselweni**	ridge/area

Table 0.1 *(cont.)*

Place name	Feature	Place name	Feature
eNsikaneni	stream	**iSiwasempila**	cliff
iNtabayamanina	hill	**uSokhwezele**	hill
iNtabayamaphiva	hill	**uSoncunda**	hill
uNtoyiyana	hill	**uSontuli**	hill/area
uNtshiyana	stream	**uTeke**	stream
iNyamakayithengwa	stream	**uThobothi**	stream
uNyonikazana	stream	**iThumbu**	stream
uQaqalwempisi (3)	hills/ridges	**iTshele likaFosingi**	hill
iSabokwe	hill	**eTshenilentombi**	area
iSalathiyela	stream	**iTshenteka**	cliff
uShoshangesisila	hill	**uTshokolwana**	hill
uSilevana	hill	**eZigubeni**	area
eSivivaneni	hill/stream	**eZikhayenizenkosi**	hill
iSiwasamagunda	cliff	**eZimbokodweni**	area
iSiwasamsasaneni	cliff	**eZimenyaneni**	hills
iSiwasamanqe	cliff	**eZintunzini**	range of hills
iSiwasamhlosheni	cliff	**eZintuthwaneni**	area
iSiwasemfene	cliff		

[a] eNgilandi was on this list and the map in error, and has been removed. This area is in the Hluhluwe sector. No member of NPB staff from the 1950s to the 1970s, including me, knew of such an area in Mfolozi GR (J. Anderson, J. Forrest, P. Hitchins, R. Porter, J. Vincent, A. Whately, *in litt.*, 2015).

Table 0.2 *Places northward of Black Mfolozi river to main road (R618) (Magqubu Ntombela)*

Place name	Feature	Place name	Feature
eBhavulomu	area	**eMpindisweni**	ridge
oBhembedwini	stream	**uMsinyane**	stream
uBhokosa	stream	**eMsokosokweni**	stream
uCaya	hill	**uMtshongweni**	stream
uDlogodlo	ridge	**uNcengeninhliziyo**	hill
uDomu	stream	**eNdlovaneni**	stream
uDonsagolo	hill	**iNdondwane**	stream/hill
eDuduseni	ridge	**iNgceba**	ridge
esiFusamvini	ridge	**oNgeni**	hill
eGobhe	stream	**eNhlonhleniyamathonga**	area
iGoqo	ridge	**uNkonyane**	hill
eGwalagwaleni	stream	**uNondubela**	ridge
uHlathikhulu	ridge	**uNonqishi**	area
uHlaza	hill	**eNqunyeni**	stream

(cont.)

Table 0.2 *(cont.)*

Place name	Feature	Place name	Feature
uHlebomunye (= Mshukulo)	area	**iNtabakamayanda**	hill
uHlekuzulu (= eNtuzuma)	ridge	**iNtabakamthwazi**	hill
iHlengwa	stream	**iNtabayamaphiva**	hill
iKhandalomuntu	ridge	**iNtabayentombi**	hill
eKushesheni	ridge	**uNxabo**	stream
Kwesemvivi	ridge	**iNyalazi**	stream
Kwesogada	ridge	**uNyongwana**	stream
eLabelweni	stream	**uPhondo**	stream
uLubisana	stream	**uQikiyana**	area/stream
eMadotsheni	stream	**eSangcobeni**	ridge/stream
uMagqayiza	area	**uSangobo**	stream
uMagula	area	**uShiyane**	ridge
eMashashangeni	ridge	**eSigoqweni**	hill
uMajojoyi	stream	**uSikhovana**	hill/stream
eMakhandeni ezindlovu	stream/ ridge	**eSivivaneni**	col on hill
eMalalaneni	stream	**eSiyembeni**	hill
aManzimhlophe	stream	**uSokosoko**	stream
uMasi	ridge	**uThekwane**	stream
uMasimba[a]	hill/stream	**iTshelamabhunu**	ridge
uMasimba omncane	hill	**eTsheni**	ridge
eMasundweni	hill	**eTshenteka**	ridge
uMatelembana	stream	**iTshevu**	stream
aMatshemnyama	ridge	**iZalani**	ridge
eMazondweni	hill	**oZengwaneni**	ridge/stream
uMchachazo	stream	**eZibozini**	stream
uMcibilindi (2)	streams	**eZihlabeni**	ridge/stream
uMcobosi	ridge	**eZiklebheni**	area
uMcumane	stream	**eZimambeni**	stream
uMfulawembuzi	stream	**eZinhlonhlwaneni**	stream
eMguthwaneni	stream	**eZinqunyeni**	stream
iMona	stream	**eZinqwambeni**	ridge
eMondini	stream	**eZinsisheni**	stream
iMpelenyane emhlophe	stream	**eZishamashameni**	ridge
iMpelenyane emnyama	stream		

[a] Ntombela suggested *Masinda* for the visitor facilities, as an inoffensive alternative to *Masimba* (dung heap) nearby. This was not traditional, as Ntombela acknowledged (I. C. Player, *in litt.*, 2014).

Table 0.3 *Places from Hluhluwe Sector southward to main road (R618) (Thembeni Mthethwa)*

Place name	Feature	Place name	Feature
uBelebane	stream	**iMpongo**	forest
eBomvini	stream	**uMthole**	hill
uCakula	stream	**uMuntulu**	area
iCalalendlu	area	**uMunywana**	stream
eCekaneni	area	**eMunywaneni**	area
iChibilamanqe	pan	**uMzini**	stream
iChibilezangoma	pan	**oNcobeni**	stream
oDakaneni	stream	**iNdabakazipeli**	ridge
eDubeni	hill	**uNdantsha**	stream
uFuzula	stream	**uNdimbili**	stream
uGontshi	hill	**iNdlunkulu**	stream
eGunjaneni	stream	**iNdodanye**	stream
uHidli	hill	**uNgalonde**	ridge
uHlathikhulu	thicket	**eNgilandi**[a]	area
uHlaza	hill/stream	**iNgqungqulu**	stream/ridge
uHlokohloko	hill/stream	**iNgwenyaneni**	stream
iHluhluwe	river	**iNhlabashana**	stream
uKubi	stream	**uNhlayinde**	hill
uMabombothelana	stream	**uNkonono**	hill
uMacabuzele	stream	**uNkwakwa**	hill
eMagangeni	ridge	**uNomageje**	stream
uMagwanxa	hill/stream	**uNombali**	ridge
eMahlabathini	area	**uNqodi**	hill
aMahlungulu	hill	**eNqoklweni**	area
uMahwanqana	ridge	**iNqumela (2)**	streams
uMakhokhoba	hill	**iNsizwa**	hill
uMalikayiko	stream	**uNtabamhlophe**	hill
aMansiya	stream	**iNzimane**	river
aManzamnyama	stream	**uQholwane**	hill
aManzibomvu	stream	**uQololenja**	hill
aMaphumulo	ridge	**eSaheni**	area
uMaqanda	stream/thicket	**uSankoya**	ridge
uMashiya	hill	**uSeme**	hill
uMatikalala	ridge	**eSikelemeni**	stream
aMatshemhlophe	hill	**iSikhalasomoya**	hill
aMatshovozo	stream	**uSiqwashu**	stream
aMawane	forest	**uSisuze**	area
aMawuzi	forest	**iSitezi**	hill

(cont.)

Table 0.3 *(cont.)*

Place name	Feature	Place name	Feature
uMbango	area	**uSithole**	hill
uMbhombe	forest	**iSivivaneni**	hill
eMcibilindini (2)	streams	**iSiwasamakhosikazi**	cliff
uMcincinya	ridge	**uSomaxekwane**	ridge
eMfukuzweni	stream	**oThiyeni**	bush
uMgovuzo	stream	**iTsheliyamfoma**	stream
uMjantshi	hill	**iTshempofu**	hill
uMkhombe[b]	hill	**iVivi**	hill/forest
uMlebezi omkhulu	stream	**uVumbe**	stream
uMlebezi omncane	stream	**uZangomfe**	hill
uMnqabatheki	ridge	**eZidonini**	area
iMpanzakazi	hill	**eZimbokodweni**	stream
iMpisaneni	stream	**eZincakeni**	hill/dam
iMpolomba	stream	**eZiqhumeni**	ridge
		eZisengeni	ridge

[a] Eastward of the confluence of the Hluhluwe and Manzibomvu rivers; omitted from the original list. It refers to where Captain H. B. Potter attempted to introduce English fallow deer and pheasant in the 1940s. They did not survive for long.

[b] Westward of the confluence of the Hluhluwe and Manzibomvu rivers. Vaughan-Kirby (1920) records that Zulu guards in Hluhluwe used this name for large male black rhino. They had never seen a white rhino, because they did not occur north of the Black Mfolozi river early in the twentieth century. However, the name could derive from an even earlier time when white rhino probably occurred there. They are there now.

Cliff *isiwa*, usually on the outside of the bend along a major perennial stream: (S) *iSiwasamagunda*, *iSiwasamsasaneni*, *iSiwasamanqe*, *iSiwasamhlosheni*, *iSiwasemfene*, *iSiwasempila*, (N) *iSiwasamakhosikazi*.

Hill (mountain) *intaba:* (S) *iNtabayamanina*, *iNtabayamaphiva* (2), (C) *iNtabakamayanda*, *iNtabayentombi*, (N) *iNtabamhlophe*.

Pan (pond) *ichibi*, with temporary water after rain, used as a wallow by elephant, rhinos, buffalo and common warthog, thus enlarging with time: (S) *iChibi elibomvu*, *iChibilembube*, *iChibilentungunono*, *iChibilenyathi* (2), *iChibilokumbiwa*, (N) *iChibilamanqe*, *iChibilezangoma*.

Stream amanzi (water), minor perennial: (C) *aManzimhlophe*, (N) *aManzibomvu*, *aManzamnyama*.

Total 27 places.

Mammals, Birds, and Reptiles Occurring in Place Names

imbube, lion; *indlovu*, elephant; *idube*, zebra; *imfene*, baboon; *uhobosha*, puff-adder; *igwalagwala*, purple-crested turaco; *ingonyama*, lion; *ingqungqulu*, bateleur eagle; *ingwe*, leopard; *ingwenya*, Nile crocodile; *inkawu*, vervet monkey; *umkhombe*, white rhino; *i(ama)nqe*, vulture(s); *inyathi*, buffalo; *iphiva*, waterbuck; *impisi*, spotted hyena; *impofu*, eland; *iseme*, Denham's bustard; *uthekwane*, hamerkop; *intungunono*, secretary bird. Total 19 taxa.

List of Place Names (June 1968)

The Zulu spelling of the place names in Tables 0.1–0.3 accords with the list produced by Magqubu Ntombela, Thembeni Mthethwa, and Reg Mayne at Hluhluwe Hilltop Camp in June 1968 (338 places, provided by John Vincent *in litt.*, 2014). This list gives an average of around one place name per 2.7 km^2 in the Park, as shown on the 1979 map (see online Supplementary Material). It was duplicated for official use by the Natal Parks Board, although not published before. These names omit the initial lower-case vowel and have the initial consonant capitalized.

Magqubu Ntombela[1] and Thembeni Mthethwa[2] provided an oral rendering, in each other's hearing, of the Zulu place names in the southern and northern sectors, respectively, of the Hluhluwe-Mfolozi Park (P. M. Hitchins *in litt.*, 2014; J. Vincent *in litt.*, 2014). This was done at a two-day meeting arranged by the Natal Parks Board (NPB) at Hluhluwe Hilltop Camp in June 1968. They were illiterate men in their 60s who spent their working lives as game rangers in the Park. The former was stationed in the Umfolozi Game Reserve and the latter in the Hluhluwe Game Reserve, as the southern and northern sectors of the Park were then known. For management purposes these sectors included the intervening unreserved State land known as 'the Corridor', with each sector extending to the Mtubatuba-Hlabisa main road (R618).

Indeed, Ntombela was born and grew to manhood in the southern Corridor, at his father's homestead on a hill (*oNgeni*) overlooking the Black Mfolozi river, at the turn of the twentieth century. His father, a member of the iNgobamakosi regiment who fought at Isandlwana in 1879 (I. C. Player, pers. comm., 1979), would have been born around 1853

[1] Hugh Dent, a fluent and literate Zulu linguist who knew him well, corrected the spelling from 'Maqubu Nthombela' that was used in earlier documents (H. R. Dent, pers. comm., 1973).

[2] Mtethwa or Mtetwa in earlier documents.

(Faye, 1923). He and Mthethwa acquired their knowledge in the traditional way by remembering precisely: (1) the teaching of their parents and other elders, and (2) the information provided by their contemporaries as well as their own observations while walking over the ground for many years. Their memories were prodigious and reliable.

The place names were transliterated by Reg Mayne, a retired high court interpreter of Zulu–English who was fluent and literate in each. He listened carefully to the spoken names and their discussion of them, in order to spell them correctly and learn their meaning where known. These he dictated to John Vincent and Peter Hitchins of the NPB scientific staff, emphasizing the importance of distinguishing the prefix from the stem of a noun. They compiled his spellings in an alphabetical list based on the first consonant, and prepared a map with these place names (P. M. Hitchins *in litt.*, 2014; J. Vincent *in litt.*, 2014). The list and map were duplicated for the use of NPB staff. This use must have been discontinued because at a meeting in 2008, staff members of Ezemvelo KZN Wildlife seemed to be unaware of either (N. Turner *in litt.*, 2014).

The list in Tables 0.1–0.3 is resurrected from Peter Hitchins' notes and a copy held by John Vincent. As an archive of indigenous knowledge obtained up to a century and more ago (mid-nineteenth century), it cannot be replicated. As such, this list can be regarded as a more reliable record of tradition than any obtained in the present century. The names in the list below are given on a map available through the online Supplementary Material of this book (made in 1979 by Hitchins and Vincent).

There is a notable difference between the orthographic convention used in this list and in the standard dictionary (Doke and Vilakazi, 1953). The latter gives *im-Folozi* as the name of the second river for which the Park is named. However, Ntombela gives *iMfolozi* in Mayne's transcription (below). Chief Mangosuthu Buthelezi also used *iMfolozi*, rather than *imFolozi*, in having the name of the game reserve corrected from Umfolozi (I. C. Player, personal communication, 2014). Both probably follow the accepted spelling convention of the time, capitalizing the initial consonant. However, the Park's legal name is Hluhluwe-Imfolozi (KZN Provincial Gazette Extraordinary, Vol. 6 No. 799, Provincial Notice No. 83, 30 August 2012). The dictionary has no entry for the Hluhluwe river. It has *um-Hluhluwe* for: (1) the thorny rope, a climbing plant (*Dalbergia armata*), and (2) the spur on a cock's leg (that the plant's thorns resemble).

Official policy in KwaZulu-Natal (KZN) is to include the whole prefix in the writing of isiZulu place names. This is not so in the Eastern Cape Province. There, official policy continues to omit the lower-case initial vowel from the written prefix in an isiXhosa place name, e.g. *Mthatha, Mzimvubu, Dutywa.*

I thank Joris Cromsigt, Hugh Dent, Peter Hitchins, Ian Player, Noleen Turner, John Vincent, C. J. (Roddy) Ward, John Ward, and John Wright for documents, information, and comment.

References

Doke, C. M. & Vilakazi, B. W. (eds) (1953) *Zulu–English dictionary*, 2nd edn. Witwatersrand University Press, Johannesburg.

Faye, C. (1923) *Zulu references*. City Press, Pietermaritzburg.

Vaughan-Kirby, F. (1920) The white rhinoceros, with special reference to its habits in Zululand. *Annals of the Durban Museum* **2**: 223–242 (footnote 6, p. 4).

Acknowledgements

One of the unique features of HiP is its well-equipped research station where external researchers can rent a room and use common facilities and interact with the researchers employed by the park. The camp has become affectionately known as 'Dungbeetle' because some of its initial infrastructures were funded by an Australian dung beetle research programme. This excellent research facility has ensured that South African and international universities have been able to run several large research programmes within the park. For many years, up to 20–30 external researchers (from BSc student to Professor) spent many months together at the station. This meant that the Dungbeetle kitchen was often filled with lively discussions of research projects, the ecology and management of the park, and other earthly matters. Often, this initiated new ideas and collaboration among projects. This open, enlightened, atmosphere at Dungbeetle has strongly contributed towards the nature of this book, indicated, for example, by co-authors of many different institutions sharing chapters. The first ideas for this book also originated, in the late 1990s, from dinner table discussions among researchers at Dungbeetle. As Park Ecologist at that time, Dave Balfour was important in these initial discussions. Much later, some of these ideas were formalized during two workshops in 2007, one in the park and one at the Society for Conservation Biology's conference in Port Elizabeth. Sue van Rensburg and Han Olff were important in driving these workshops. During the more recent years, Dave Druce facilitated the book process on behalf of the park's management authority Ezemvelo KZN Wildlife. We are grateful for the Dungbeetle spirit and the many people that have helped creating and maintaining it.

As reviewers for each chapter, we sought internationally recognized experts within the field matching the chapter. The book has benefited hugely from their critical assessment of chapters. They include: Keryn Adcock, Alan Andersen, Michael Anderson, Jane Carruthers, Johan du Toit, Richard Emslie, Jim Feely, Sam Ferreira, Hervé Fritz, Navashni Govender, Danny Govender, Niall Hanan, Gareth Hempson, Ricardo Holdo, Andrew Illius, Marietjie Landman, Caroline Lehman, Donal McCracken, Joseph Ogutu, Craig Packer, Owen Price, Rob Pringle, Dave

Richardson, Bob Scholes, Göran Spong, Michael Usher, Sue van Rensburg, Nikki Stevens, Kari Veblen, Freek Venter, Tony Whateley, and John Wright. We especially acknowledge Roger Porter, who kindly agreed to review several chapters. The contents of many of the chapters in this book are the result of close interactions between researchers and conservation management staff. Although too many to mention by name, the park's current and historical conservation managers and section rangers deserve a big thank you for their openness towards research and their active involvement in many of the research projects. We pay special tribute to three 'old-timers' who passed away during the preparation of this book for their foundational contributions to conservation in HiP: Ian Player, Jim Feely, and Roddy Ward. The book has also built upon much of the work of previous researchers and scientific staff. Finally, we thank the former Natal Parks, Game and Fish Preservation Board and Ezemvelo KZN Wildlife for having been so facilitative towards research and hope that this generous attitude will continue.

Preamble

This preamble is based on an interview of Laurence M. Kruger with Abednig Mkhwanazi in June 2015. Abednig was born in the area that is now the corridor section of the park and later became a much-valued research technician based at Hluhluwe Research centre until his retirement in February 2015. During his long career he contributed to most of the chapters in this book, either by contributing to data collection or by transferring his huge knowledge on field methods, plant and animal species or functioning of the ecosystems to other field staff or students from all levels (from BSc to Professor!). This extract of the interview has been edited by Laurence M. Kruger and Sally Archibald.

In the 40 years I have been working in the park I have seen many changes. Compared to the 70s, tree density in the park is greater. In '76 when I started working at Mbulunga hill, near Mpila, it was open grassland. But you look now towards Mpapha it is thick with woody plants. We also see a great deal of change in animal numbers, especially outside the park. As a boy of 15 years, animals were walking everywhere, and people were used to animals and there were rules from the chief that prevented the hunting of animals. People thought the park was for the animals and for them. Now that there is a fence between the people and the park people kill animals inside the park for meat, but they don't understand how to live with animals any more. If you take down the fence, then there will be chaos.

The relationship between research and management in HiP hasn't always been good, but it has improved with time. Management needs scientists to interact with them, providing input into management practices. When students come, a big challenge is that their training has been mostly theoretical, but we at scientific services have practical knowledge and experience, and so we are working together. For instance, when working with William Bond, who came to see how we identify the seedlings of *Acacia*, including *A. nilotica* and *A. robusta*, I taught him how to identify the seedlings by tasting them!

Regarding the future of the park, a big issue is the relationship with the people outside. People outside have goats and donkeys and the predators eat the animals. The park is still rich in water and grass, and we don't have access. Regarding managers, scientists, and local communities I will say to them: each must understand and respect the other. We still have a long way to go in resolving these challenges.

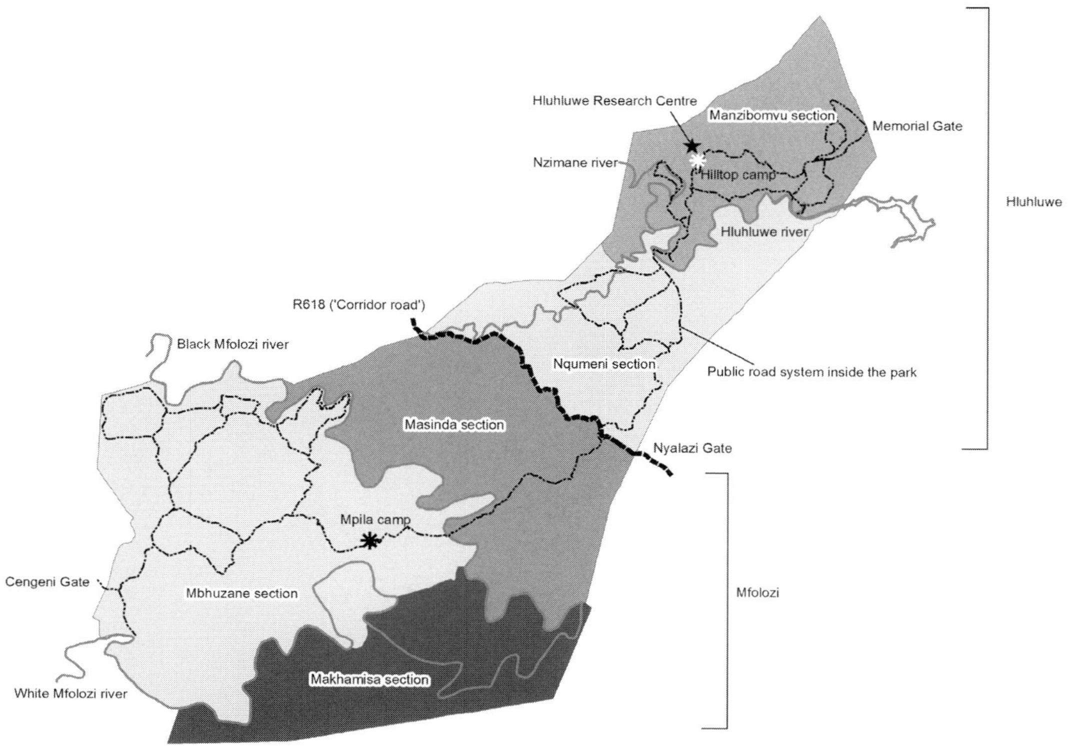

Basic map of current-day Hluhluwe-iMfolozi Park.

Part I
Setting the Scene

1 · *Anthropogenic Influences in Hluhluwe-iMfolozi Park: From Early Times to Recent Management*

MARISKA TE BEEST, NORMAN OWEN-SMITH, ROGER PORTER, AND JIM M. FEELY

1.1 Introduction

Early humans (*Homo* spp.) have been an integral component of African savannas since their origination around 2 million years ago. These early humans influenced their environment by harvesting edible plants, hunting large animals, and at some stage through igniting fires, during the prolonged period while their tools remained constructed of stone. People with implements and weapons made of iron immigrated into southern Africa from the north nearly 2000 years ago, absorbing some of the hunter-gatherers, and displacing wild ungulates from where they grazed their herds of domestic sheep and cattle. Over 500 years ago, people in ships travelling from Europe towards East Asia set foot on South African shores. They established temporary settlements that soon became permanent and spread to become initially Dutch and later British colonies. Firearms were brought and an expanding trade in ivory and other wildlife products developed. Eventually the disappearance of the wild animals prompted legislation to establish 'game reserves' where hunting would be prohibited. Two of these game reserves became consolidated to form the Hluhluwe-iMfolozi Park (HiP).

The initial history of these game reserves was turbulent, because local white farmers fought to have them deproclaimed to eliminate

Conserving Africa's Mega-Diversity in the Anthropocene, ed. Joris P. G. M. Cromsigt, Sally Archibald and Norman Owen-Smith. Published by Cambridge University Press.

the wildlife that formed a reservoir for cattle diseases. After a series of campaigns aimed at the eradication of tsetse flies (*Glossina* spp.), which transmitted blood parasites (*Trypanosoma* spp.) that infected cattle, a provincial authority was established to administer and eventually consolidate the game reserves. Management philosophies evolved from simply protecting the surviving wild animals to restoring the former fauna. This proved so successful that culling was introduced to alleviate perceived overgrazing. Eventually a more scientifically informed approach to management was adopted, aimed at fostering the ecological processes that had formerly operated on a vaster scale. HiP currently persists as a fenced island surrounded by increasingly dense human settlements.

It is the purpose of this chapter to describe these changing anthropogenic influences on the ecology of HiP as context for the chapters that follow. As will become evident, both the vegetation (Chapter 3) and wildlife (Chapter 4) have undergone continual flux. Expanding human populations and consequent land transformation in the surrounding region increasingly threaten conservation objectives within the protected area.

1.2 Archaeological History: Middle to Later Stone Age

Humans living as hunter-gatherers with tools made of stone were present in most parts of South Africa from far back in time, and had become anatomically modern around 120,000 years ago (Mitchell, 2002). Their presence in HiP is confirmed by stone implements or rock paintings recorded at more than 65 sites (Penner, 1970). It is uncertain until what time they inhabited the region that now includes HiP, but as recently as 1593 survivors of the wrecked Portuguese ship *Santo Alberto* met people armed with spears and arrows, who were not farmers, about 40 km northwest of HiP (Vernon, 2013). There are no historical reports of hunter-gatherers, i.e. people of Khoi-San ancestry, in this region of KwaZulu-Natal during the nineteenth century. Nevertheless, the click consonants that are typical of Khoi-San languages became incorporated into Zulu and related Nguni languages spoken by the people with Iron Age culture who displaced and absorbed these earlier inhabitants from around 500 AD.

It is unclear whether hunting by Stone Age people affected wildlife populations to any great extent. Humans were thinly scattered and their weapons were of short range with limited power. Probably more important ecologically would have been the practice of these people to use fire to attract ungulates for hunting (Deacon and Deacon, 1999). As a consequence, humans would probably have changed the fire regime throughout the summer rainfall region of southern Africa and elsewhere (see Chapter 10). More-or-less random ignition by lightning in spring or early summer probably gave way to regular veld burning in autumn or winter, once the field layer became dry enough to burn. The changed fire regime probably commenced well before the appearance of anatomically modern humans in South Africa over 120,000 years ago and was perpetuated from then into modern times (Deacon and Deacon, 1999; Kingdon, 2003).

Some indication of the predominant animal species that Stone Age hunters killed is provided by archaeological excavations conducted at Sibudu Cave, located alongside the Tongati river approximately 150 km south of HiP (Plug, 2004; Clark and Plug, 2008). Sedimentary layers there span the Middle Stone Age from 77,000 to 38,000 years ago as well as more recent layers from 900 to 1000 AD, and contain abundant bones of large mammals, particularly Burchell's zebra (*Equus burchelli*), hartebeest (*Alcelaphus* sp.), African buffalo (*Syncerus caffer*) and the extinct giant buffalo (*Pelerovis antiquus*) (see Chapter 4 for a detailed account). The people inhabiting this cave gathered plant material for bedding as well as for food, and thus had impacts on vegetation besides their use of fire (Wadley *et al.*, 2011).

1.3 Early to Late Iron Age

Early farmers with spears, axes, and hoes made of iron arrived in South Africa from the north nearly 2000 years ago (Hall, 1987; Huffman, 2007). Initially, their geographical distribution was confined to savanna and forest regions where fuel wood for their iron smelters was available (Feely, 2004). Two periods associated with metal smelting have been distinguished in South Africa as the Early Iron Age and the Late Iron Age. The Early Iron Age commenced around 300 AD and ended during the eleventh century. Settlements were clustered into villages, separated by half a kilometre or more, usually occupied for two or more generations, even for a century or longer (Maggs, 1984; Hall, 1987; Huffman, 2007). Indeed, the KwaGandaganda site near present Durban was continuously occupied for four centuries (Whitelaw, 1994). Besides constructing dwellings, clearing fields, and grazing livestock (initially sheep and soon after goats and cattle as well), these Iron Age people needed to gather huge amounts of wood for smelting iron ore (Feely, 1980).

Some of the earliest Iron Age sites, dated to the third–fifth centuries AD, occur along the coastal strip of KwaZulu-Natal, from the northern border with Mozambique southward to near the Mzimkhulu river mouth (Maggs, 1984). One example exists near Lake Mphangazi north of St Lucia estuary about 30 km east of HiP (Hall, 1981). Between the sixth and eighth centuries AD, farming settlements in KwaZulu-Natal expanded inland along the entrenched valleys of perennial streams (Maggs, 1984). In HiP, a seventh-century site from this period has been identified close to the Hluhluwe river (Hall, 1979a,b, 1981). Two other Early Iron Age sites have been found along the southern bank of the Black Mfolozi river (J. M. Feely, unpublished records, KwaZulu-Natal Museum). All three of these sites are located on fertile, valley bottom soils close to rivers, as is typical of this period elsewhere in south-eastern Africa (Hall, 1981; Maggs, 1984).

The Early Iron Age was succeeded by the Late Iron Age during the eleventh century AD (Maggs, 1984; Huffman, 2007), probably indicating the arrival of new settlers with a distinct culture. No sites between the tenth and fourteenth centuries have been firmly identified in KwaZulu-Natal, suggesting that farming people may have moved away from this region in response to unfavourable climatic conditions during this period (Prins, 1996). An archaeological survey conducted by Penner identified 134 sites representing the Late Iron Age within HiP, mapped by Hall (1979a,b; Figure 1.1). Artefacts from this period are relatively more obvious than those at sites from earlier in the Iron Age, which may have been buried by soil and revealed only after subsequent erosion or the activities of animals (J. M. Feely, personal observation, 1973–1982). Hence, mapped Late Iron Age sites are probably biased towards the youngest sites from the eighteenth and nineteenth centuries. The most important change occurring in KwaZulu-Natal during the Late Iron Age was in the size and location pattern of settlements. They were generally small and occupied by a single family for perhaps only one generation and were situated on the slopes and crests of hills and ridges (Maggs, 1984). The change in the location of settlements probably reflects Late Iron Age farmers wanting to avoid increased parasite infection risks for their cattle near rivers (Feely, 2004). Within HiP, Late Iron Age sites are commonly present near the 280-m contour, adjoining dolerite outcrops (Hall, 1981). The numerous Zulu place names for every hill, stream, or other place of note within HiP (see the Appendix in the Preface) attest to how thoroughly settled or traversed the region was by these Late Iron Age inhabitants.

The Iron Age peoples harvested wood as fuel for cooking, smelting, and working metals, mainly iron, and for the construction of dwellings,

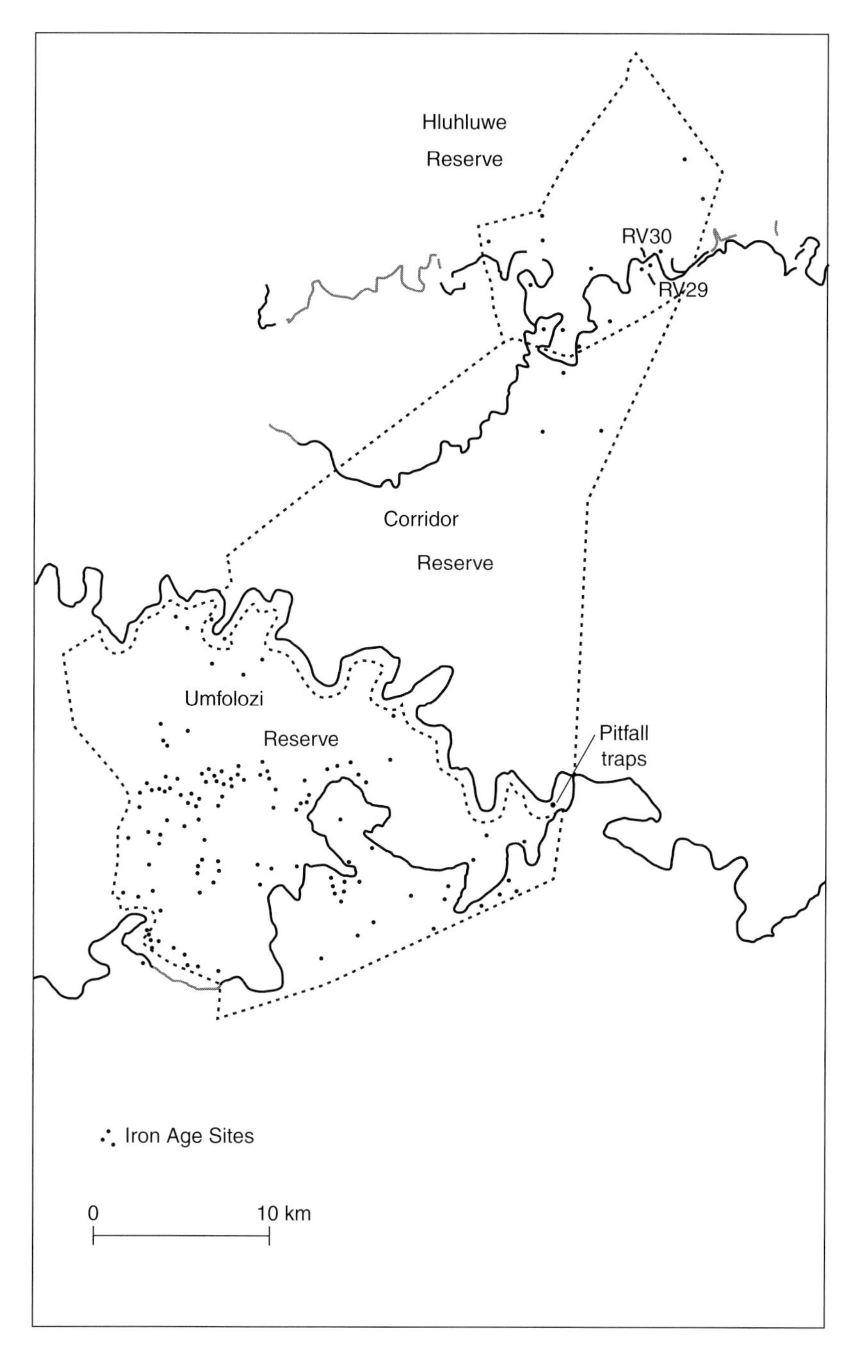

Figure 1.1 Positions of known Iron Age sites within HiP. RV29 and RV30 refer to two sites in the Hluhluwe river valley that were originally recorded by Penner (1970) and investigated and described in detail by Hall (1979b). The pitfall traps refer to a third site that has been investigated in detail by Hall (1977). It consists of a line of pitfall traps which lie near the confluence of the two Mfolozi rivers. (Reprinted with permission from Ezemvelo KZN Wildlife; Hall, 1979b.)

as well as grass for thatching (Maggs, 1984; Hall, 1987; Huffman, 2007). Staple crops grown were the grasses sorghum (*Sorghum bicolor*) and pearl millet (*Pennisetum glaucum*), and after the mid-seventeenth century also maize (*Zea mays*). Accumulations of cattle dung within the livestock enclosures and ash from cooking fires enriched mineral nutrients in soils around these settlements. Early African farmers would also have gathered plants for medicines, fruits, bark, and other natural resources, as rural people do today (van Wyk and Gericke, 2000). Fruits of marula trees (*Sclerocarya birrea*) were evidently esteemed as early as the first millennium AD in the nearby Thukela valley (Maggs, 1984). Pitfall traps for the hunting of large game remain evident near the confluence of the Mfolozi rivers within HiP (Figure 1.1). These probably date to the reign of King Shaka in the 1820s (Hall, 1977).

Among the Late Iron Age sites mapped by Penner (1970), 15% showed signs of being used either for smelting iron ore (Hall, 1980) or as forges for iron-working (Hall, 1979b, 1981; Huffman, 2007; J. M. Feely, unpublished records, KwaZulu-Natal Museum). Hardwoods such as tamboti (*Spirostachys africana*) and red bushwillow (*Combretum apiculatum*) would have been selectively felled to produce charcoal for smelting and forging. At a smelting site near the Hluhluwe river in HiP, clearance apparently initiated a succession over 200–300 years towards domination by magic guarri (*Euclea divinorum*) by the late 1970s (Hall, 1980, 1981, 1984).

Grazing and browsing by herds of domestic livestock kept by the Iron Age settlers would have had additional impacts on vegetation (Hall, 1987; Huffman, 2007). Besides selective grazing and browsing in the vicinity of settlements, a wider impact would have been the burning, in most years, of the grass layer (Hall, 1981). Burning would probably have been carried out in autumn or winter, once grasses were dry enough to burn. This would have reduced the incidence of lightning-caused fires during the following spring and early summer.

Natural vegetation would have been cleared for the cultivation of grain crops, although such fields are no longer evident (Hall, 1979b, 1981, 1984; Feely, 1980). Dwelling sites are indicated by the remains of pottery and grinding stones. Circular depressions in the ground, 1–2 m in diameter, indicate collapsed grain pits. These were dug beneath cattle kraals centrally placed in the dwelling area. Trees such as marula, jacket-plum (*Pappea capensis*), weeping boer-bean (*Schotia brachypetala*) and buffalo thorn (*Ziziphus mucronata*) left to provide fruit or shade, or for spiritual reasons, may still remain standing (Feely, 1980).

In Mfolozi, former Iron Age settlements show a distinct grass community characterized by bushveld signal grass (*Urochloa mosambicensis*) on the locally compacted soils (Hall, 1981, 1984; J. M. Feely, personal observation, 1973–1982). These sites are favoured by wild grazers including white rhinos (see also Chapter 6). Many of these sites could represent homesteads destroyed a century earlier, in 1883, during the southward invasion by Mandlakazi (Ndwandwe) people, under Zibhebhu kaMaphitha, from north of the Black Mfolozi river (Laband, 1995).

1.4 Early Historical Period 1790–1887

The historical record spans the last part of the Late Iron Age. The early history of KwaZulu-Natal has been recorded beginning sketchily from the sixteenth century in the form of reports by survivors of shipwrecks (Vernon, 2013), and expanded through the eighteenth to nineteenth centuries following the arrival of white explorers, hunters and later settlers. Towards the end of the eighteenth century, the region of KwaZulu-Natal north of the Thukela river, labelled Zululand, was partitioned among distinct chiefdoms who warred with one another over land and livestock. The territory north of the Black Mfolozi and combined Mfolozi rivers was the domain of the Ndwandwe, while the Mthethwa ruled to the south (Laband, 1995; Wright, 2008). The Zulu were a small group under the hegemony of the Mthethwa, occupying land to the west of current-day HiP in the basin of the White Mfolozi river. By the early nineteenth century, the Mthethwa had largely abandoned the low-lying land between the Mfolozi rivers, due either to the incidence of nagana (trypanosomiasis), transmitted by tsetse flies to their cattle (McCracken, 2008), or to tensions with the Ndwandwe (Wright, 2008). At that time the Mthethwa leader, Dingiswayo kaJobe, occupied a site south of the White Mfolozi river and east of uDadethu ('our sister') pan. He is said to have named this feature for his sister after she was killed by a crocodile while fetching water there (Magqubu Ntombela, personal communication to J. M. Feely in 1960). This would have happened before Dingiswayo's death in 1817 (Laband, 1995).

The Mthethwa and other groups became consolidated into the Zulu Kingdom after Shaka rose to power in 1819. Shaka's military forces drove out the remaining occupants between the Mfolozi rivers, leaving behind broken pottery and grinding stones. According to oral tradition, the hunting of wild animals thereafter became restricted seasonally. Species such as buffalo, greater kudu (*Tragelaphus strepsiceros*) and waterbuck (*Kobus ellipsiprymnus*) were largely protected, and white rhino (*Ceratotherum simum*) were not regarded as edible (McCracken, 2008). Following Shaka's conquest of the Mthethwa, a great game drive took place between the Mfolozi rivers, with animals funnelled into concealed pits dug just above the confluence (Hall, 1977). Elephant hunts directed from uNqolothi hill overlooking the White Mfolozi river have been described (McCracken, 2008), although Shaka's royal hunting ground is reputed to have been in the Mbhekamuzi Valley, to the west of the present-day park boundary (Vincent, 1970). The killing of wild animals was restricted mainly to intermittent ceremonial hunts, such as 'the washing of the spears' following the death of a chief. Possibly the last such hunt took place in the Somkhele area a little to the east of HiP's present Nyalazi Gate in 1955, after the death of Chief Matubatuba. Some 300 men armed with spears and accompanied by many dogs took part (J. M. Feely, personal observation, May 1955). Killing was accomplished by spear, snare or pit trap before guns became available (McCracken, 2008).

During Shaka's rule, the first white settlers arrived at Port Natal (later renamed Durban), and were granted permission by Shaka to stay. In 1824, Messrs Farewell and Fynn met with Shaka, requesting permission to trade with his people in ivory (Laband, 1995). Trading in wildlife products was already ongoing by that time, with ivory being exported through Delagoa Bay (now known as Maputo) in southern Mozambique. Fynn remarked on the abundance of game in the Zululand region coexisting alongside the people (Fynn, 1950). Elephant and buffalo were targeted especially by the white hunters and the African hunters whom they employed. By 1832, hunting had greatly reduced the populations of these and other big game around Port Natal, but wildlife still abounded to the north of the Thukela river (Herman and Kirby, 1970; McCracken, 2008). Delegorgue described seeing buffalo, greater kudu, eland (*Taurotragus oryx*) and zebra, along with white rhino, elephant (*Loxodonta africana*), wild dog (*Lycaon pictus*), and spotted hyena (*Crocuta crocuta*) between the White and the Black Mfolozi rivers while hunting there in 1840 (Delegorgue, 1847). This area was apparently free of settlements at that time. Hunting expanded from the 1830s onwards, but wild ungulates remained abundant in wooded lowlands where malaria and tsetse fly inhibited occupation by people and their livestock into the 1850s.

Dutch-speaking farmers who had trekked from the Cape to avoid British rule (Boer Voortrekkers) had established a republic called Natalia at Port Natal in 1839. They fought the British in the Battle of Congella in 1840. In 1843, the area to the south of the Thukela river as far as the Mzimkhulu river was formally annexed by Britain and became the colony of Natal in May 1844 (Laband, 1995). Until 1856 it was administered from the Cape Colony. The area north of the Thukela river remained the independent kingdom of Zululand under Shaka's successors, the kings Dingane, Mpande, and Cetshwayo.

During the 1870s, guns became more widely available and the abundance of wildlife declined due to the escalation in hunting (McCracken, 2008). The decimation of cattle by an outbreak of lung-sickness in 1874/5 (Laband, 1995) also forced people to rely more on wild ungulates for food. Ivory exports from Delagoa Bay and later Durban peaked in 1854 with an annual total of 85,000 kg and then declined during the 1880s. Hide exports from Durban peaked in 1872, representing 417,000 animals during that year (McCracken, 2008), but ceased in 1885. A substantial portion of these hides would have originated from inland parts of southern Africa rather than from Zululand (Boshoff and Kerley, 2013). Rhino horn exports peaked in 1884 at 679 horns, but declined after 1888 (McCracken, 2008).

1.5 Initiation of Game Protection: 1887–1897

The Anglo-Zulu war of 1879 ended with British victory and led eventually to the annexation of most of Zululand under British colonial rule in 1887 (Laband, 1995). The first game law came into effect in Natal in 1884 and was extended to Zululand in 1890 (Brooks, 2001). Wild animals were categorized into species designated as 'royal game', given maximum protection (e.g. black (*Diceros bicornis*) and white rhino, elephant, buffalo, kudu), 'closed season' (e.g. bushbuck (*Tragelaphus scriptus*), blue duiker (*Cephalophus monticola*), oribi (*Ourebia ourebi*), and steenbuck (*Raphicerus campestris*)), and 'not listed' (e.g. red duiker (*Cephalophus natalensis*) and mountain reedbuck (*Redunca fulvorufula*)) (McCracken, 2008). In subsequent proclamations (1893, 1897) many species were downgraded from 'royal game' to 'closed season', including black and white rhino, which reflected the growing concerns with the disease nagana killing large numbers of cattle (Brooks, 2001; McCracken, 2008).

The Zulu people recognized the association between infection of their cattle with nagana, or animal African trypanosomiasis, and contacts with wild ungulates. They attributed an increase in the incidence of nagana to an increase in large game brought about by the protection afforded by the game laws (Pringle, 1982). In 1894 Dr David Bruce, who was stationed in Ubombo in northern Zululand, experimentally confirmed the role of the tsetse fly in the transmission of the trypanosome parasites from infected wild ungulates to domestic livestock. With the arrival of Marshal Clarke as the new resident commissioner in Zululand in 1893, the concerns of the Zulu people regarding nagana were acknowledged. The new game law of 1893 allowed Zulu residents affected by nagana to reduce the numbers of game in their areas. To prevent the total destruction of game, Marshal Clarke wrote to the governor that 'in light of the relaxed game laws it might be a good idea to create game reserves in areas already infested with nagana' (Brooks, 2001).

By that time it was feared that white rhinos were on the brink of extinction. Following the shooting of six white rhinos near the confluence of the two Mfolozi rivers, in February 1895 Mr C. D. Guise wrote a letter to the Secretary for Zululand, for consideration by the Colonial Governor, requesting that 'a particular range of country in Zululand which embraces the habitat of the white rhinoceros should be beaconed off as a game reserve and no shooting or destruction of game be allowed therein'. He also requested that the white rhino be returned to the list of royal game and that no permits for their hunting be issued (Pringle, 1982; McCracken, 2008). Stirred by this letter, Commissioner Clarke submitted proposals for the establishment of game reserves within Zululand to the governor, who was sympathetic.

The Zululand Government Notice dated 30 April 1895 proclaimed five game reserves named as follows: Hluhluur Valley Reserve, Umfolosi Junction Reserve, St Lucia Reserve, Umdhletshe Reserve, and Reserve No. 5 between the Pongolo and Mkuze rivers to the east of the Lebombo mountains. Within them, the 'killing of game will be altogether prohibited' (Ellis, 1994). With the impending consolidation of the Natal and Zululand colonies, the status of four of these reserves (Umdhletshe, Hluhluwe, St Lucia, and Umfolosi) was reaffirmed by a proclamation dated 27 April 1897 (Pringle, 1982; McCracken, 2008). Maps showing the boundaries of the separate Hluhluwe and Umfolozi game reserves as originally proclaimed are reproduced in Figure 1.2 (Ellis, 1994). As noted in this figure, the boundaries of the original Hluhluwe and Umfolozi game reserves were defined somewhat vaguely using hills, homesteads, rivers, and footpaths. For instance, the Umfolozi GR was described as 'The country between the Black and White Umfolozi rivers from the junction of the rivers to the Mandhlagazi [*sic*] footpath' (Ellis, 1994; McCracken, 2008). The Mandlakazi footpath marked the route that was supposedly followed in August 1883 when the Mandlakazi people from north of the Black Mfolozi river invaded the people living to the south and destroyed their

A

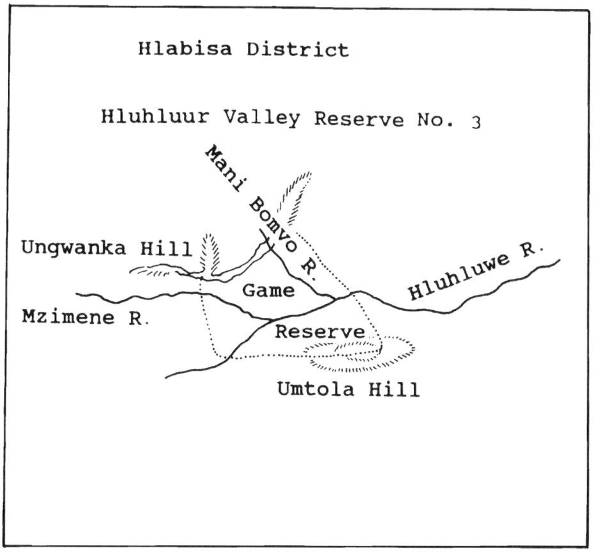

BOUNDARIES

A straight line from the highest point of the Zankomfe ridge to the Mpanzakazi hill; from thence to the present sites of the kraals of Umdimdwane, Mantunjana, Saziwayo, and Umswazi; from the latter kraal to the nearest point of the Mzinene stream; thence to the Mehlwana hill, south of the Hluhluwe river; thence to the Mtolo hill; from thence in a direct line with the same hill to the Hluhluwe river; and from there to the highest point of the Zankomfe hill.

B

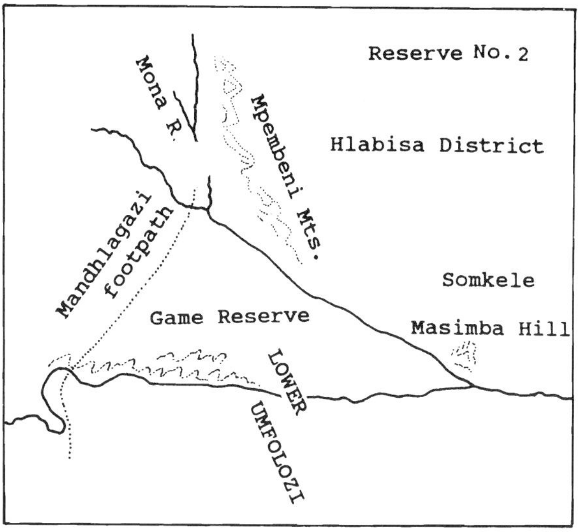

BOUNDARIES

The country between the Black and White Umfolozi rivers from their junction to the Mandhlagazi.

Figure 1.2 Original maps and description of the location of the game reserves at the time of proclamation in 1895. (A) Hluhluur Valley Reserve (Hluhluwe GR). (B) Umfolosi Junction Reserve (Umfolozi GR). (Reprinted with permission from the Natal Society Foundation (natalia.org.za); Ellis, 1994.)

homesteads. This path defining the original western boundary did not remain evident very long.

At the time of their proclamation the Hluhluwe and Umfolozi game reserves were largely uninhabited. However, it is doubtful whether the vicinity was ever completely uninhabited (McCracken, 2008). The Mthethwa people living to the south of the White Mfolozi river had vacated the region between the Mfolozi rivers in 1883 following raids by the Mandlakazi people, perhaps also prompted by a nagana outbreak among their cattle (Vincent, 1970, McCracken, 2008). Some people still occupied the high-lying area to the west of Umfolozi GR, which was incorporated into HiP during the 1960s. Human density in the region surrounding the two reserves was relatively low – about 2.5 people per km^2 in 1895, compared with almost 6 people per km^2 in Zululand as a whole (McCracken, 2008).

1.6 The Difficult Early Years: 1898–1952

Important contributions to the protection of game in these early years were made by local magistrates who did their best to enforce the game laws and ensure the protected status of the game reserves, in collaboration with local African chiefs, by restricting the issuing of hunting permits (McCracken, 2008). African people were generally supportive of the protection afforded to the wild ungulates, having lived alongside wild animals since early times. Moreover, during this time the game reserves were still remote and not readily accessible by white hunters. As late as 1916, only Hluhluwe GR could be reached by wheeled transport. Finally, the presence of tsetse flies probably kept livestock out of the reserves to a large extent.

Between 1895 and 1904, southern Africa was ravaged by the rinderpest epizootic (a viral disease transmitted from cattle to wild ungulates), which decimated buffalo, eland, kudu, and wildebeest (*Connochaetes taurinus*) populations as well as domestic cattle (McCracken, 2008). East Coast fever (theileriosis) followed among cattle, and people lost around 80% of their livestock. After this reduction in wild ungulates, the occurrence of nagana declined in Zululand, but resurged again following the recovery of wild ungulates. By 1907, the incidence of nagana had almost regained its former levels (Pringle, 1982; McCracken, 2008).

After the incorporation of Zululand into the colony of Natal in 1897, the report of a Delimitation Commission was eventually finalized in 1905, opening about a third of Zululand to white settlement (Ellis, 1994).

The remaining area, apart from the game reserves and 'Crown' land, was designated as tribal lands under communal tenure. The resident commissioner of Zululand at that time, C. R. Saunders, played an influential role in protecting the game reserves from land alienation (Brooks, 2001). A new reserve, Hlabisa, was proclaimed in 1905 to encompass the land between the Hluhluwe and Umfolozi GRs as far east as the shores of Lake St Lucia (Pringle, 1982). Part of it was later to become the Corridor Game Reserve. It was abolished in 1907 following complaints from transport riders that nagana killed the oxen pulling their wagons through this region. The Umdhletshe Game Reserve was also abolished in that year for the same reason.

In 1904, the Natal Game Protection Association was established to lobby for the protection of wildlife. In 1909 Dr Ernest Warren, the Director of the Natal Museum, changed its name to the Natal and Zululand Protection Association. Warren played an important role in conserving wildlife in Zululand by influencing public opinion to oppose the killing of wild animals (Brooks, 2001).

In 1910, Natal became a province of the newly established Union of South Africa. In the short term, not much changed. Like the old colonial administration, the new provincial government strongly supported the game reserves. Several members of the Natal Provincial Council were sport-hunters who had an interest in preserving the game. In 1911, the Provincial Council appointed Mr Frederick Vaughan-Kirby as Game Conservator for Zululand (Brooks, 2001). He became a strong advocate for conservation, emphasizing how few white rhinos remained in Umfolozi GR. White farmers who had settled in regions of Zululand became an influential counter-lobby. They resented both the sport-hunters, who belonged mostly to a privileged, often urban, elite, and the game reserves (Brooks, 2001). Their cattle were vulnerable to infection with nagana through contact with wild ungulates remaining outside the game reserves. Accordingly, tsetse flies switched from being the saviours of the wildlife to becoming the cause of much conflict (Pringle, 1982). The official strategy for resolving the 'nagana problem' became the 'elimination of all game' (McCracken, 2008).

Pressure increased after 1918, following the opening of farms in the Ntambanana district to the south of the Umfolozi GR for settlement by former World War I servicemen (Vincent, 1970; McCracken, 2008). The Umfolozi GR was abolished by provincial notice in 1920, but reinstated in 1930. In 1929, a campaign was mounted to eliminate wild ungulates from the area between Umfolozi GR and the Ntambanana farms, by either shooting them or driving them into the game reserve (Mentis, 1970; McCracken, 2008) and over 25,000 animals were killed between 1929 and 1930 (the species breakdown is given in Chapter 4). This slaughter failed to prevent cattle from being infected with nagana. In 1932, administration of Umfolozi GR (but not Hluhluwe GR) was transferred from the province to the national Department of Veterinary Services (McCracken, 2008). Nevertheless, field rangers remained stationed there to prevent any shooting of white rhinos.

In 1921, the entomologist R. H. T. P. Harris had been assigned to Umfolozi GR to undertake research on the tsetse fly problem (Brooks, 2001). Recognizing that these flies found their animal hosts by sight rather than smell, he invented the Harris Fly Trap. This was a V-shaped structure, which was open beneath with a gauze trap above. Flies entering the trap from below moved up towards the light and into the gauze trap, where they died. To prevent damage to these traps, fires were suppressed during the two decades during which they were deployed. Millions of flies were destroyed, reducing their numbers but not eliminating them. In 1931, the Minister of Agriculture stated that 'The only solution to the problem is to eliminate the host of the fly, which can only be done by eliminating the game reserves. Hluhluwe and Mkuzi must be abolished. The white rhino can be preserved at Umfolozi, but the rest of the game there must be reduced . . . ' (quoted by McCracken, 2008). The Zululand Game Reserves and Parks Board, the Wildlife Society of South Africa and individual conservationists and scientists challenged the threat that this standpoint posed to the wildlife of Zululand. Nevertheless, a second operation to eliminate most wild ungulates within Umfolozi GR as well as in the adjoining lands was launched in 1942 and continued until 1950, under the management of the veterinary authority (see Chapter 4 for detailed account). Wildlife within Hluhluwe GR retained its protected status, but significant numbers were killed in buffer zones surrounding it (see Chapter 4).

Eventually, a veterinary scientist conceded that 'The reduction of the thicket-inhabiting species (of ungulates) to the level where they could no longer support a population of this species of tsetse fly was ultimately found to be so formidable a task as to be almost impossible of achievement' (R. du Toit, cited by Vincent, 1970). A different intervention was then launched: aerial spraying with insecticides. This drew on the finding that tsetse flies bred only in confined, shaded sites. Hence, a 3-km wide strip was cleared of all woody vegetation along parts of the perimeter of the game reserves, to confine and prevent the shade-dependent flies

Figure 1.3 Actions aimed at eradicating tsetse flies from the Hluhluwe-iMfolozi Park during the 1930s, 1940s, and early 1950s. (A) Insecticides being sprayed from aircraft; (B) 3-km wide strip cleared of all woody vegetation along the western boundary of Umfolozi Game Reserve to prevent dispersal by surviving flies. (Photographs taken by an unknown photographer, figure made by C. Staver and J. Graf.)

from dispersing (see also Chapter 3). Areas of dense woodland within the reserves were also cleared. People settled in parts of the Corridor between the two game reserves were removed in the 1940s, supposedly temporarily while the aerial spraying took place. This led to much discontent when they were not allowed back at a later stage (Brooks, 2001). Aerial spraying was undertaken between 1947 and 1951 initially with dichlorodiphenyltrichloroethane (DDT), and later with benzene hexachloride (BHC), throughout Umfolozi GR, the Corridor region and low-lying parts of Hluhluwe GR (Figure 1.3). This finally eradicated the tsetse fly species mainly responsible for the transmission of nagana from the Umfolozi region, although pockets of forest-inhabiting flies persisted in Hluhluwe GR and towards the coast. Effects of the insecticide spraying on other insects and birds were not investigated.

In 1929 Captain Harold Potter, who succeeded Vaughn-Kirby as Game Conservator, established his headquarters within Hluhluwe GR. He appointed field rangers and enforced the protection of the wildlife. The first rest huts for visitors were built in 1934, and a road network was established. In an attempt to promote the increase of ungulate populations, Potter poisoned spotted hyenas. Lion (*Panthera leo*), cheetah (*Acinonyx jubatus*), and African wild dog were extinct in the region and leopard (*Panthera pardus*) were scarce (see Chapter 12). Potter brought in nyala (*Tragelaphus angasi*) and impala (*Aepyceros melampus*) from the Mkhuze region on the eastern coastal plain, although neither antelope had been recorded historically within HiP (see Chapter 4 for more details on introductions).

Meanwhile, adjustments took place to the boundaries of HiP, which as originally proclaimed excluded a large area of prime wildlife habitat along the southern bank of the Hluhluwe river. Hunters entered this region during the 1930s and shot animals coming to drink. In 1939, the boundary of Hluhluwe GR was extended to incorporate the entire river (marked as 'original boundary' in Figure 1.4). A proposal to move it further to include the corridor between the two reserves was rejected, because the Minister of Lands was prepared to do this only on condition that Mkuzi GR was abolished, which was opposed by the conservation lobby. The matter was resolved when it was decided to extend the southern boundary of Hluhluwe GR to a position well short of the Mtubatuba–Hlabisa road. Further extension would be considered later when nagana and tsetse fly were under control (Vincent and Porter, 1979). In the early 1960s there was a further adjustment of the boundary which was moved south in compensation for the area that was to be lost to the Hluhluwe dam (marked as 'present boundary' in Figure 1.4). In 1941, a five-strand barbed wire fence was completed along the boundary of Hluhluwe GR and the northern edge of the Corridor, leaving a gap to allow animals to move between both areas (Vincent and Porter, 1979). Due to increased poaching, a section of this fence was electrified in 1950. Another fence was constructed to the south of the White Mfolozi river in 1952 to stop white rhinos from straying towards the Ntambanana farms, enclosing a buffer zone that eventually became incorporated within the game reserve.

1.7 Consolidation of the Hluhluwe-iMfolozi Park after 1952

The Zululand Game Reserves and Parks Committee had been established in 1937 to oversee conservation in the game reserves, with William Power as its chairman. During the 1940s, moves were made towards ceding control of the provincial game reserves to the National Parks Board headquartered in Pretoria. This strengthened lobbying for unified control of conservation in the province, including inland and coastal fishing, led by Power together with the Administrator of Natal, Mr Douglas Mitchell. This led to the establishment of the Natal Parks, Game and Fish Preservation Board in December 1947.

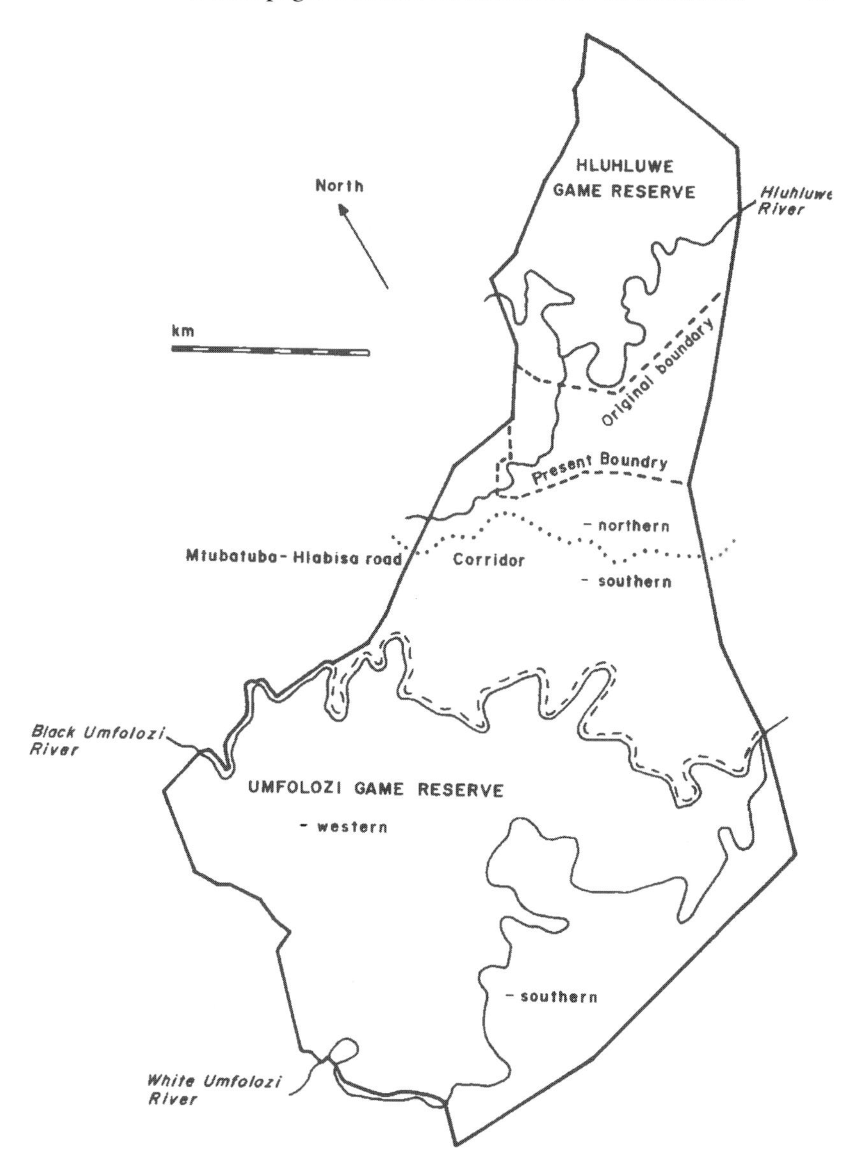

Figure 1.4 Map of the game reserves in 1983, showing the division between Hluhluwe GR, Umfolozi GR, and the Corridor, with major rivers. For Hluhluwe GR the proclaimed boundary after the 1939 extension ('original boundary') and the later 1960s extension ('present boundary') are shown. (Drawn by Ms A. Ludbrook and reprinted from Brooks and Macdonald, 1983.)

On 1 January 1953 control of Umfolozi GR passed back to the province (Vincent, 1970). The corridor between Umfolozi and Hluhluwe GRs remained State land, although fenced on the north to prevent rhinos from moving out. Following the handing back of Umfolozi GR the major concern was for the protection of the white rhino, particularly the relatively small population present in the former hunting area to the south of the reserve boundary. Poaching was rife during the whole of the 1950s both within Umfolozi GR and in adjacent areas. In addition, incursions of cattle into southern Umfolozi GR became a major problem. In an attempt to intensify security a ranger outpost was built south of the White Mfolozi river in the 'Southern Buffer Zone' and game guard outposts were strategically placed near the boundaries. With Umfolozi GR reverting back to a protected area in 1953 there was confusion regarding the hunting of wild animals among local communities given that they had been allowed to kill animals during the nagana control period but were now being arrested and charged for poaching. Resentment and hostility grew among these people and management staff undertook meetings with the tribal leaders in an effort to explain the change in policy in order to get their cooperation and reduce the incidence of poaching.

In 1954, a botanical ecologist, Mr C. J. (Roddy) Ward, was appointed research officer and based in Hluhluwe GR. The section of Umfolozi GR to the south of the tourist road, including the White Mfolozi river basin, became designated as a 'wilderness area' free of roads and buildings. This idea was instigated by Ranger Jim Feely, inspired by the concept after reading the principles of the US Wilderness Society, and implemented vigorously by the ranger-in-charge, Ian Player (Player, 1997). Tourists were allowed to enter only on foot on conducted wilderness trails, camping in tents. This practice is maintained today. In 1958, huts for the accommodation of visitors travelling in motor vehicles were established in Umfolozi GR on Mpila hill.

During the 1950s and 1960s poaching remained a serious problem, particularly by young men hunting and snaring warthog (Annual Reports, 1951–1960). This trend became more prevalent with the use of dogs by poachers. Towards the end of the 1950s gangs of poachers, usually armed with .303 rifles, became more frequent, and night poachers operated using spotlights. In 1961 the first black rhino was snared and killed by a poacher who was subsequently arrested, convicted and sentenced to a year's imprisonment or fine of R200. Strategic placing of guard camps in the more troubled areas, the increased use of mounted patrols, and possibly the 'restraining influence of the presence of lions' since the early 1960s, led to a decline in poaching. In subsequent years from 1967 to 1970, poaching was considered to be largely under control (Annual Reports, 1969–1973). In addition to poaching of animals, certain plant species used in traditional medicine were also gathered inside the Park. *Warburgia salutaris* was particularly sought after and is now extremely rare as a result of overharvesting. Bark, roots, corms, bulbs, and leaves of several plant species were collected. After 2000, the anti-poaching system was improved by the establishment of Anti-Poaching Units that operated secretly and independently of other management staff. Intelligence has proved to be essential in anti-poaching operations and a Rhino Security Intervention Strategy was developed making use of CyberTrackers (software installed in hand-held computers), which enabled efficient analysis of patrol data to make them more strategic and effective.

Influxes of homesteads into the State land adjoining the western boundary of Umfolozi GR took place from 1955 to 1961 (Annual Reports, 1956–1959). These were people relocated by the government from the Msinga district in the Thukela river basin to the area bordering this State land. This settlement of the western State land caused much tension, with repeated attacks on field rangers, leaving one game guard killed, and led to increased poaching, veld burning, and cattle problems. In 1959 a Boundary Commission made recommendations to the national government, under whom unreserved State land fell, to solve the position of the boundaries (Figure 1.5 shows the boundaries as they were in 1960). Settlement was finally reached in 1962 and the western and southern boundaries of Umfolozi GR were officially extended to encompass portions of the State land that had separated it from the adjoining tribal land (reflected by the 'desirable boundary' in Figure 1.5). People were forcefully evicted from occupying the State land along the western boundary in 1962 (Annual Reports, 1960–1963; Vincent, 1970). With the final fixing of the reserve's southern and western boundaries, a start was made on erecting a fence. Between 1962 and 1965, a 2.8-m high fence consisting of four to five strands of heavy cable was constructed along the new boundaries. The status of the Corridor remained unresolved. It was officially State land, but full of wildlife. Protracted negotiations with the Veterinary Department, who insisted on the construction of a game-proof fence to ensure that diseases carried by buffalo (Corridor disease, which is a form of East Coast fever or theileriosis) and wildebeest (bovine malignant catarrhal fever or snotsiekte) were confined to the game reserves,

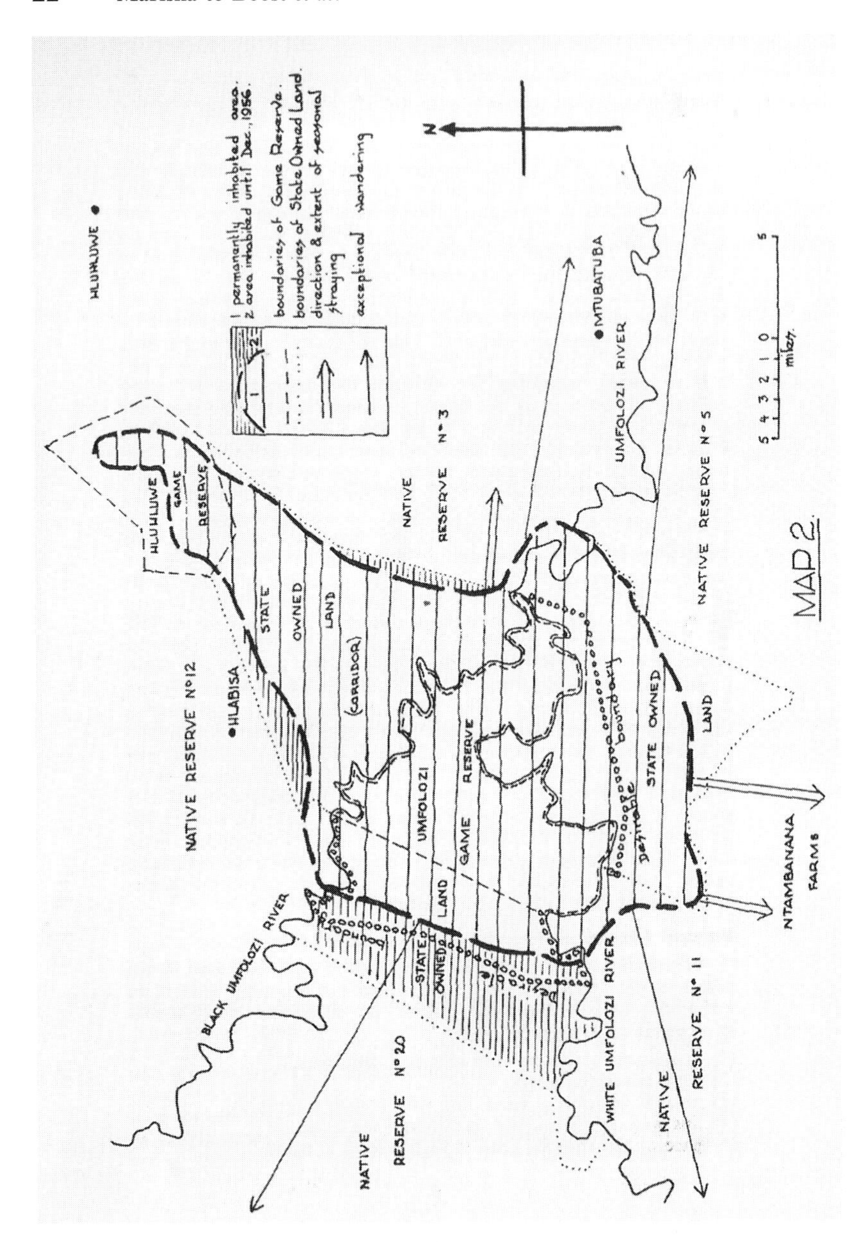

Figure 1.5 Map of the game reserves, State land, and native reserve boundaries as they were in 1960. This map was originally drawn up to illustrate the distribution of white rhino, which is represented by the thick black dashed line (area permanently inhabited by white rhino). The boundary of Hluhluwe GR is shown by the thin dashed line, and the boundary of the State land by the thin dotted line. The desirable boundary of Umfolozi GR is shown with open circles. (Reprinted with permission from Ezemvelo KZN Wildlife; Player and Feely, 1960.)

continued until an agreement was finally reached in 1968. Fencing then commenced on the eastern Corridor boundary to confine rhino to the reserve and stop encroachment by people. The veterinary game-proof fence was completed at great expense in 1970 and was successful in preventing invasion by cattle, and disease transmission to domestic animals, as well as protecting people's crops from the depredations of buffalo and elephant. This fence was consolidated in 1973 by a rhino- and lion-proof fence completely enclosing the borders of the Corridor and Hluhluwe GR as well (Figure 1.4 shows the boundaries as they were in the 1980s).

In February 1988 a land surveyor was appointed to survey the boundaries of the tribal authority areas that adjoined Hluhluwe GR. It was found that the boundaries between the tribal authorities and Hluhluwe GR were vague and that the only legal boundary was the one given in the 1897 proclamation (Isherwood, 2002). However, local communities had established their homesteads and croplands inside the proclaimed game reserve boundary. The 1970 veterinary game-proof fence that took cognizance of the occupation by the people was considered to be in the ideal position and had been accepted by local community leaders, traditional authorities, and all other government parties. A recommendation that the two areas now occupied by communities be excised from the Hluhluwe GR was agreed upon in 1997 after negotiations with the neighbouring tribal authority.

In 1989, the Corridor Game Reserve was formally proclaimed, joining Hluhluwe GR and Umfolozi GR into a unified protected area, divided only by a pre-existing dirt road through the Corridor joining the towns of Mtubatuba and Hlabisa. This road was upgraded and tarred around 2003–2004. The extent of the land originally comprising Hluhluwe, Umfolozi, and Corridor game reserves was 934 km^2. However, after the adjustments of the final boundaries decided upon in 1997, the consolidated 'Hluhluwe-Imfolozi Park' [*sic*] as legally promulgated in 2012 enclosed a reduced fenced area of 897 km^2.

Meanwhile, the land surrounding HiP under communal tenure has become increasingly densely settled with dispersed homesteads, following Zulu custom. The associated land transformation from rangelands to agricultural fields means that the protected area has become effectively an ecological island in a matrix of human-modified landscapes (Figure 1.6), with implications for animal movements (see Chapters 4, 11, 12, and 14), invasive alien plants (see Chapter 15), fire management (Chapter 10), and the social and economic benefits the park is expected to provide.

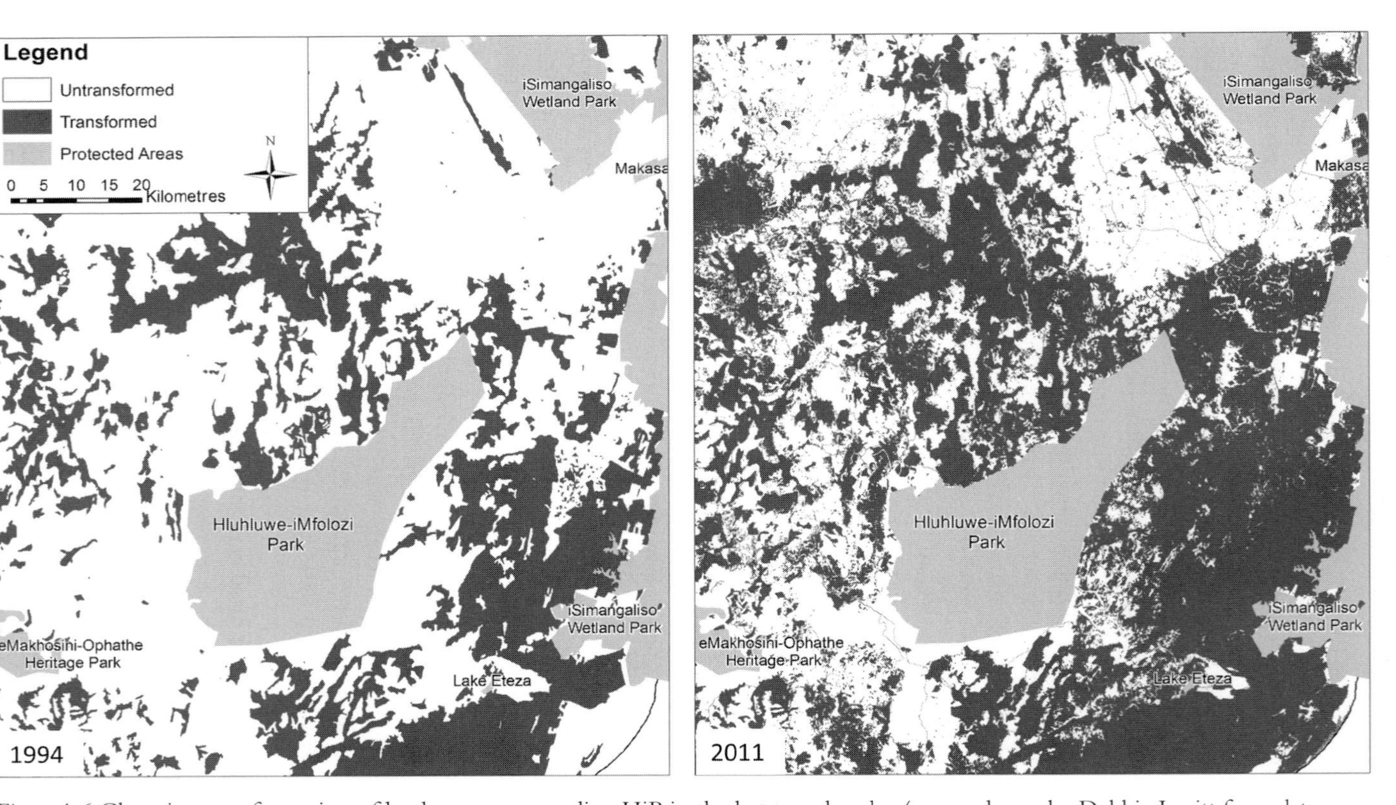

Figure 1.6 Changing transformation of landscapes surrounding HiP in the last two decades (map redrawn by Debbie Jewitt from data presented in Jewitt *et al.*, 2015).

1.8 Management History

The Natal Parks Board (as it became known) was the official management authority for all of the game and nature reserves in the Province from its inception in 1947 until December 1997, when a new authority, the KwaZulu-Natal Nature Conservation Board, was formed following South Africa's first democratic elections in April 1994. The organization operates as Ezemvelo KZN Wildlife, incorporating the Zulu word for conservation.

Management policies for the park have undergone changes over time. During the 1950s they were stated for Hluhluwe GR as being 'to maintain a balance among the animals and between them and the soil, the vegetation and the water supplies . . . interpreted as maintaining the habitats as they occurred in the days prior to the restriction of game movements' (Ward, 1961 as cited in Brooks and Macdonald, 1983). With the appointment of qualified natural scientists, the first one appointed for Hluhluwe GR in 1953, and many research and monitoring projects undertaken, management interventions became better informed.

As a result of effective wildlife protection, by 1953 animal numbers within Hluhluwe GR had reached levels raising concerns about degradation of soils and vegetation and the emaciated appearance of impalas and warthogs (Cowles, 1959). Predation was solely by relatively few spotted hyenas and leopards, until a male lion settled in the Umfolozi GR in 1958, followed by two lionesses a few years later (see Chapter 12). Soil erosion caused by overgrazing, trampling, and poorly positioned tracks and roads was noted in 1955 in Hluhluwe GR (Annual Report, 1955–56). Areas perceived to be overgrazed were also recorded in Umfolozi GR, although only white rhinos and warthogs were abundant at that time. Various reclamation methods were used, including stone packs, gabion structures and thorn scrub packs, to rehabilitate gullies formed along roads, streams, and fence lines. In addition, animal removals were initiated in Hluhluwe GR and the northern Corridor in 1954 and extended to Umfolozi in 1959, aimed at alleviating grazing pressure by species besides white rhinos (see Chapter 4 for details). Removals initially tracked rainfall variation, increasing in dry years and decreasing in wet years (Brooks and Macdonald, 1983).

In 1955 a controlled burning programme was instituted and annual firebreaks were cut to protect infrastructure and to prevent fires from entering HiP. Fires were generally ignited after the first spring rains in early to mid-September (see Chapter 10 for details on history of fire management). During the same period, the encroachment of woody plants into grassland became a recognized problem (see Chapter 3), particularly in Hluhluwe GR as it reduced visibility of wildlife for visitors. Management-related factors that may have contributed to bush encroachment might have been the removal of large numbers of impala and nyala from the late 1960s to 1980s, and possibly the elimination of invertebrate seed predators as a result of the aerial spraying of insecticide during the nagana campaign, resulting in higher seed viability of *Acacia* and other tree species. From 1959 to 1962 bush-clearing operations were undertaken, especially in the northern Hluhluwe section. Later, the removal of bush was associated with gully rehabilitation activities, to make scrub packs (*Acacia* and *Dichrostachys* spp.) in eroded areas.

In 1961 the Department of Water Affairs started work on the construction of the much-disputed dam on the Hluhluwe river downstream of HiP to supply water to farms and Hluhluwe village. The dam had a full level, extending into Hluhluwe GR, but confined largely within the banks of the river and its main tributaries (Isherwood, 2002). In February 1972 the dam flooded extensive areas of *Ficus sycamorus–Schotia brachypetala* riparian forest, which subsequently died (Porter, 1981). In recent years (from 2009), a number of boreholes in the Hluhluwe river upstream from HiP have led to increasingly low water levels in Hluhluwe river and the dam.

The growth of the white rhino population under effective protection raised concerns about overgrazing and the risk of a disease outbreak. Pioneering trials with the use of chemical immobilization led to the initiation of a live capture and translocation programme in 1967, which was greatly expanded from 1970 onwards (Brooks and Macdonald, 1983). This programme, initiated by the park warden, Ian Player, became known as 'Operation Rhino' and is one of the greatest conservation success stories of the last century. By translocating rhinos from Umfolozi GR to other reserves, such as Kruger National Park, the Southern white rhino population increased 20-fold in the last 30 years or so (see Chapter 11). In 1970 an effective method of live capture was developed for other ungulates, driving herds into temporary enclosures formed by plastic sheeting, which enabled populations of, particularly, wildebeest and zebra to be reduced in subsequent years (Annual Report, 1970–1971).

During the late 1970s, the management policy was to hold herbivore populations below the estimated carrying capacity of the park, guided by agricultural norms. This policy was challenged following animal removals so excessive that few animals died during the extreme drought conditions that prevailed through 1981–1983 (Walker *et al.*, 1987). Thereafter, a 'process-based' philosophy was adopted, allowing animal populations and the incidence of fires to fluctuate in response to rainfall variation to the extent that they might have done in the past, before the park became fenced (Ezemvelo KZN Wildlife, 2011). During drought years, animal removals were curtailed to allow for natural mortality. From 1999, disease management within the park has focused on restricting the incidence of bovine tuberculosis in buffalo herds, acquired from surrounding cattle (Chapter 13). Elephants, which had been absent for a century, were brought back in 1981 (Chapter 14). The first alien invasive plant survey was conducted in 1978 (Macdonald, 1983). *Chromolaena odorata* was identified as the plant of greatest concern, which unfortunately proved true (Chapter 15). Given the concerns regarding invasive alien plants, a list of species that were prohibited from being brought into HiP was included in the Management Plan (Ezemvelo KZN Wildlife, 2011).

The first formal management plan was produced in 1972. Currently such plans must comply with the requirements of the National Environment Management: Protected Areas Act of 2003 and undergo periodic review and updating every 5 years. The most recent integrated management plan for HiP (Ezemvelo KZN Wildlife, 2011) stated the following objectives.

1. To contribute to the achievement of provincial and national nature conservation objectives and targets, as a component of the national protected area system, through protection of a representative sample of the indigenous ecosystems, communities, ecotones, and representative landscapes of the area, their indigenous biodiversity, and the ecological and evolutionary processes that generate and maintain this diversity.
2. To protect and conserve species of conservation significance (e.g. endangered, rare, and endemic plant and animal species) indigenous to the area.
3. To conserve the ecological integrity and the wild character of the park.
4. To conserve the integrity of the iMfolozi Wilderness Area.
5. To safeguard the archaeological, historical, paleontological, and living cultural heritage of the area.
6. To promote awareness and appreciation of the natural environment, scenic beauty, and outstanding aesthetic value of the area.
7. To provide controlled access by the public to the area and its resources.

8. To contribute to local, regional, and national economies through the provision of ecosystem services, eco-cultural tourism, and the sustainable use of natural resources.
9. To provide a major destination for eco-cultural tourism in SA.
10. To provide opportunities for management-orientated and other forms of research, monitoring, education, interpretation, and awareness programmes that contribute to improved understanding and awareness of the values of the area and excellence in the management of the park.

Management actions implemented towards achieving these objectives include:

1. protection of animals and plants from poachers;
2. arresting and rehabilitating areas of active soil erosion;
3. controlled use of fire to remove moribund pasture and suppress bush encroachment (Chapter 10);
4. eradication of alien invasive plants (Chapter 15);
5. animal population management, particularly of rhino and elephant (Chapters 11 and 14);
6. re-introduction and management of predators (Chapter 12); and
7. disease control programmes (Chapter 13).

How successfully these objectives can be achieved in the future remains uncertain. With lands surrounding the protected area mostly densely occupied by people (Figure 1.6), HiP has become a fenced island in a matrix of human-modified landscapes. Close interactions with, and socioeconomic development of, the park's neighbouring communities are thus perhaps the most urgent issues to safeguard the future of HiP. Climate change, shrub encroachment, invasive species, and disease are all issues that need to continue to be addressed. Mining activities on the boundaries of HiP have increased enormously. However, most disturbing at the current time is the strong rise in poaching of rhinos prompted by the enormous price fetched by their horns in the Far East. This threatens to reverse one of the most renowned conservation successes of the twentieth century.

1.9 Summary

The area that became the Hluhluwe-iMfolozi Park was not a wilderness free of human influences. Stone Age hunter-gatherers were present from probably as far back as the origins of the earliest humans, killing animals, and igniting fires. Immigrants with implements made of iron harvested wood for smelting iron ore, constructed dwellings, cleared land for crops, and grazed livestock. Nevertheless, a rich abundance of wild ungulates continued to persist alongside these farmers and pastoralists, gaining protection from the local presence of tsetse flies and the disease that these flies transmitted from wild ungulates to cattle. White settlers with guns decimated the wildlife, assisted by African hunters. Rinderpest, spread from Eurasia into Africa via cattle, contributed further to the reduction in wild ungulates. White farmers near the reserves mounted campaigns to eradicate wild ungulates outside the game reserves and later within them, to wipe out the tsetse flies and the blood parasites that they carried. Aerial spraying with insecticides eventually eliminated tsetse flies, and eradicated nagana. Protected, but without effective predators, herbivore populations grew to levels causing perceived overgrazing of vegetation and associated soil erosion. Ungulate populations became restricted by live capture and sales, while large predators and elephants were restored. Through much of this history, the white rhino has been pivotal, initially for the proclamation of the game reserves and continuing protection afforded to the land, subsequently for management interventions and successful conservation, and currently once again because of their vulnerability to poaching due to the monetary value of their horns.

1.10 Acknowledgements

For comment and information we thank André Boshoff, Peter Hitchins, Tim Maggs, Eugene Moll, Terry Oatley, Ken Tinley, Gillian Vernon, and John Vincent. We thank Debbie Jewitt for creating Figure 1.6.

1.11 References

Annual Reports. Reports of the Natal Parks, Game and Fish Preservation Board to the Administrator of Natal. Unpublished reports, Natal Parks Board, Pietermaritzburg.

Boshoff, A. F. & Kerley, G. I. H. (2013) *Historical incidence of the larger mammals in the Free State Province (South Africa) and Lesotho*. Centre for African Conservation Ecology, Nelson Mandela Metropolitan University, Port Elizabeth.

Brooks, P. M. & Macdonald, I. A. W. (1983) The Hluhluwe-Umfolozi Reserve: an ecological case history. In: *Management of large mammals in African conservation areas* (ed. R. N. Owen-Smith), pp. 51–77. Haum, Pretoria.

Brooks, S. J. (2001) A critical historical geography of the Umfolozi and Hluhluwe Game Reserves, Zululand, 1887–1947. PhD thesis, Queens University, Canada.

Clark, J. L. & Plug, I. (2008) Animal exploitation strategies during the South African Middle Stone Age: Howieson's Poort and post-Howieson's Poort fauna from Sibudu Cave. *Journal of Human Evolution* **54**: 886–898.

Cowles, R. B. (1959) *Zulu journal. Field notes of a naturalist in South Africa.* University of California Press, Berkeley, CA.

Deacon, H. J. & Deacon, J. (1999) *Human beginnings in South Africa: uncovering the secrets of the Stone Age.* David Philip, Cape Town.

Delegorgue, A. (1847) *Voyage dans l'Afrique Australe*, reproduced as *Adulphe Delegorgue's travels in Southern Africa*, Vol. 1 (F. Webb. (transl.), S. J. Alexander & C. de B. Webb (eds), 1990). University of Natal Press, Pietermaritzburg.

Ellis, B. (1994) Game conservation in Zululand, 1824–1947. *Natalia: Journal of the Natal Society* **23/24**: 27–44. Available at natalia.org.za.

Ezemvelo KZN Wildlife (2011) *Integrated management plan: Hluhluwe-iMfolozi Park, South Africa.* Ezemvelo KZN Wildlife, Pietermaritzburg.

Feely, J. M. (1980) Did Iron Age man have a role in the history of Zululand's wilderness landscapes? *South African Journal of Science* **76**: 150–152.

Feely, J. M. (2004) Prehistoric use of woodland and forest by farming peoples in South Africa. In: *Indigenous forests and woodlands in South Africa: policy, people and practice* (eds. M. J. Lawes, H. A. C. Eeley, C. M. Shackleton, & B. G. S. Geach), pp. 284–286. University of KwaZulu-Natal Press, Pietermaritzburg.

Fynn, H. F. (1950) *The diary of Henry Francis Fynn*, compiled from original sources and edited by J. Stuart and D. McK. Malcolm. Shuter & Shooter, Pietermaritzburg.

Hall, M. (1977) Shaka's pitfall traps: hunting technique in the Zulu Kingdom. *Annals of the Natal Museum* **23**: 1–12.

Hall, M. (1979a) A list of known Iron Age archaeological sites in the Umfolozi, Corridor and Hluhluwe Reserves. Unpublished report, Natal Parks Board, Pietermaritzburg.

Hall, M. (1979b) The Umfolozi, Hluhluwe and Corridor reserves during the Iron Age. *The Lammergeyer* **27**: 28–40.

Hall, M. (1980) An iron-smelting site in the Hluhluwe Game Reserve, Zululand. *Annals of the Natal Museum* **24**: 165–175.

Hall, M. (1981) *Settlement patterns in the Iron Age of Zululand: an ecological interpretation.* Cambridge monographs in African archaeology, Volume 5. British Archaeological Reports, Oxford.

Hall, M. (1984) Prehistoric farming in the Mfolozi and Hluhluwe valleys of southeast Africa: an archaeo-botanical survey. *Journal of Archaeological Science* **11**: 223–235.

Hall, M. (1987) *The changing past: farmers, kings and traders in southern Africa, 200–1860.* David Phillip, Cape Town.

Herman, L. & Kirby, P. R. (1970) *Travels and adventure in eastern Africa: descriptive of the Zoolus, their manners, customs, with a sketch of Natal, by Nathaniel Isaacs.* Struik, Cape Town.

Huffman, T. N. (2007) *Handbook to the Iron Age: the archaeology of pre-colonial farming societies in southern Africa.* University of KwaZulu-Natal Press, Pietermaritzburg.

Isherwood, H. B. (2002) Umfolozi-Hluhluwe Game Reserve and adjoining State land. Unpublished report to Director: Survey Services.

Jewitt, D., Goodman, P. S., Erasmus, B. F. N., O'Connor, T. G., & Witkowski, E. T. F. (2015) Systematic land-cover change in KwaZulu-Natal, South Africa: implications for biodiversity. *South African Journal of Science* **111**: 1–9.

Kingdon, J. (2003) *Lowly origin: where, when and why our ancestors first stood up*. Princeton University Press, Princeton, NJ.

Laband, J. (1995) *Rope of sand: the rise and fall of the Zulu Kingdom in the nineteenth century*. Jonathan Ball, Johannesburg.

Macdonald, I. A. W. (1983) Alien trees, shrubs and creepers invading indigenous vegetation in the Hluhluwe-Umfolozi Game Reserve Complex in Natal. *Bothalia* **14**: 949–959.

Maggs, T. (1984) The Iron Age sequence south of the Zambezi. In: *Southern African prehistory and paleoenvironments* (ed. R. G. Klein), pp. 329–359. Balkema, Rotterdam.

McCracken, D. P. (2008) *Saving the Zululand wilderness. An early struggle for nature conservation*. Jacana Media, Johannesburg.

Mentis, M. T. (1970) Estimates of natural biomasses of large herbivores in the Umfolozi Game Reserve area. *Mammalia* **34**: 363–393.

Mitchell, P. (2002) *The archaeology of southern Africa*. Cambridge University Press, Cambridge.

Penner, D. (1970) Archaeological survey in Zululand game reserves. Unpublished report, Natal Parks Board, Pietermaritzburg.

Player, I. C. & Feely, J. M. (1960) A preliminary report on the square-lipped rhinoceros (*Ceratotherium simum simum*). *The Lammergeyer* **1**: 3–24.

Player, I. C. (1997) *Zululand wilderness: shadow and soul*. David Philip, Cape Town.

Plug, I. (2004) Resource exploitation: animal use during the Middle Stone Age at Sibudu Cave, KwaZulu-Natal. *South African Journal of Science* **100**: 151–158.

Porter, R. N. (1981) A preliminary impact assessment of the environmental effects of proposed dams in the Mfolozi catchment, Natal. Unpublished report, Natal Parks Board, Pietmaritzburg.

Pringle, J. A. (1982) *The conservationists and the killers: the story of game protection and the Wildlife Society of Southern Africa*. T. V. Bulpin and Books of Africa, Cape Town.

Prins, F. E. (1996) Aspects of Iron Age ecology in the Eastern Cape and KwaZulu/Natal. In: *The growth of farming communities in Africa from the equator southwards* (ed. J. E. G. Sutton), pp. 71–90. *Azania* special volume 12.

Van Wyk, B. E. & Gericke, N. (2000) *People's plants: a guide to useful plants of southern Africa*. Briza Publications, Pretoria.

Vernon, G. N. (2013) *Even the cows were amazed: shipwreck survivors in south-east Africa 1552–1782*. Jacana Press, Johannesburg.

Vincent, J. (1970) The history of Umfolozi Game Reserve, Zululand, as it relates to management. *Lammergeyer* **11**: 7–49.

Vincent, J. & Porter, R. N. (1979) The boundaries of Hluhluwe–Corridor–Umfolozi (Central Complex) and the background to their present situation. Unpublished report, Natal Parks Board, Pietermaritzburg.

Wadley, L., Sievers, C., Bamford, M., *et al.* (2011) Middle Stone Age bedding construction and settlement patterns at Sibudu, South Africa. *Science* **334**: 1388–1391.

Walker, B. H., Emslie, R. H., Owen-Smith, N., & Scholes, R. J. (1987) To cull or not to cull: lessons from a southern African drought. *Journal of Applied Ecology* **24**: 381–401.

Ward, C. J. (1961) Burning as it affects veld management policy of the Hluhluwe Game Reserve. Unpublished report, Natal Parks Board, Pietermaritzburg.

Whitelaw, G. (1994) KwaGandaganda: settlement patterns in the Natal Early Iron Age. *Natal Museum Journal of Humanities* **6**: 1–64.

Wright, J. (2008) Rediscovering the Ndwandwe kingdom. In: *Five hundred years rediscovered: southern African precedents and prospects* (eds. N. Swanepoel, A. Esterhuysen, & P. Bonner), pp. 217–235. Wits University Press, Johannesburg.

2 · *The Abiotic Template for the Hluhluwe-iMfolozi Park's Landscape Heterogeneity*

RUTH A. HOWISON, HAN OLFF, NORMAN OWEN-SMITH, JORIS P.G.M. CROMSIGT, AND SALLY ARCHIBALD

2.1 Introduction

The Hluhluwe-iMfolozi Park (HiP) is situated in the transitional region between the coastal lowlands and dissected interior plateau of northern KwaZulu-Natal Province in South Africa (Figure 2.1). The terrain ranges from high rolling hills in the north-east to lowlands bordering the two Mfolozi rivers in the south-west (Figure 2.2). In this chapter, we outline how this landscape setting has contributed to an exceptional heterogeneity in topography, geology, soils, and rainfall within its 950 km^2 extent. This heterogeneous biophysical template forms a basis for the diverse vegetation and associated fauna conserved within the park, which rival the diversity of much larger protected savannas such as Serengeti and Kruger National Park.

2.2 Landscape Context

HiP lies between coordinates 28°00″ and 28°26″S and 31°43″ and 31°09″E, and on the basis of its latitude falls just south of the tropics. However, the continuation of the Mozambique coastal plain lying to its east is warmed by the south-flowing Agulhas current, forming a biogeographic corridor that extends the distribution of plant and animal species with tropical affinities into this region. The local topography has been shaped by erosion cycles following uplift of the southern African

Conserving Africa's Mega-Diversity in the Anthropocene, ed. Joris P. G. M. Cromsigt, Sally Archibald and Norman Owen-Smith. Published by Cambridge University Press.

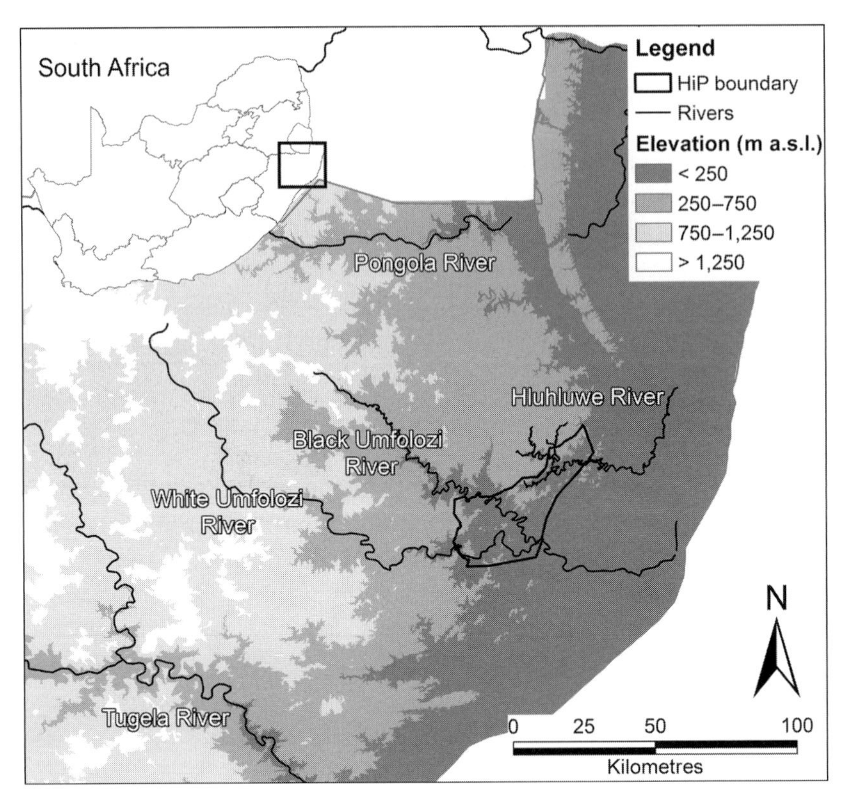

Figure 2.1 Regional location of Hluhluwe-iMfolozi Park (black polygon) within the northern part of the KwaZulu-Natal Province of South Africa, showing the increase in elevation from the coast to the interior and the three major rivers flowing through the park.

interior in two main periods: the first around 20 million years ago during the Miocene, and the second around 5 million years ago during the Pliocene (Partridge, 1998). This uplift led to the down-tilting of geological strata towards the continental margin by 10–15°. HiP lies on the east flank of this geological monocline (Figure 2.1) and features land surfaces shaped during Miocene, Pliocene, and Quaternary times. Its landscapes vary from steeply undulating hills intersected by narrow river valleys in the north, through the rolling uplands of the central area, to the gently undulating river basins in the south (Figures 2.3–2.9). Three major rivers traverse the park. The Black and the White Mfolozi rivers in the south have distant catchments in the interior plateau (Figure 2.1),

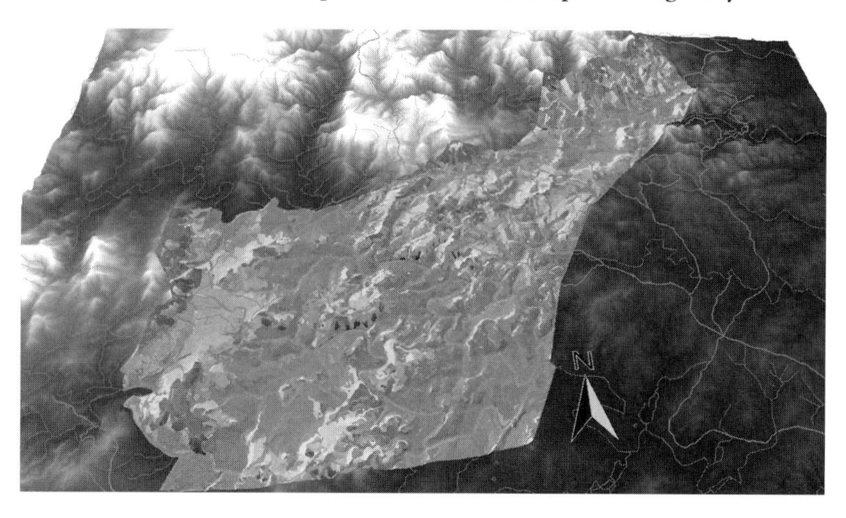

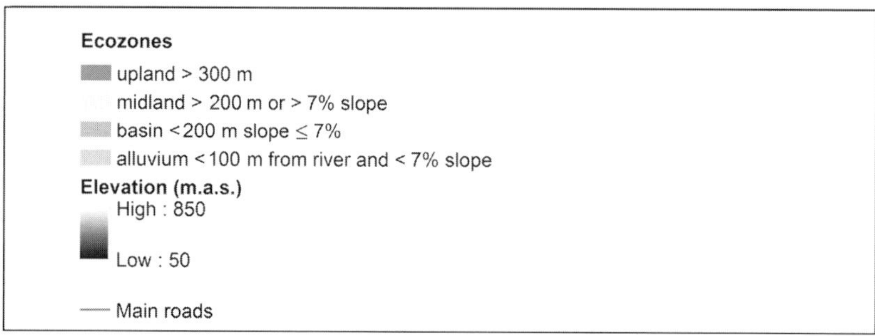

Figure 2.2 A three-dimensional view of model of Hluhluwe-iMfolozi Park showing the regional setting with respect to elevation in detail, and the main elevation-based landscape zones. (For the colour version, please refer to the plate section. In some formats this figure will only appear in black and white.)

while the Hluhluwe river and its major tributaries, the Nzimane and Manzibomvu, in the north, have their catchments a short distance beyond the park boundary. All major confluences of these rivers are inside the park boundaries (Figure 2.1). The valleys of the two Mfolozi rivers are separated by a wedge-shaped watershed, broadening from their confluence in the east towards around 20 km in width on the western boundary (Figure 2.6). The White Mfolozi river gets its name because the shallow water flowing over its sandy substrates is typically clear, in contrast with the deeper murky water of the Black Mfolozi river. The altitude within the park varies from 45 m above sea

Figure 2.3 Rolling hills with mosaic of grassland, thicket, and forest in northern Hluhluwe (photo: Norman Owen-Smith). (For the colour version, please refer to the plate section. In some formats this figure will only appear in black and white.)

Figure 2.4 Upland 'scarp forest' alternated with fire-dominated grassland patches in the northern part of the park (photo: Norman Owen-Smith). (For the colour version, please refer to the plate section. In some formats this figure will only appear in black and white.)

Figure 2.5 Rolling fire-dominated upland grassy hills in the central (corridor) part of the park, with trees only occurring along the drainages (photo: Han Olff). (For the colour version, please refer to the plate section. In some formats this figure will only appear in black and white.)

Figure 2.6 Basin of the White Mfolozi river in the southern part of the park looking towards the hills on the western boundary (photo: Norman Owen-Smith, taken in 1970). (For the colour version, please refer to the plate section. In some formats this figure will only appear in black and white.)

Figure 2.7 Knob thorn savanna in Mfolozi (photo: Norman Owen-Smith). (For the colour version, please refer to the plate section. In some formats this figure will only appear in black and white.)

Figure 2.8 Umbrella thorn savanna in western Mfolozi with grazing lawn grassland (photo: Norman Owen-Smith). (For the colour version, please refer to the plate section. In some formats this figure will only appear in black and white.)

Figure 2.9 Black Mfolozi river. Much of the riverine woodland was removed by floods following cyclone Demoina in February 1984 (photo: Norman Owen-Smith). (For the colour version, please refer to the plate section. In some formats this figure will only appear in black and white.)

level (asl) below the confluence of the two Mfolozi rivers to 750 m asl in the north west where the park boundary abuts the interior plateau. The central part of the park, traversed by the road between Mtubatuba and Hlabisa (also known as 'the Corridor Road'), forms an elevated grassy upland separating the main river basins in the south and the north of the park.

2.3 Geology and Soils

The topography formed by rivers and uplands (with main variation north to south, Figure 2.2) dissects the landscape perpendicular to the geomorphic tilting (with main variation west to east). This has exposed a diversity of geological substrates within the broader region, which are mostly represented within the boundaries of HiP (Downing, 1980; King, 1982). Most widespread are shale, siltstone, and sandstone of the Ecca Group (63% of HiP; Figure 2.10A). They are a component of the Karoo Supergroup formed by lacustrine deposits within the interior of the super-continent known as Gondwana during Permian times. Ecca

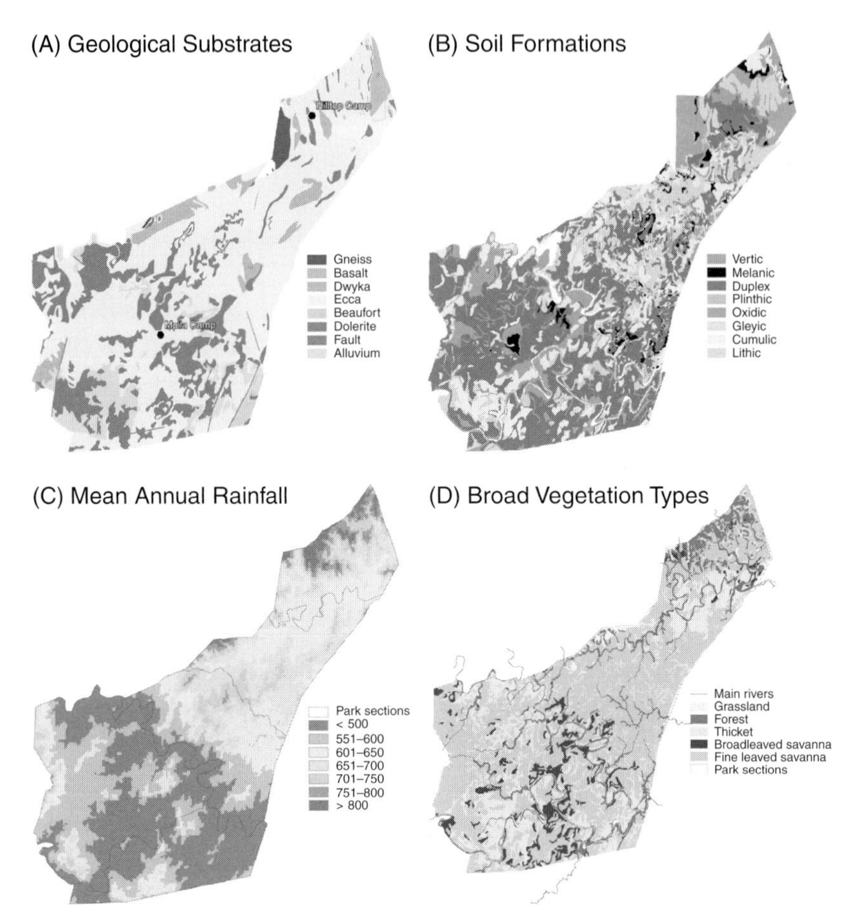

Figure 2.10 Landscape-scale heterogeneity within HiP in (A) geological substrates, (B) soil formations, (C) mean annual rainfall (from Kriging of records from 17 recording stations during 2001–2007), and (D) broad vegetation types (adapted from Whateley and Porter, 1983 – see Chapter 8 for a more detailed classification of this map). (For the colour version, please refer to the plate section. In some formats this figure will only appear in black and white.)

sandstones and shales underlie much of interior South Africa and include coal beds that are generally extensively mined (but not in the park). Several dykes and sills of dolerite intrude through the Karoo sediments and cap eminent hills in the Mfolozi region (21% of HiP). They represent feeders to the basalt that covered much of the land surface of southern Africa around the time of the break-up of Gondwana during early Tertiary times. A small region of basalt abuts Ecca sediments in the

Manzibomvu valley in northern Hluhluwe as a result of faulting. Breccias filling the fault zone persist as ridges of hard white stone on hills east of the northern road entrance. Dwyka tillite, a basal member of the Karoo Supergroup formed from glacial moraine, is exposed in scattered regions at different elevations throughout the park (5% of HiP). Sandstone of the Natal Group, which underlies the Karoo formations, forms conspicuous cliffs where the White Mfolozi river enters the park in the south-west. Granitic gneiss of the Archaean Complex is exposed in highlands in the northwest where the Nzimane river enters HiP, shown by rocky ridges with little domes and bare rock faces. Alluvial deposits occur along all main river valleys, most extensively in a basin adjoining the upper section of the White Mfolozi valley (7% of HiP). Terraced topography has resulted from differential erosion, with harder sandstone or dolerite persisting as rocky scarplets and softer shale strata forming gulleys and terraces.

Different geological substrates tend to produce different soils, but soil types are also dependent on temperature, rainfall, landscape position and corresponding drainage and soil movements (Fey, 2010; Figure 2.10B). Because of the complex topography and geology, the catenary gradients that are typical of African plateau landscapes are not readily evident in the park. Much of the soil in lower-lying regions is transported and mixed downslope. Lowland soils associated with Ecca shales are clay-rich and have a distinct and fairly hard B horizon aligning them with the Duplex soil group (Table 2.1A). This makes both water infiltration and capillary uplift difficult, causing these soils to dry first at the onset of the dry season. The pH of the surface horizon is typically slightly acid to neutral (5.5–7), and calcium carbonate nodules are commonly present in the subsoil. Duplex soils are susceptible to becoming sodic and may show surface crusting. However, calcium carbonate hardpans are not formed. The surface soil texture is sandier where sandstones and granites occur near the surface. In regions underlain by dolerite, the predominant soils are oxisols typified by their enrichment in iron oxides (Table 2.1B). Although clay-rich, these soils are well aerated and well drained. Their B horizon is uniformly coloured, either from dark red or yellow oxides of iron. Structured oxisols have a high cation exchange capacity and consequently are relatively fertile, given sufficient water availability. In some upland regions shallow lithic soils predominate, forming a soil layer of a few decimetres overtopping hard or weathered rocky layers. In footslopes and basins, deep cumulic soils develop from the deposition of unconsolidated colluvium or alluvium of mixed upslope origin. In areas with poor drainage and

Table 2.1A *Grouping of soil forms adapted from Fey (2010), based on the presence of specific diagnostic horizons or materials, and related to South African soil forms (SCWG, 1991)*

Soil group	Description	Identification	South African soil forms	Proportion of area (%)
Duplex	Marked clay enrichment	Pedocutanic or prismacutanic B	Sepane, Sterkspruit, Swartland	41
Lithic	Young soil on weathered rock	Lithocutanic B or hard rock	Cartref, Glenrosa, Mispah	24
Oxidic	Iron enrichment (residual), uniform colour	Red apedal, yellow–brown apedal or red structured B	Clovelly, Hutton, Shortlands	14
Cumulic	Young soil in unconsoli-dated sediment (colluvial, alluvial, aeolian)	Neocutanic or neocarbonate B, regic sand, deep E, or stratified alluvium	Alluvia, Augrabies, Dundee, Fernwood, Namib, Oakleaf, Tukulu	12
Gleyic	Reduction (aquic subsoil or wetland)	Stagnant wet G	Katspruit, Kroonstad	7
Melanic	Black clay, high base status	Melanic A	Bonheim, Inhoek, Mayo, Milkwood	2
Vertic	Swelling, cracking clay	Vertic A	Arcadia, Rensburg	< 1
Plinthic	Iron enrichment (absolute), mottling or cementation	Soft or hard plinthic B	Avalon, Bainsvlei, Dresden, Wasbank, Westleigh	< 1

sufficient rainfall, gleyic soils are found in small wetland pockets. Vertic or melanic clays occur is isolated patches, notably in the Manzibomvu valley where basalt occurs.

2.4 Climate and Hydrology

HiP falls within the summer rainfall region of South Africa, but is subject to a coastal influence. Annual rainfall has averaged 968 mm ± 36 SE

Table 2.1B *Features of a dolerite-derived oxisol sampled near Hilltop camp in Hluhluwe (from Cramer and Bond, 2013)*

Feature	Unit	Measure
pH		4.94 ± 0.11
C	%	2.8 ± 0.2
Total N	%	0.19 ± 0.01
Total P	$mg\ kg^{-1}$	151 ± 7
Available P	$mg\ kg^{-1}$	10.9 ± 2
K	$mg\ kg^{-1}$	173 ± 15
Cu	$mg\ kg^{-1}$	4.16 ± 0.16
Zn	$mg\ kg^{-1}$	1.43 ± 0.16
Mn	$mg\ kg^{-1}$	17 ± 4
B	$mg\ kg^{-1}$	0.31 ± 0.02
Fe	$mg\ kg^{-1}$	165 ± 15
Exchangeable cations		
Na	$cmol\ H^+\ kg$	0.37 ± 0.02
K	$cmol\ H^+\ kg$	0.44 ± 0.04
Ca	$cmol\ H^+\ kg$	39.2 ± 0.9
Mg	$cmol\ H^+\ kg$	4.1 ± 0.2
$\Delta^{15}N$	‰	5.9 ± 0.1

(calculated from monthly rainfall collected 1933–2012) at Hilltop Tourist Camp in northern Hluhluwe at an altitude of 436 m, and 691 mm ± 30 SE (from monthly rainfall collected 1960–2012) at Mpila Tourist Camp in Mfolozi at an altitude of 261 m. However, local rainfall is strongly dependent on elevation (Balfour and Howison, 2001). Spatial extrapolation between various recording stations indicates a rainfall range from around 1000 mm on the highest hills in the north to less than 550 mm in low-lying regions of Mfolozi (Figure 2.10C). The uplands in the far north, the middle of the park, and the far south produce local rainfall shadows that make the valleys between them much drier than expected from the coastal location. Approximately 75% of the annual total falls during the six summer months (October–March) and 25% during the dry winter months (April–September; Figure 2.11). Nevertheless, none of the dry season months receives less than 20 mm of rainfall on average. Much of the summer rainfall is associated with frontal systems drifting north-eastwards from the Western Cape region, drawing in moist air from the Indian Ocean to the east. Rainfall patterns over eastern South Africa have apparently exhibited a cyclic component with a period of approximately 18 years during the past century (Nevill, 1908; Tyson and Dyer,

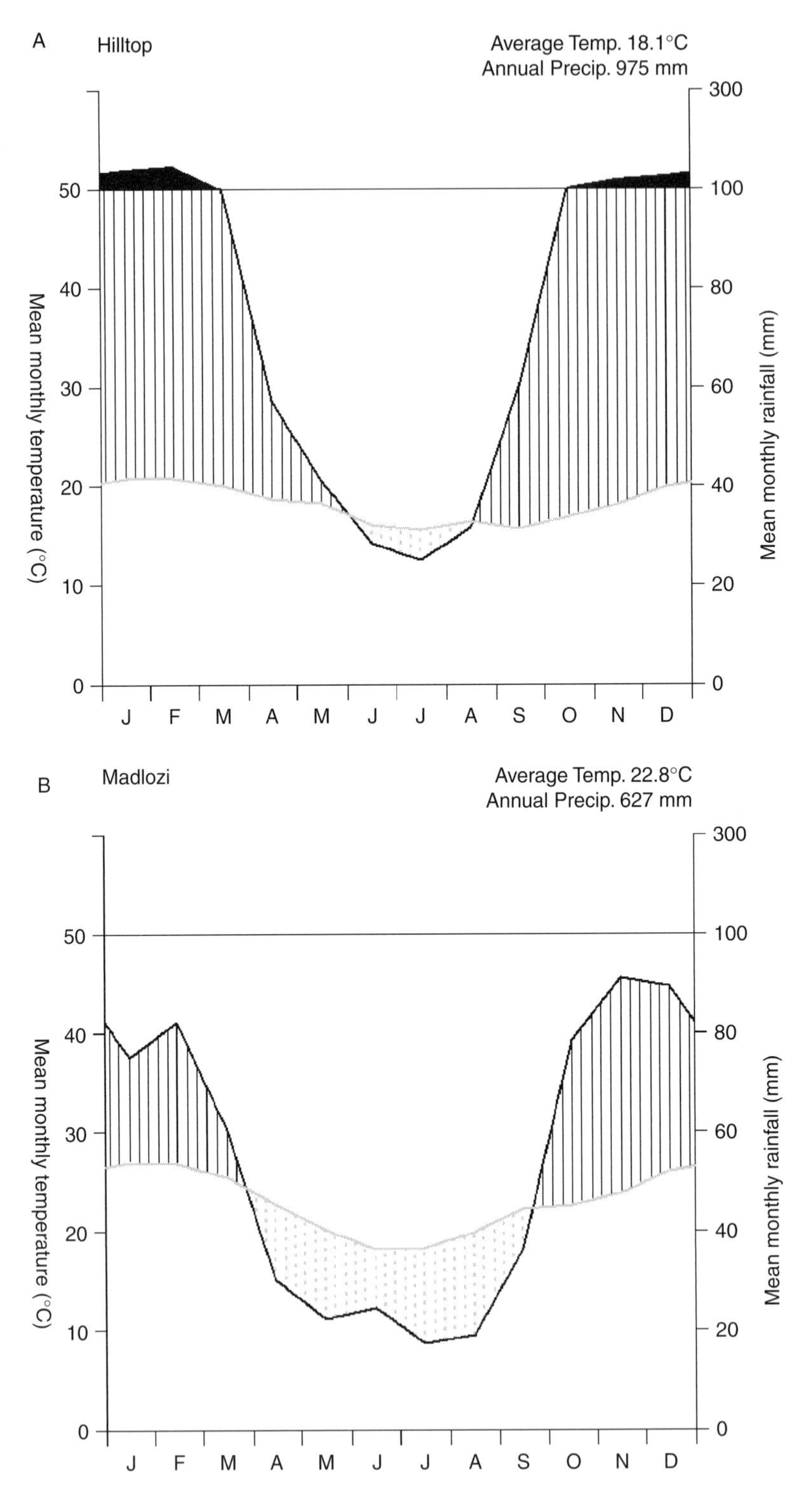

Figure 2.11 Walter climate diagrams showing seasonal rainfall and temperature variation for (A) the northern Hluhluwe region of the park (Hilltop), and (B) the south-western Mfolozi region (Madlozi).

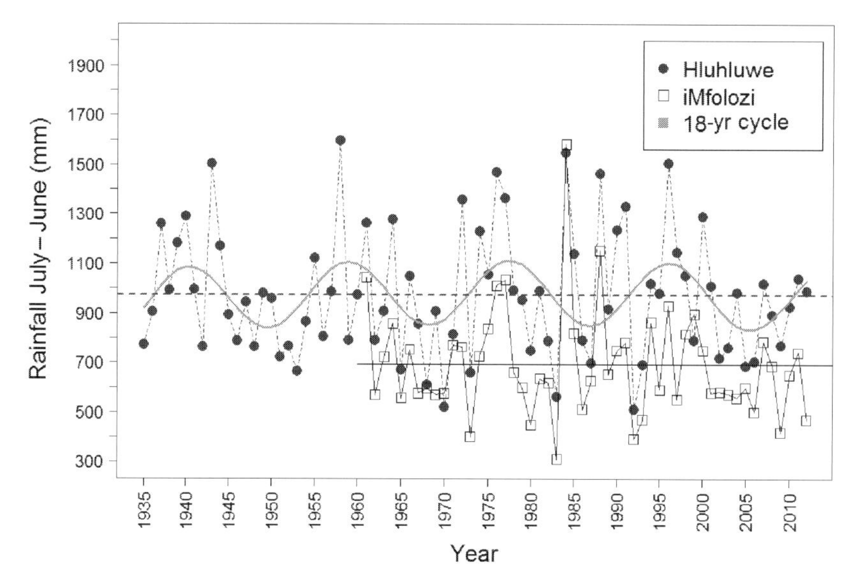

Figure 2.12 Long-term variation in annual rainfall total (July–June) recorded at Hilltop (Hluhluwe) and Mpila (Mfolozi) camps in comparison with mean values and the phase of a fitted 18-year oscillation in rainfall.

1980). Rainfall within HiP seemed to be consistent with this pattern prior to 1984, and again after 1994 (Figure 2.12). The contribution of the El Niño Southern Oscillation (ENSO), generated by temperature anomalies in the Pacific Ocean, to this variation in rainfall is not consistent, and additional influences from the Indian Ocean seem to operate (Nicholson and Kim, 1997). The highest annual rainfall total exceeded 1500 mm at both Hilltop and Mpila, associated with cyclone Demoina in February 1984. Other notable floods associated with intense rainfall occurred in July 1962 and November 2000. Extreme droughts were experienced during 1968/69, 1972/73, 1982/83 and 1991/92. Periods of persistently low rainfall occurred through 1945–1953, 1979–1983, and 2001–2012.

At Hilltop Camp, temperatures were recorded between 1973 and 1995. During this period, the maximum daily temperature averaged 24.8°C ± 0.25 SE in January and the minimum 11.7°C ± 0.24 SE in September. The highest temperature recorded was 34.6°C in May 1986 and the lowest 3.8°C in September 1994. There has been no discernible trend in temperatures over this ~20-year period covered by the data from Hilltop. For lower-lying western Mfolozi, temperature records were only kept over the period 1969–1971 (N. Owen-Smith, unpublished). The mean daily maximum temperature was 32.8°C ± 0.8 SE in January and the mean minimum 10.7°C ± 1.0 SE in July. In Mfolozi, the highest maximum temperature during 1969–1971 was 43.9°C in October 1969, and the lowest minimum temperature was 6.7°C recorded in both June and July 1971. A Walter diagram of relationships between temperature and rainfall shows that the Hluhluwe region (Figure 2.11A) is seldom in water deficit, whereas the Mfolozi region (Figure 2.11B) shows a more 'savanna-type' period of water deficit.

The two Mfolozi rivers and the Hluhluwe river were formerly perennial, but all three have recently stopped flowing occasionally in the dry season during very dry years, although pools remain in their channels. Small lakes adjoining the White Mfolozi river retain water year-round in most years. A dam constructed across the Hluhluwe river just outside the park impounds water back into the park for about 2 km. A few springs located between the two Mfolozi rivers ooze trickles of somewhat saline water. In addition, numerous pan depressions retain water for varying periods into the dry season. Three pans with hides for tourists constructed alongside them have had water pumped into them for varying periods. The furthest distance from perennial surface water is about 10 km in the west of Mfolozi. Cultivation, creation of boreholes, and heavy stocking with domestic livestock in upstream catchments outside the park have contributed to reduced water quality in the rivers passing through the park.

2.5 Vegetation

Savannas are defined as systems with a continuous layer of C4 grasses and discontinuous tree layer (Mucina and Rutherford, 2006), but tree cover may vary locally from sparse to almost closed-canopy woodland, depending on soil moisture conditions and consumer impacts of fire and herbivory. Moreover, a basic distinction has been recognized between mesic/dystrophic savannas characterized by broadleaved trees predominantly in the Caesalpinioideae, and arid/eutrophic savannas characterized by trees with fine leaves mainly representing the Mimosoideae (Huntley and Walker, 1982). Below around 650 mm rainfall, available moisture generally restricts woody canopy cover, while above this threshold fire and herbivory become more important in preventing canopy closure (Sankaran *et al.*, 2005). Around 1000 mm in annual rainfall, African savanna vegetation grades into either more closed woodland, or a mosaic of forest and grassland dependent on position in the landscape.

HiP is located in the transitional region between the predominantly savanna vegetation in the coastal lowlands to the east and open grassland plus patches of scarp forest in the upland plateau to the north-west. The prevalent vegetation types within this region were labelled Zululand Thorn Veld and Tropical Bush and Savanna by Acocks (1988), and Zululand Lowveld and Zululand Sourveld by Mucina and Rutherford (2006). At a finer scale, 20 vegetation communities were originally distinguished within HiP from patterns shown in aerial photographs, supported by ground surveys of woody species associations (Whateley and Porter, 1983; Table 2.2). For our mapping, these communities have been grouped into five functionally distinct vegetation types (Figure 2.10D) which are strongly associated with the landscape zones shown in Figure 2.2. However, there is evidence of substantial change in the extent of some of these formations, reflecting the dynamic nature of these mixed tree–grass systems (Chapter 3). Chapters 7 and 8 address the mechanisms involved further.

Most widespread in HiP are various forms of fine-leaved thorn savanna, with specific communities distinguished by predominance of different species of *Acacia* (see also Chapter 8). Knob thorn (*A. nigrescens*) is typically associated with clay soils derived from dolerite; red thorn (*A. gerrardii*) with clay soils derived from shale; and monkey thorn (*A. burkei*) with sandy loam soils on sandstone. Umbrella thorn (*A. tortilis*) and scented thorn (*A. nilotica*) occur more widely, with *A. tortilis* most common in the semi-arid, herbivore-dominated Mfolozi section of the park (Chapter 8). Sweet thorn (*A. karroo*) and sickle bush (*Dichrostachys cinerea*) have become abundant in Hluhluwe and in the central Corridor section of the park and form dense stands replacing previously open savanna grassland (see section on bush encroachment in Chapter 7). Whateley and Porter (1983) termed these encroached savanna patches 'induced thicket' (see Table 2.2). Encroached savanna dominated by *A. caffra* and *A. davyi* was likewise labelled 'thicket' due to their dense woody state (Table 2.2). More recently, it has been argued that the term thicket should be reserved for distinct vegetation types of broadleaved plants that, in contrast to savanna, lack a grassy component and have very different functional characteristics (see Mucina and Rutherford, 2006 and Chapter 8). Throughout HiP, one can find such dense vegetation patches in catchments of perennial and seasonal rivers and elsewhere in bottomlands dominated in particular by tamboti (*Spirostachys africana*), common guarrie (*Euclea divinorum*) and sea guarrie (*E. racemosa*). Although categorized as (savanna) woodland types by Whateley and Porter (1983), recent work on the

Table 2.2 *Whateley and Porter's (1983) classification of the vegetation communities of Hluhluwe-iMfolozi Park grouped according to six physiognomically distinct habitat types and proportional contribution to landscape position*

Vegetation community	Description	Proportion of HiP	Habitat type	Combined proportion of HiP (%)
Acacia nigrescens woodland	Found on hillsides with a dolerite substrate	12.3	Fine-leaved savanna	55.81
Acacia tortilis woodland	Occurs throughout Mfolozi, usually on east-facing slopes	10.3	Fine-leaved savanna	
Acacia gerrardii woodland	Covers gently undulating areas on shale	9.8	Fine-leaved savanna	
Acacia karroo–Dichrostachys cinerea induced thicket	Found on steeply undulating hillsides particularly in the Corridor and Hluhluwe	8.4	Fine-leaved savanna	
Acacia burkei woodland	Occurs both on moderately steep hillsides and flat riverine terraces	5.4	Fine-leaved savanna	
Acacia karroo woodland	Spread throughout NE Hluhluwe especially in river valleys	3.2	Fine-leaved savanna	
Acacia nilotica woodland	Occurs mostly south of Hluhluwe river	3.0	Fine-leaved savanna	
Acacia caffra thicket	Dense encroached savanna found on hillsides and ridges throughout HiP	3.4	Fine-leaved savanna	
Acacia davyi thicket	Dense encroached savanna. Covers only a few hectares in Hluhluwe	0.01	Fine-leaved savanna	

Table 2.2 (cont.)

Vegetation community	Description	Proportion of HiP	Habitat type	Combined proportion of HiP (%)
Combretum molle woodland	Occurs mainly north of Hluhluwe river on steep rocky slopes	0.5	Broad-leaved savanna	0.7
Combretum apiculatum woodland	Restricted to rocky hillsides in Mfolozi	0.2	Broad-leaved savanna	
Celtis africana–Harpephyllum caffrum forest	Occurs on high hills in NW Hluhluwe	5.1	Forest	9.7
Ficus sycamorus–Schotia brachypetala forest	Riverine community confined to banks of larger rivers and their major tributaries	4.6	Forest	
Spirostachys africana woodland	Covers bottomlands in Mfolozi, particularly in catchments of seasonal watercourses	27.0	Thicket	32.6
Euclea divinorum woodland	Occurs on gentle sloping ground in Hluhluwe river valley	1.7	Thicket	
Celtis africana–Euclea racemosa forest	Found in lower altitudes under drier conditions	0.3	Thicket	
Spirostachys africana–Euclea racemosa riverine forest	Occurs as a narrow strip along seasonal water courses	3.6	Thicket	
Themeda triandra grassland	Limited to tops of high hills and ridges	1.4	Grassland	1.42
Sporobulus africanus–Cyperus textilis sedge grassland	Present in valley bottoms in northern Hluhluwe	0.02	Grassland	

functional characteristics of these vegetation states suggests that they should be allied with the thicket biome (see Chapter 8). Accordingly, two distinct 'thicket' types have been recognized in HiP: broadleaf thicket (typically *Spirostachys africana–Euclea racemosa–A. grandicornuta* associations), and seral 'thicket' (which is a form of encroached savanna typified by *Dichrostachys cinerea*, *A. karroo* and *A. caffra*). The different application of the term thicket has led to various interpretations of the map developed by Whateley and Porter (1983), which complicates assessments of past vegetation change (see Chapter 3 and Chapter 8 for further discussion). In this chapter, and for most of the book, we have reserved the term thicket for vegetation patches that represent the broadleaf thicket biome (see division in six distinct habitat types in Table 2.2 and Figure 2.10D).

The dystrophic component of savannas (Huntley and Walker, 1982) is not widespread in HiP due to the predominance of clay-rich soils. However, on shallow rocky or sandy soils, broadleaf wooded savanna characterized by bush willows (*Combretum* spp.) occurs. Large marula (*Sclerocarya birrea*) trees are present quite widely in both the fine-leaved and broadleaved savanna. Adjoining the two Mfolozi rivers and parts of the Hluhluwe river is a narrow riparian fringe typified by sycamore fig (*Ficus sycamorus*) and winter thorn (*Acacia robusta*) trees plus local patches of *S. africana* woodland. Nestled on south-facing hillslopes at higher elevations in Hluhluwe are upland scarp forests (Figure 2.4) typified by white stinkwood (*Celtis africana*) and African wild plum (*Harpephyllum caffrum*) trees. Open grassland is restricted to the crests of the highest hills and uplands (Figure 2.5).

In the herbaceous layer, the main division is between areas dominated by red grass (*Themeda triandra*) on clay soils and those supporting a mixture of love grass (*Eragrostis superba* and *Eragrostis curvula*) and dropseed (*Sporobolus pyramidalis* and *S. africanus*) on more sandy soils. Interspersed amid the bunch grasslands are grazing lawns typified by low-growing, commonly stoloniferous grasses due to grazing and soil compaction (see Chapter 6). Guinea grass (*Panicum maximum*) occurs under the shade canopy of trees. In woodlands more shade-tolerant grasses prevail – needle grass (*Enteropogon monostachyus*) in Mfolozi and LM grass (*Dactyloctenium australe*) in Hluhluwe.

2.6 Inter-Relationships and Comparisons

The notable features of the landscape heterogeneity within HiP are (1) the varied topography; (2) the diverse geology, but with substrates yielding clay-rich and hence relatively fertile soils predominating; (3) the spatially compressed rainfall gradient spanning the range from that supporting semi-arid savanna to the transition into forest; (4) the diverse interspersion of woody vegetation communities, but with fine-leaved acacia trees associated with eutrophic conditions most prevalent; and (5) the predominance of palatable grasses such as *Themeda triandra*. A cross-sectional north–south transect of the park (Figure 2.13) depicts how these different features are inter-related.

Despite lying outside the tropics, HiP shares some notable affinities with Tanzania's Serengeti ecosystem (Sinclair and Norton-Griffiths, 1979). As in Serengeti, the predominant vegetation within HiP is fine-leaved thorn savanna with iconic umbrella thorn trees prominent, grading into open grassland with forest patches in the wetter north, with red grass widespread in the grass layer. However, within HiP, the range in annual rainfall from under 600 mm to around 1000 mm is compressed within 30 km, while in Serengeti it is spread over 100 km (Sinclair and Norton-Griffiths, 1979). Rainfall is concentrated within a single summer season within HiP, whereas in Serengeti, rainfall tends to be partitioned between early short rains and later more prolonged rains following the passage of the Intertropical Convergence Zone. In both areas, topographically related rainfall shadows are responsible for strong gradients in precipitation that are steeper than those governed by regional atmospheric circulation differences. The fertility of soils within the Serengeti region is derived from very recent volcanic layers of soda ash overlying Archaean igneous and metamorphic rocks due to the presence of still active volcanos nearby, while in HiP soil fertility is influenced by older volcanic intrusions of dolerite that were feeders to more ancient flood basalts, along with clay soils derived from fine-grained sediments. Serengeti is situated within an interior plateau region, meaning that gradients in soils and associated vegetation are prolonged and typically follow a catenary sequence from shallower soils on drier uplands to deeper alluvial soils in moister bottomlands. HiP presents a more complex, finer-scale mosaic of geology, soils and vegetation because of its dissected topography in the transitional region between the coastal and interior plateaus. Because of its lower altitude, summer temperatures within HiP are somewhat hotter than the prevalent range in Serengeti, intensifying evaporative losses of soil moisture. There are only local forest patches in northern Serengeti developed where there is protection from fire, but closed-canopy thickets – analogous to HiP's broadleaved thickets – are prominent there on dry hillslopes. Serengeti's open short-grass plain underlain by volcanic ash,

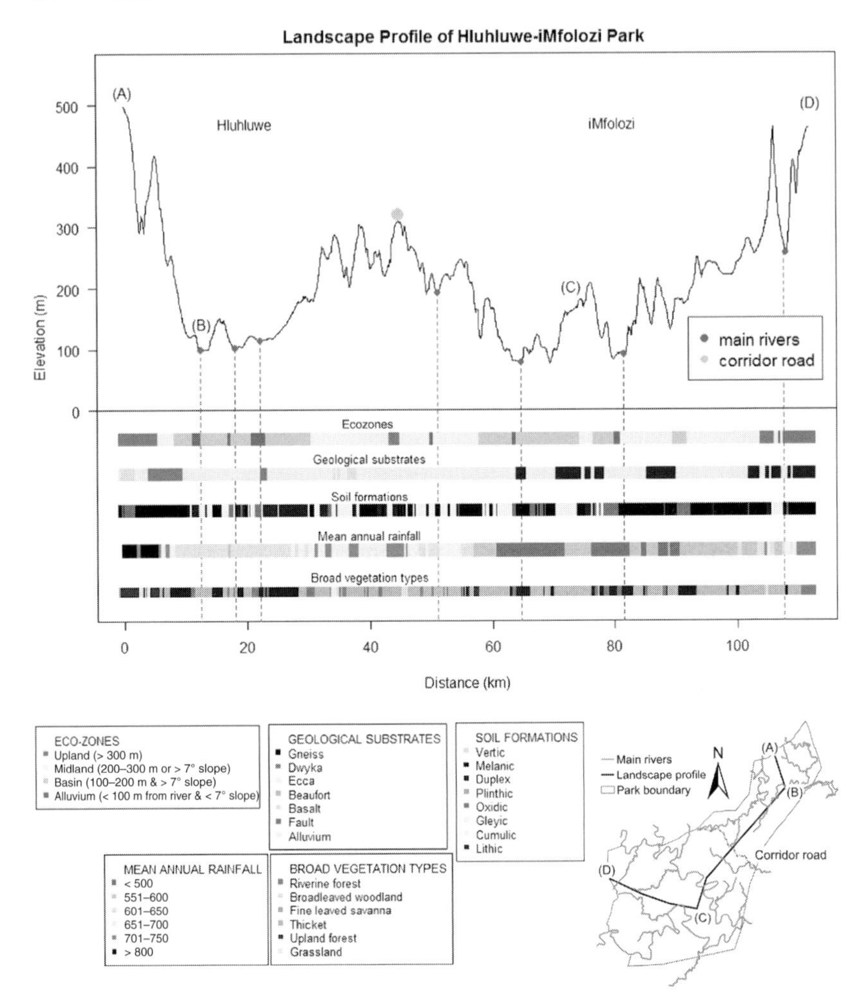

Figure 2.13 Landscape profile of Hluhluwe-iMfolozi Park, showing the general undulating profile from high elevation in the north through the basin of the Hluhluwe river, rising again through the central Corridor and through to the undulating basin of the Black and White Mfolozi rivers, rising again in the south to an elevation almost equal to that of the upland forests in the north. The five coloured bars beneath the graph show the proportion of the landscape within the four ecozones and the different geological substrates, soil formations, mean annual rainfall regimes, and the broad vegetation types. (For the colour version, please refer to the plate section. In some formats this figure will only appear in black and white.)

which draws enormous concentrations of migratory grazers during the wet season, has no counterpart within HiP. Overall, HiP contains a greater diversity of land forms, geology, soils, and vegetation within its 950 km^2 area than is represented within the 25,000 km^2 extent of the Serengeti ecosystem. This makes it an exceptionally productive place to study the key drivers of savanna ecosystem structure and functioning within a relatively compact geographic area.

The other African savanna ecosystem that is particularly well known because of its long history of research is South Africa's Kruger National Park (du Toit *et al.*, 2003). Its 20,000 km^2 area is situated in the gently undulating region between a high interior escarpment and the Lebombo range of hills separating it from Mozambique's coastal plain. The Lebombo hills, formed by rhyolite extrusions associated with the flood basalts, continue into northern KwaZulu-Natal to the east of HiP. The basaltic plain bordering these hills on the west extends from the Kruger Park through Swaziland into northern KwaZulu-Natal, but only marginally enters HiP. Because of the tilting of strata, Karoo sediments form only a narrow band running through central Kruger Park separating the basalt from the granitic gneiss in the west (Venter, 1986). Kruger's altitude is similar to that of HiP, ranging from 250 m in the north-east to 800 m in the south-west, and while the Mfolozi region of HiP is similarly hot, northern Hluhluwe is much cooler. Kruger Park is also somewhat drier than HiP, with a rainfall range from 750 mm year^{-1} in the south-west to 400 mm year^{-1} in the north extending over a distance of 250 km, and wet season rains commence only in October or November. Mixed thornveld including umbrella thorn and scented thorn similar in its features to the thorn savanna prevalent in HiP is restricted within Kruger largely to gabbro sills intruding within the granite (Venter *et al.*, 2003). Knob thorn–marula parkland, associated with doleritic soils in HiP, occurs widely in Kruger Park on the south-eastern basalts. On less-fertile sandy soils formed from granitic gneiss in the south-west, the predominant vegetation is broadleaf savanna typified by species of *Combretum* and *Terminalia*. In the hot and dry northern half of Kruger Park, broadleaved mopane trees (*Colophospermum mopane*) prevail on all soil substrates.

The template in geology, soils and vegetation within HiP described in this chapter is greatly enhanced by additional heterogeneity imposed by the effects of fire (Chapter 10), grazing (Chapter 6), and invertebrate activity (termites, dung beetles; Chapter 9) on vegetation and soils. The relatively fertile soils and hence comparatively nutritious grasses and trees contribute to the abundance of grazing and browsing herbivores and the high herbivore biomass levels attained within HiP, despite the lack of the migratory opportunities underlying large herbivore concentrations in Serengeti and elsewhere.

2.7 References

Acocks, J. P. H. (1988) *Veld types of South Africa*. Botanical Research Institute, South Africa.

Balfour, D. A. & Howison, O. E. (2001) Spatial and temporal variation in a mesic savanna fire regime: responses to variation in annual rainfall. *African Journal of Range & Forage Science* **19**: 43–51.

Cramer, M. D. & Bond, W. J. (2013) N-fertilization does not alleviate grass competition induced reduction of growth of African savanna species. *Plant and Soil* **366**: 563–574.

Downing, B. H. (1980) Relationships between rock substrate, landform and soil in Umfolozi Game Reserve, as explained for conservation purposes. *Lammergeyer* **30**: 32–48.

du Toit, J. T., Rogers, K. F., & Biggs, H. C. (2003) *The Kruger experience: ecology and management of savanna heterogeneity*. Island Press, Washington.

Fey, M. (2010) *Soils of South Africa*. Cambridge University Press, Cambridge.

Huntley, B. J. & Walker, B. H. (1982) *Ecology of tropical savannas*. Springer-Verlag, Berlin.

King, L. (1982) *The Natal monocline: explaining the origin and scenery of Natal, South Africa*. University of Natal Press, Pietermaritzburg.

Mucina, L. & Rutherford, M. C. (2006) *The vegetation of South Africa, Lesotho and Swaziland*. South African National Biodiversity Institute, Pretoria.

Nevill, E. M. (1908) The rainfall in Natal. *Agricultural Journal of Natal* **11**: 1531–1533.

Nicholson, S. E. & Kim, J. (1997) The relationship of the El Nino–Southern oscillation to African rainfall. *International Journal of Climatology* **17**: 117–135.

Partridge, T. C. (1998) Of diamonds, dinosaurs and diastrophism; 150 million years of landscape evolution in southern Africa. *South African Journal of Geology* **101**: 167–184.

Sankaran, M., Hanan, N. P., Scholes, R. J., *et al.* (2005) Determinants of woody cover in savannas. *Nature* **438**: 846–849.

SCWG (1991) *Soil classification: a taxonomic system for South Africa*. SIRI: Department of Agricultural Development, Pretoria.

Sinclair, A. R. E. & Norton-Griffiths, M. (1979) *Serengeti, dynamics of an ecosystem*. The University of Chicago Press, Chicago.

Tyson, P. D. & Dyer, T. G. J. (1980) The likelihood of droughts in the 80s in South Africa. *South African Journal of Science* **76**: 340–341.

Venter, F. J. (1986) Soil patterns associated with the major geological units of the Kruger National Park. *Koedoe* **29**: 125–138.

Venter, F. J., Scholes, R. J., & Eckhardt, H. (2003) The abiotic template and its associated vegetation pattern. In: *The Kruger experience: ecology and management of savanna heterogeneity* (eds. J. T. du Toit, K. F. Rogers, & H. C. Biggs), pp. 83–129. Island Press, Washington.

Whateley, A. & Porter, R. N. (1983) The woody vegetation communities of the Hluhluwe–Corridor–Umfolozi Game Reserve Complex. *Bothalia* **14**: 745–758.

3 · *Long-Term Vegetation Dynamics within the Hluhluwe iMfolozi Park*

A. CARLA STAVER, HEATH BECKETT, AND JAN A. GRAF

3.1 Introduction

Long-term vegetation dynamics present a long-standing challenge for ecology. Historical and paleo-analyses of vegetation change can provide ecological insights into the stability and determinants of vegetation dynamics that are lacking from analyses of extant vegetation patterns (Prins and van der Jeugd, 1993; Willis *et al.*, 1999; Rees, 2001; Jeffers *et al.*, 2011). Long-term data on pre-industrial/colonial ecosystem processes can also help guide management decision making by informing desired and potential outcomes of management actions (Gillson and Duffin, 2007; Gillson, 2008; Willis *et al.*, 2010). In this chapter, we consolidate the available evidence of vegetation transformations in HiP and of its relation with drivers of vegetation change. For the long-term dynamics we rely on early historical descriptions of vegetation, anecdotes of animal sightings/hunting, fixed-point and aerial vegetation photographs, and paleo-analyses of soil carbon. We evaluate whether changes in vegetation in HiP have been widespread and generalized, whether they are local in response to heterogeneous human, wildlife, or fire impacts, and, to the extent possible, whether vegetation has been temporally stable or dynamic.

3.2 General Models of Long-Term Vegetation Dynamics in Savannas

Both bottom-up (water and soil nutrients) and top-down processes (fire and herbivory) shape vegetation dynamics in savanna (Scholes and Archer,

Conserving Africa's Mega-Diversity in the Anthropocene, ed. Joris P. G. M. Cromsigt, Sally Archibald and Norman Owen-Smith. Published by Cambridge University Press.

1997). However, there remains uncharacteristic disagreement in the ecological literature about the temporal dynamics that characterize savanna vegetation. The implications for the management of savanna ecosystems are fundamental, because dynamics determine the extent to which management interventions can actually impact vegetation change. We identify four models that describe temporal vegetation dynamics and its drivers in savannas.

The classic model for biome dynamics and biome distributions globally is that climate and soils determine vegetation structure (Holdridge, 1947; Whittaker, 1975). This model is clearly useful to some extent. Tree cover tends generally to increase with increasing rainfall (Staver *et al.*, 2011a,b), decreasing rainfall seasonality (Staver *et al.*, 2011b), and decreasing rainfall episodicity (Good and Caylor, 2011). Meanwhile, the impacts of soils on tree cover, while difficult to generalize, can clearly be substantial (Williams *et al.*, 1996). Explaining tree–grass coexistence in this classic model – wherein mean climate and soils determine biome distributions – presents a challenge; the oldest explanation, termed the Walter hypothesis, invokes strong separation in rooting niche, and therefore water and nutrient use, between trees and grasses to explain coexistence (Walter, 1971; Walker *et al.*, 1981). Recent experimental work has shown that trees and grasses do compete for water and nutrients, but such root–niche separation in trees and grasses is probably insufficient for equilibrium tree–grass coexistence (Cramer *et al.*, 2007; Riginos, 2009; February *et al.*, 2013). Even early formalizations of this classic biome-distribution model concede that climate and soils are probably not perfectly predictive of vegetation structure (Whittaker, 1975), and evidence for drivers other than climate and soils in determining savanna structure and distribution is extensive.

Many, even most, savannas have fewer trees than local climate could potentially support (Bond *et al.*, 2005; Sankaran *et al.*, 2005; Staver *et al.*, 2011a,b), especially where mean annual rainfall exceeds 650–1000 mm. Much of this work is not explicit about vegetation dynamics in response to disturbance, however. Are savannas characterized by succession towards a climate-defined climax, where tree–grass coexistence is transient? The frequent conclusion that disturbances force savannas away from a climate equilibrium (Scholes and Walker, 1993; Sankaran *et al.*, 2005) suggests that they could be unstable and transient, even successional, systems that would tend towards equilibrium if undisturbed by fire or disturbances.

Alternatively, ecosystem dynamics in savanna may not be transient and successional at all. The transition from savanna to forest is abrupt, and vegetation types intermediate to savanna and forest are not widespread

(Hirota *et al.*, 2011; Staver *et al.*, 2011a,b). Positive feedbacks with fire maintain savanna as a stable alternative to forest (Bond *et al.*, 2005; Staver *et al.*, 2011b); abundant grass in savanna promotes fire spread, which in turn prevents tree establishment and favours grass. This feedback potentially stabilizes savanna dynamics and prevents continuous, autogenic succession towards forest (Beckage and Ellingwood, 2008; Staver *et al.*, 2011a). Biogeographic patterns of savanna tree evolution and functional traits tend to support this view that savanna represents a stable and qualitatively different biome from forest. Savanna trees are characterized by functional traits that are distinct from forest tree functional traits, especially with respect to fire versus shade tolerance (Hoffmann *et al.*, 2003, 2009), and phylogenetic evidence suggests that, in South America at least, savanna trees probably diversified within stable savannas, rather than dispersing from other fire-prone biomes (Simon *et al.*, 2009). This modern evidence does tend to suggest that savanna and forest distributions are subject to alternative stable-state dynamics, but good long-term data are mostly lacking. Savanna and forest may represent alternative stable states only on short time scales, but slow time-scale interactions at the boundary between savanna and forest – such as rare forest fires or forest expansion via dispersal – may play a fundamental role in determining biome distributions. In the modern context, some authors have argued that these have only become important during the last century or so, as climate and land-use change have radically changed fire management (Nepstad *et al.*, 2004; Silva *et al.*, 2008; Brando *et al.*, 2011).

Some work in savannas supports an alternative fourth class of model for savanna dynamics, which explicitly incorporates variability in drivers of savanna vegetation structure to explain tree–grass coexistence. Generally, temporal variability, in which some periods favour populations of one type and other periods another, promotes coexistence, especially of long-lived organisms like trees and grasses (Hutchinson, 1961; Chesson and Warner, 1981). Indeed, savannas often occur in temporally variable areas (Scholes and Archer, 1997; Sankaran *et al.*, 2004), and functional coexistence may result from variability in climate (Jeltsch *et al.*, 2000; Scanlon *et al.*, 2005; Good and Caylor, 2011) or from stochastic fire events (Higgins *et al.*, 2000). Rare, stochastic events like floods or ungulate disease outbreaks can also have catastrophic, long-lasting effects on vegetation (Prins and van der Jeugd, 1993; Gillson, 2006), which are challenging to study on short time scales. Where variability drives tree–grass coexistence, coexistence may be theoretically stable, but vegetation structure can nonetheless fluctuate widely as drivers do.

These four models for ecosystem dynamics in savanna generate distinct predictions for the long-term dynamics of vegetation within savanna and at the savanna–forest boundary. The first predicts that vegetation should change only as climate does (clearly poorly supported). The second predicts that vegetation should tend towards some climate-determined climax vegetation, except following disturbances; in a long-term vegetation record, this might manifest as substantial variability in woody cover through time, potentially characterized by alternating slow increases and rapid decreases in woody cover. The third predicts that vegetation should be relatively stable through time, punctuated by rapid changes during transitions among stable states. Finally, the fourth again suggests that vegetation is temporally variable, although not trending towards any obvious climax.

These predictions have not received much direct evaluation, in part because data are scarce and difficult to evaluate quantitatively. However, even a qualitative treatment of temporal vegetation change – in this case in Hluhluwe-iMfolozi Park (HiP) – may provide insights into vegetation dynamics within savanna, as well as stability dynamics at the boundary between savanna and forest.

3.3 A Long-Term Vegetation History of Hluhluwe-iMfolozi Park

HiP is climatically, edaphically, topographically, and even biogeographically dynamic, and includes an area where savanna and forest co-occur. Rainfall varies broadly with elevation, from about 975 mm mean annual rainfall (MAR) in the north (elevation 580 m), characterized by savanna–forest and savanna–thicket mosaic, to < 600 mm MAR in the south (elevation 50 m), dominated by savanna (Balfour and Howison, 2002; see Chapter 2 for more about the HiP landscape). Currently, HiP's forests border both grassland and mesic savanna; given the lack of a clear distinction between grassy biomes, we henceforth refer to these as 'savannas'.

The area has likely been subject to significant climatic variability during the Holocene and Pleistocene. Climatic changes in the savanna regions of South Africa over the past ~200,000 years have been driven by a combination of variability in the Earth's axial precession, which affects insolation on a ~23 ky cycle, in the southern displacement of the intertropical convergence zone, and in oceanic circulation that affects the formation of weather systems (Kristen *et al.*, 2007). More locally, recent paleoclimatic work (Baker *et al.*, 2014) from nearby Mfabeni peatland in iSimangaliso Wetland Park has identified five distinct climatic stages during the past ~50,000 years: (1) a cool and wet period lasting from 47 to 32.2 kya transitioned to (2) a dry and windy period from 32.2 to 27.6 kya, which was followed by (3) another cool and wet period from 27.6 to 20.3 kya. This was punctuated by an abrupt transition to (4) dry and cool conditions, which lasted until ~15 kya, when precipitation began to gradually increase to (5) the current climate regime, dating back 10,400 years, which has been relatively wet but has been characterized by variable temperatures.

Concurrent with substantial changes in climate, the area in and around HiP has a long history of human occupation (see Chapter 1 for more detail), resulting in direct effects on long-term vegetation dynamics and on the ecological processes that determine vegetation structure, including herbivory and fire. The first evidence of sedentary agriculture in the HiP region dates from approximately 200–300 AD, when the first Bantu-speaking peoples immigrated to the area at the beginning of the Iron Age (Hall, 1978, 1987), clearing land for cereal farming and cattle, sheep, and goat pasture (van der Merwe, 1980; Hall, 1987; Maggs and Whitelaw, 1991), and for smelting iron, which relied on fuelwood harvest and may have decreased woody cover (Goucher, 1981; Hall, 1984; Feely, 1985). Around 1000 AD, an abrupt break in the nature and content of archaeological sites (Maggs and Whitelaw, 1991) suggests that settlements may have shrunk significantly in the area (Maggs, 1980). During the last 200 years, HiP has experienced a long series of major events, many of anthropogenic origin, that have driven ecological and vegetation change (Table 3.1).

Sociopolitical consolidation of the fragmented and culturally diverse Nguni peoples into the Zulu Kingdom began during the 1700s and concluded between 1818 and 1828, during the reign of Shaka kaSenzangakhona. Hunting pits near the confluence of the White and Black iMfolozi rivers suggest that this area was used intensively, possibly by royal directive (but see Brooks and Macdonald, 1983). Emerging infectious diseases (Lincoln, 1995), including *nagana*, transmitted from wild ungulates to cattle by tsetse flies, and malaria affecting humans (Foster, 1955; Hall, 1978), likely kept the HiP area sparsely inhabited throughout the late eighteenth and early nineteenth centuries. Both tsetse flies and *nagana* persisted into the twentieth century in HiP (Henkel, 1937), and the early history of the HiP area is characterized by tension between highly politicized attempts to eradicate the disease versus efforts to protect wildlife (Lincoln, 1995; Chapter 1). Hluhluwe and Umfolozi Game

Table 3.1 *Known major event-driven ecological and vegetation change in HiP*

Date	Event	Ecological impacts
1830–1900	Intensive hunting of white and black rhino, elephant, buffalo, and other herbivores	Reduction in herbivore populations (Chapter 4). Evidence variously of tree establishment events (Prins and van der Jeugd, 1993) and increases in fire frequency (Holdo *et al.*, 2009) from East Africa; effects in HiP unclear (Staver *et al.*, 2011c). Possible decrease in the extent of grazing lawns (Bond *et al.*, 2001; Waldram *et al.*, 2008; Chapter 6)
1890s	Rinderpest epidemic	
1920s–1950s	Major herbivore culling for *nagana* control	
1930s+	Road construction	Fragmentation results in local changes in fire regimes, soil moisture and drainage; establishment of forest/thicket patches
1940s	Bush clearing along western iMfolozi boundary to contain *nagana*	Effects on tsetse populations and *nagana* prevalence inconclusive (Chapter 13), Transformation of woody savanna and thicket into open 'grassland' (see Figure 3.1B)
1942–1946	Bush clearing along eastern/northern Hluhluwe boundary to contain *nagana*	
1947–1952	DDT and BHC aerial spraying (see Figure 3.1A)	Tsetse (and other invertebrate?) extinction (Chapter 13). Cascading effects unknown but likely included high mortality of invertebrates and, possibly, small vertebrates, leading to reduced seed predation and increased woody plant establishment (Chapter 9)
1940s	People and cattle removed from Corridor	Possible recolonization of kraal sites by savanna trees and shrubs. However, old kraals also turned into grazing lawns (see Chapter 6)
1960s–1977	Fence construction	Restriction of animal movement between interior and exterior of park
1967	Hluhluwe dam construction	Downstream from park; impacts on riverine forest and associated vegetation
1918, 1925, 1957, 1963, 1984, and 1987	Exceptional floods (Tropical Cyclone Domoina in 1984)	Removal of riverine forest, changes in river geomorphology, and possible expansion/contraction of *Phragmites* reedbeds
1990s and 2000s	Large-scale invasion by South American shrub *Chromolaena odorata*, especially in Hluhluwe	Potential negative effects on native flora and fauna (Chapter 15). Clearing programmes have led to successful control of *C. odorata*, but also induced scarp and riverine forest fires (Chapter 15)
Early 1920s, 1944–1955, late 1960s, 1973, 1984, 1992, 2009	Major droughts	Herbivore mortality, especially of white rhino. Probable effects on grass productivity; effects on woody vegetation unknown (see Chapter 2)
1999–2008	Hluhluwe forest and *Euclea* thicket fires	Fires burn extensive areas of marginal scarp and riverine forest and *Euclea* thicket in northern Hluhluwe; vegetation trajectories unknown

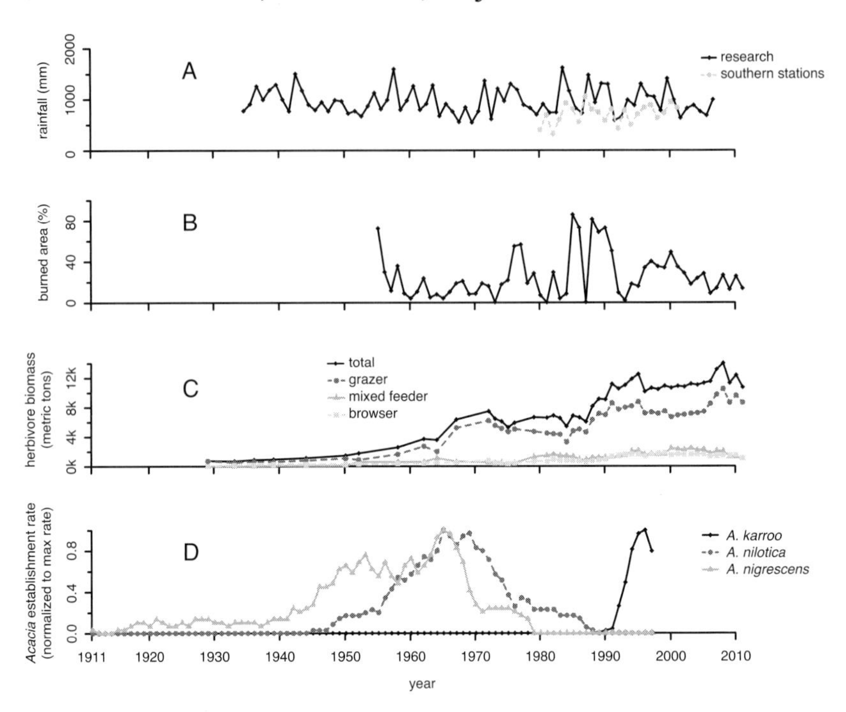

Figure 3.1 Long-term annual rainfall (calculated July–June) (A), % burned area (B), herbivore biomass (C), and acacia establishment rates (D) in HiP. Data for A–C were derived from park monitoring records; data for D were adapted by normalizing establishment rates and applying a five-year moving average to tree-ring data from Staver *et al.* (2011b).

Reserves were proclaimed at the same time in 1895, concurrent with the beginning of European settlement and agriculture in Northern Zululand. Despite the conservation mandate of the reserves, pressure to eradicate tsetse flies and *nagana* mounted (Lincoln, 1995), and early efforts included large-scale eradication of herbivore populations (for more details on culling campaigns, see Chapter 4). Tsetse flies and *nagana* persisted, until aerial spraying with insecticides was deployed during the late 1940s (Figure 1.3). In association with the use of these chemicals, a 3-km wide strip was cleared of all woody vegetation along the boundary of the Umfolozi GR (Figure 1.3) to confine tsetse flies to within the reserve (Downing, 1979). By 1952, when *nagana* was effectively eliminated, game populations in protected areas in KZN had been decimated (see Figure 3.1C and Chapter 4), and the majority of game species had mostly been eliminated from unprotected areas of the province.

During the early history of the reserves until 1952, fire application was haphazard, although burning to prevent bush encroachment later became an important management goal (Chapter 10 presents more information about historical fire regimes). Fires were set to stimulate forage production, but fire suppression was also common. Fire management was rationalized with the appointment of the first research officer to the Hluhluwe GR in 1952 (Brooks and Macdonald, 1983). He reported severe vegetation and soil degradation attributed to overburning plus heavy grazing and browsing pressure, exacerbated by the effects of a long drought, within Hluhluwe GR. In Umfolozi GR, veld condition was still considered 'good' because following the *nagana* campaigns few animals were present apart from a few hundred rhinos (Brooks and Macdonald, 1983).

Nonetheless, the system was viewed as a stable, equilibrium entity; preservation of extant systems remained the conservation goal. Game culling continued in an attempt to ameliorate overgrazing, but herbivore populations nonetheless recovered substantially from *nagana*-era eradication campaigns (see Figure 3.1C and Chapter 4). Overgrazing became an increasing concern in Umfolozi GR as white rhino numbers along with other herbivore populations grew (Mentis, 1968; although areas interpreted as 'overgrazed' may have been what we now consider grazing lawns – see Chapter 6). Most areas in the Hluhluwe Reserve, including areas of relatively low rainfall, burned > 10 times between 1956 and 1996, during which time areas in Umfolozi GR burned only between one and eight times (Balfour and Howison, 2002). Except for decadal-scale fluctuations in fire frequency, there are no obvious long-term trends in burned area across HiP (see Figure 3.1B and Chapter 10).

HiP has undergone an explicit change in management policy in the last decade to 'process-based' management of herbivory and fire (Ezemvelo KZN Wildlife 2002) in line with a change in the global conservation paradigm (Pickett *et al.*, 1992). The park's current management plan acknowledges that the park is a product of a highly varied human and ecological history (see Table 3.1) and that ecological processes, including herbivory and fire, do not function as they did in the pre-colonial landscape. However, quantitative data that address the roles of fire, herbivory, and climate variability in shaping vegetation change during the last century in Hluhluwe iMfolozi Park are sparse. Here, we combine qualitative and anecdotal data with quantitative assessments for a more detailed view of long-term vegetation dynamics in HiP. We focus on variation in savanna versus forest distributions in northern Hluhluwe on millennial time scales, and on changes in woody cover in HiP's savannas during the last ~100 years.

3.3.1 Long-Term Savanna and Forest Distributions

Globally, the response of tropical savanna/grassland and forest distributions to glacial–interglacial climate cycles is a matter of significant debate (Colinvaux *et al.*, 2000; Mayle, 2000; Mayle *et al.*, 2007); maximum forest distributions probably occurred during hotter, wetter periods, and forest minima during cooler, drier periods (Maley, 2011). On long time scales, the vegetation dynamics in the HiP area have not been documented systematically. At Sibudu Cave north of Durban, an area currently within the forest biome, where stone tools and human remains indicate that the area was heavily utilized by nomadic hunter-gatherers (Wadley and Jacobs, 2004), animal and plant remains dating from 61,000 to 26,000 years ago indicate a savanna environment supporting both browsers and grazers (Plug, 2004; Wadley, 2004). This period may have been somewhat drier than the area is currently (Baker *et al.*, 2014), but Stone Age inhabitants also used fire, traces of which can be found in the cave as charcoal (Wadley and Jacobs, 2004), which may have promoted an expansion of savanna in the region.

Carbon isotope depth profiles from HiP indicate that forests are currently more widespread, or distributed differently, than they were in the recent past (~1000 years ago; Figure 3.2; West *et al.*, 2000; see also Gillson, 2015). Carbon isotopes in soils can be used to examine historical shifts between savanna/grassland and forest (1) because trees and tropical grasses generate litter with distinct carbon isotopic signatures (mean $\delta^{13}C$ is ~27‰ for C3 trees and ~13‰ for C4 grasses (Boutton, 1996)), and (2) because organic carbon at depth in the soil profile is older than soil carbon in shallow soil layers. The $\delta^{13}C$ signature of organic carbon in a soil profile therefore yields a rough proxy record for the contributions of trees and grass to soil carbon; mostly tree litter indicates that forest occupied the site of the profile, while mostly grass litter indicates savanna or grassland. Direct interpretation of the $\delta^{13}C$ signature of organic matter is somewhat complicated in soils, however, because organic carbon reaches deeper parts of the soil profile not by sedimentation or progressive deposition, as is the case in lakes, but by vertical transport. Thus, organic carbon at any given depth represents a time-integrated mixture of carbon; deeper carbon is older because it has been transported down through the soil profile for longer. Thus, both $\delta^{13}C$ measurements and

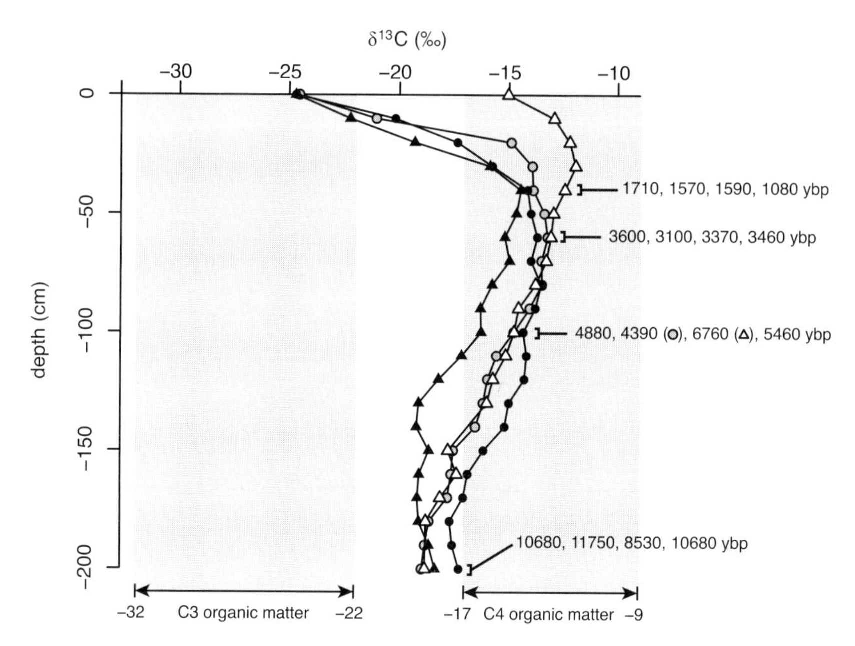

Figure 3.2 Bulk soil organic matter carbon isotopic signature with depth at four sites in the savanna–forest mosaic in northern HiP, collected in March 2010. Dates represent mean carbon ages estimated using radiocarbon dating. Additional data, examining only the top 50 cm of the soil profile, show similar trends (West *et al.*, 2000).

the $\delta^{14}C$ measurements used to calibrate carbon age at depth represent an aggregate reading through time.

Nonetheless, carbon isotope profiles in soils do indicate clearly those areas that are currently forests were grasslands ~1000 years ago (see Figure 3.2). Forest distributions have likely increased since then to their current distribution, which, at rough temporal resolution, represents a recent maximum. Human impacts on forest distributions may have been substantial, or some non-anthropogenic ecological processes (e.g. variation in climate) may have acted to restrict forest distributions at earlier times.

In parts of the hilly north of HiP, forests are currently dominated by shade-intolerant species that are not regenerating within forests. This mismatch between the species composition of forest canopy trees and regenerating forest trees also suggests a successional shift, at least locally (West, 1999). Anecdotal evidence suggests that people farmed small areas within HiP's forests up until the early twentieth century (Henkel, 1937), which

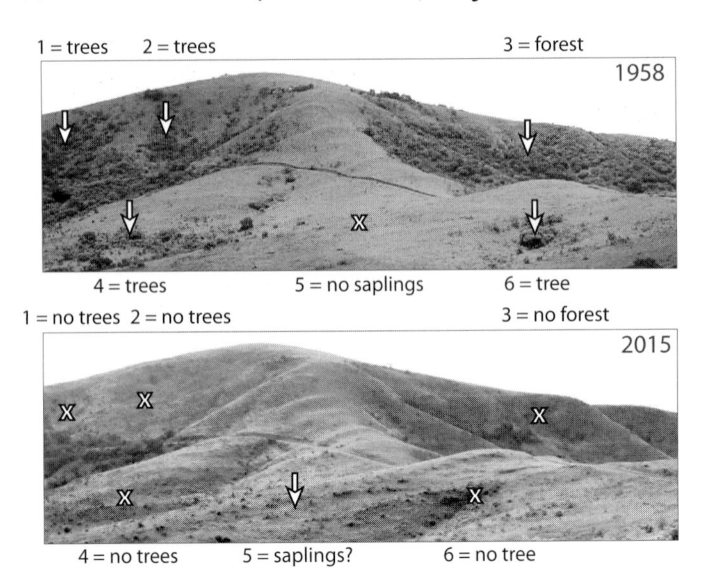

Figure 3.3 Hlaza hill in Hluhluwe photographed in September 1958 (a few months after the fire) and again in 2015 (several years since the fire). Forest cover has decreased (1–4, 6) while sapling density (5) has increased. Photo from 1958 was taken by Norman Owen-Smith, and the 2015 photo was taken by A. Carla Staver.

may explain the prevalence of shade-intolerant forest trees in localized areas (West, 1999). This may provide some explanation for why forests are more extensive currently than they were ~1000 years ago. Archaeological evidence suggests that human impacts may have become intensive as long as ~1400 years ago, with potential impacts on forest extent.

Fire spreading from savanna into forest might also result in major effects on forest community composition. Unlike large savanna trees (see Chapter 8), forest trees in HiP are highly fire-sensitive. Intense, stand-replacing fires have burned into marginal forests repeatedly during the last 10 years. Early-successional trees usually replace late-successional trees following stand-replacing fires (Heath Beckett, unpublished), but repeated recent fires have eliminated some forest patches locally (see Figure 3.3). The historical frequency of stand-replacing disturbances in HiP's forests is unclear; recent intense forest fires may be a by-product of a recent *Chromolaena odorata* invasion (te Beest *et al.*, 2012; Chapter 10, Chapter 15), but the prevalence of forest-margin fires globally suggests that drivers may not necessarily be local (Nepstad *et al.*, 2004).

Some authors have convincingly argued that progressive fractionation with depth is not particularly strong and that soil carbon even at depth is reasonably reflective of the vegetation history of a site (Mariotti and Balesdent, 1990; Ambrose and Sikes, 1991; Silva *et al.*, 2008). If so, the most recent minimum distribution of forest in HiP at a mean carbon age of ~1000 years was preceded by another period when either forests were more extensive or savannas woodier. Biogeographic models from KwaZulu Natal more broadly tend to confirm that forest distributions are experiencing a current local maximum and experienced another maximum during the Holocene altithermal (~7 kya) and that forest distributions were severely constricted during the last glacial maximum (~18 kya) (Eeley *et al.*, 1999; Lawes *et al.*, 2007).

3.3.2 Long-Term Vegetation Structure within Savanna

Quantitative data on vegetation change during the last century in HiP are sparse. The overwhelming weight of anecdotal evidence and a few quantitative studies point (a) to an increase in tree and shrub densities, and (b) to ongoing changes in species composition of tree and shrub communities. From 1840 to 1842, Louis Adulphe Delegorgue, a French naturalist and big game hunter, visited the area which would become HiP. In his diaries, he recorded a 'vast stretch of barren country' covered in 'pasturage' (as opposed to woody species; possibly grazing lawns) in the area from Nqolothi hill in south-western Mfolozi (Delegorgue, 1847). This area is currently covered by open and closed woodland and thicket. He describes the banks of the White Mfolozi river as open country, offering good visibility for hunting elephants, and laments a lack of trees when he encounters a lion. Nonetheless, he mentions numerous 'mimosas' (acacias) dotting the landscape, as well as juveniles 'which were hidden in the grass, [and] tore at our legs and so hindered our progress'; notes numerous thick bush and forested patches, especially on steep mountain slopes; and describes the confluence of the White and Black Mfolozi rivers as a 'wooded region, which abounds in game of all kinds'. Delegorgue describes rivers lined with 'immense wild fig trees', 'kruyz berries' (*Grewia occidentalis*) and tamboti trees, a pattern also noted by Aitken and Gale (1921) during a botanical expedition to Zululand: 'The trees grow closest together along the country lying just beneath the western mountain boundary and near the river. All the trees found in the open veld are found growing along the rivers.' Trees currently abound in Mfolozi, where a century ago they were either scarce or highly localized. Historical accounts of animal species seem to confirm this trend: hunting accounts indicate that buffalo, wildebeest, zebra, eland, waterbuck, and

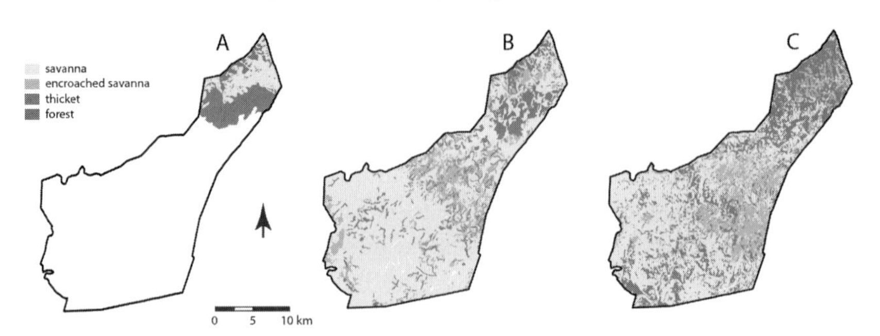

Figure 3.4 Vegetation maps for (A) 1937 (redrawn from Henkel, 1937), (B) 1979 (redrawn from data collected by Whateley and Porter, 1983), and (C) 2001 (redrawn from data in Dora, 2004). (For the colour version, please refer to the plate section. In some formats this figure will only appear in black and white.)

reedbuck were common, consistent with a regional prevalence of grassland or open savanna (McCracken, 2008), and thicket-inhabiting impala and nyala were not recorded inland of the coastal plain. Nevertheless, the localized presence of closed woodland patches amid the grassland is indicated by the abundance of bushbuck and duiker shot alongside numerous zebra during the anti-nagana campaigns of the 1920s and 1930s (Chapter 4).

An early vegetation map of Hluhluwe documents vegetation comprising predominantly open savanna parkland, mixed with pockets of forest in the north-eastern part of the reserve (Henkel, 1937; see Figure 3.4A). Multiple authors during the early part of the twentieth century repeatedly made mention, even then, of changes in vegetation (Attwell, 1948; Foster, 1955; Cowles, 1959; Deane, 1966), specifically in increases in wooded vegetation and the spread of *Acacia karroo* and other 'undesirable plants' (Attwell, 1948). Aitken and Gale (1921) present an account of a resident of the Hluhluwe valley who 'pointed to a hillside studded with thorn trees, and volunteered the information that in his childhood there had not been one tree there'. Foster recorded from several elderly people in the Mfolozi area that they could not recall any areas encroached by woody vegetation in their youth, estimated at the mid-nineteenth century (Foster, 1955).

Henkel's 1937 vegetation map shows that thicket vegetation was widespread in the lower-lying areas of Hluhluwe at that time (Figure 3.4). Comparison with subsequent maps highlights a problem with evaluating subjective historical data: definitions of open savanna, encroached savanna, and thicket (but probably not forest) may have changed as substantially as vegetation structure has changed. This is a particular problem with thicket. A number of HiP map authors have used the term 'thicket' to refer to savanna encroached by acacias (including Whateley and Porter, 1983; see also Skowno *et al.*, 1999). Other authors, however, reserve the term 'thicket' for the densely wooded formations of broadleaved trees with sparse grass cover found throughout South Africa and in South America (Mucina and Rutherford, 2006). There is increasing support that this broadleaved thicket (aka dry forest) is an entirely different non-savanna biome (Parr *et al.*, 2012; see also Chapters 2, 7, and 8). In this chapter, we have opted to reserve the use of the term 'thicket' for the densely woody formations with sparse grass cover, in line with the idea that broadleaved thicket represents a qualitatively different biome, and have interpreted historical classifications within that context.

It does appear that woody thickening within savanna continued, or even accelerated, in the mid- and late twentieth century, and there is evidence of forest/thicket expansion in Mfolozi (Figure 3.4B,C). Changes in the distributions of forest, thicket, and savanna in Hluhluwe are hard to disentangle from variations in the mapping resolution and changes in terminology, but Dora's map (Figure 3.4C; Dora, 2004) clearly shows that by 2001 this entire northern section was much woodier than before. Quantitative evaluations of vegetation change during the twentieth century, with a focus on mesic Hluhluwe, confirm that woody cover has increased there (Downing, 1980). Between 1937 and 1974, forest area had increased by ~5% and grassland decreased by ~8% (Brooks and Macdonald, 1983; Watson, 1995). More recent work, comparing sites within HiP to adjacent communal and commercial lands, shows that savannas have generally become woodier (Wigley *et al.*, 2010): between 1937 and 2004, tree cover increased from 14% to 58% in northern Hluhluwe, from 3% to 50% in nearby commercial rangelands, and from 6% to 25% in nearby communal rangelands. Differences between the park and areas under communal management emphasize the role of management in reducing woody densities in savanna.

Because woody encroachment appears widespread, it may be partially the result of global drivers, including changing rainfall distributions (despite little evidence of systemic changes in rainfall in the park – see Figure 3.1A) and increasing atmospheric CO_2 levels (Buitenwerf *et al.*, 2011; Bond and Midgley, 2012; Chapter 7). Qualitative examination of repeated fixed-point photography suggests some caution, however;

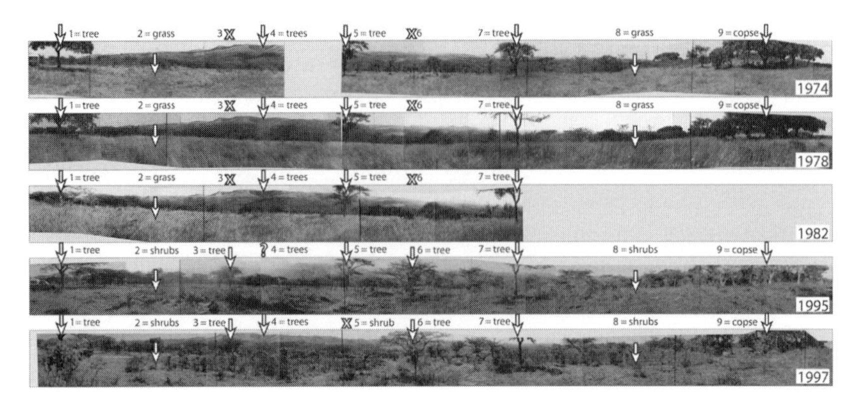

Figure 3.5 Repeat fixed-point photography from Hluhluwe (–28.18663 S, 31.95618 E). Trees have died (at 5) and shrubs have encroached (at 2, 5, and 8), but trees have also established readily over the same time period (at 3 and 6). Photos were taken by staff from Hluhluwe Research Centre, Natal Parks Board/Ezemvelo KZN Wildlife.

woody encroachment has been neither uniform nor ubiquitous, and has been largely restricted to shrubs rather than large trees, even within HiP (see Figures 3.3, 3.5, and 3.6). Discussion sessions at an unpublished workshop on 'Vegetation Dynamics in the Hluhluwe–Corridor–Umfolozi Complex', August 1979, which also used fixed-point photography, were similarly inconclusive, but suggest that drier

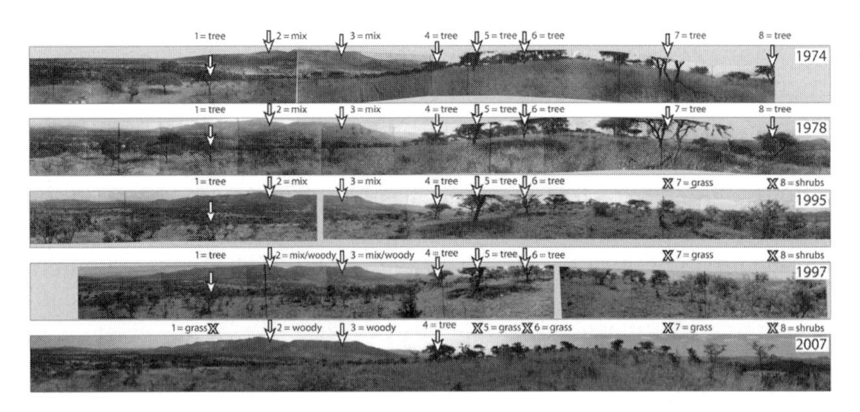

Figure 3.6 Repeat fixed-point photography from Mfolozi (–28.29271 S, 31.7512 E; B). Density of large trees has decreased (tree mortality at 1, 5, 6, 7, and 8) and shrub cover has increased (woody encroachment at 2, 3, and 8), although there is little evidence of tree establishment. Photos were taken by staff from Hluhluwe Research Centre, Natal Parks Board/Ezemvelo KZN Wildlife.

Mfolozi has experienced less woody encroachment than mesic Hluhluwe. In general, woody encroachment has been more severe in southern Africa's mesic savannas than in arid savannas (although even semi-arid savannas like those in Mfolozi have experienced substantial encroachment) (Stevens, 2014; see also O'Connor *et al.*, 2014; Russell and Ward, 2014).

An examination of rates of recruitment of three prevalent acacia species in the park from tree rings suggested that recruitment has been highly variable, potentially tied to variation in fire frequency and herbivore populations, and that species composition may be changing drastically (see Figure 3.1D) (Staver *et al.*, 2011c). The oldest *Acacia karroo* sampled became established in 1992 (possibly because their lifespans are limited to ~30 years), the oldest *A. nilotica* in 1946 and the oldest *Acacia nigrescens* in 1909, while recruitment of *A. nigrescens* appears to have stopped in ~1977 and *A. nilotica* in 1985. However, tree-ring work found little evidence for the highly episodic tree establishment events driven by large-scale herbivore population crashes documented elsewhere in Africa (Prins and van der Jeugd, 1993). A study examining community structure found evidence that HiP is currently undergoing major shifts in woody vegetation communities from herbivore-tolerant towards fire-tolerant species (Bond *et al.*, 2001). Elephant effects on trees and shrubs are potentially exacerbating these changes.

The riverine woodland or forest bordering the Mfolozi rivers, characterized by *Ficus sycamorus* and *Schotia brachypetala* trees, has been scoured by repeated floods in recent historical times (Table 3.1). However, the flood associated with the Tropical Cyclone Domoina in late January 1984 was more than twice the magnitude of any previous flood over the preceding century, with a peak flow rate of 16,000 m^3/s and a rise in water level of 20–40 m above normal river levels (Kemper, 1991). Torrential rainfall in the catchment region amounted to 450 mm over 3 days. During this flood, 80–90% of large fig trees were removed from the banks of the Black Mfolozi river, and overall 96% of the riverine vegetation along this river suffered 75–100% damage (Wills, 1984). The magnitude of this flood might have been exacerbated by overgrazing and clearing of riverbank vegetation in the catchment outside the park, because an earlier flood in March 1925 had much more rainfall but less effect on river flow. Major floods seemed to be increasing in frequency, with peak flow rates exceeding 2000 m^3/s being recorded also in 1957, 1963, 1977, and 1987. However, in the time between these earlier floods, the riverine woodland largely recovered to its former structure. Recovery from the 1984 flood has not been monitored, but is likely to have been slow.

3.4 Conclusions

Despite a paucity of quantitative data on long-term vegetation change in HiP, a number of main conclusions emerge from the available records.

1. Anthropogenic impacts have been prevalent in and around HiP for 1000+ years, and humans have been present for 170–80 kya, such that defining 'natural' vegetation dynamics in the modern context is an artificial exercise. This may be true in savannas and forests throughout Africa and, on more recent temporal scales, globally. Excluding humans and their domesticated animals from parks may have substantial, and historically atypical, effects on vegetation dynamics.
2. Forests were more restricted in their distribution ~1000 years ago than they are now, and savanna–forest boundaries appear to have shifted substantially. These shifts in forest–savanna (or –grassland) boundaries suggest that the view prevalent in the tropical forest literature (Nepstad *et al.*, 2004; Brando *et al.*, 2011) that fire-driven forest-to-savanna transitions are a recent phenomenon, belies a long history of dynamic vegetation change at the interface of savanna and forest.
3. Open savanna in HiP has become widely encroached by woody shrubs especially in more mesic Hluhluwe. This dates from the 1850s and has accelerated through the present (Wigley *et al.*, 2010). The global ubiquity of this phenomenon (Eldridge *et al.*, 2011) implicates some global-scale forcing, such as increases in atmospheric CO_2 (Bond and Midgley, 2012), in addition to changes in fire and herbivore management or reductions in fuelwood harvesting by humans.
4. There is evidence that tree establishment has been highly variable over the last century, possibly occurring during periods with exceptionally low fire frequency or intensity, low herbivore pressure, and/or insect seed predators and reflects probable species turnover within the tree community. Determining whether this pattern is characteristic of tree-layer dynamics in savannas more generally remains a challenge.

What insight do these findings give into the long-term dynamics that characterize savanna ecosystems? More direct evaluations of long-term vegetation dynamics in Hluhluwe-iMfolozi Park and elsewhere will be critical to answering that question well, and to translating insights into well-informed management goals and implementation. However, general conclusions are taking shape on the basis of anecdotal and scattered quantitative work.

First, savanna and forest form distinct biomes, with well-defined distributions and a discontinuous transition, consistent with the hypothesis that savanna and forest represent alternative stable states (Model 3). However, their distributions are dynamic; the boundary between savanna and forest has shifted over millennial time scales. Human influence has been extensive in HiP, and may have influenced long-term savanna and forest distributions. The relative roles of climate, fire, and direct human interventions remain an open question, and better estimates of the timing and rate of transition between savannas and forests would contribute significantly to understanding the distribution especially of mesic savannas.

Second, we found no evidence that savannas are prone, on decadal time scales, to widespread succession towards a climatic maximum, perhaps because savannas are sufficiently fire- and herbivory-prone to prevent succession; fire and herbivory are ubiquitous in HiP. Nevertheless, during the past century, woody species composition has changed, and woody cover has increased. Again, however, all our evidence comes from a century or more of significant human impacts on the HiP area (Table 3.1). Were human activities responsible for relatively low tree densities within savanna prior to human exclusion from parks, or has global CO_2 change driven widespread increases in woody density? Are woody establishment dynamics in savanna stochastic, or at least variable? Without baseline information, we cannot evaluate whether savannas are stable and relatively constant (Model 3), or whether vegetation in savannas is stochastically variable (Model 4). The question may sound academic, given HiP's anthropogenic context, but from a management perspective, will variability within savanna result in a catastrophic transition to another biome? Is it possible to manage for specific vegetation outcomes, and how?

Fortunately, the critical need for quantitative long-term data is well understood in the global literature (Gillson, 2008; Willis and MacDonald, 2011), and work is underway in HiP at century and millennial scales to directly address it. However, answering these questions is not trivial. Savanna and forest ecosystem dynamics are intrinsically complex, and long-term changes within HiP have occurred within the context of historically unprecedented changes in atmospheric CO_2 levels, herbivore populations, and fire management. Determining how these will impact what we need from management, all within the context of a management culture and broader social context that has and continues to change significantly, will be one of the most relevant challenges research can address, both in Hluhluwe-iMfolozi Park and elsewhere.

3.5 References

Aitken, R. D. & Gale, G. W. (1921) *Botanical survey of Natal and Zululand*. Botanical survey of South Africa, Memoir No. 2. The Government Printing and Stationery Office, Pretoria.

Ambrose, S. & Sikes, N. (1991) Soil carbon isotope evidence for Holocene habitat change in the Kenya Rift Valley. *Science* **253**: 1402–1405.

Attwell, R. I. G. (1948) Last strongholds of rhinoceros. *African Wild Life* **2**: 34–52.

Baker, A., Routh, J., Blaauw, M., & Roychoudhury, A. N. (2014) Geochemical records of palaeoenvironmental controls on peat forming processes in the Mfabeni peatland, KwaZulu Natal, South Africa since the Late Pleistocene. *Palaeogeography, Palaeoclimatology, Palaeoecology* **395**: 95–106.

Balfour, D. & Howison, O. (2002) Spatial and temporal variation in a mesic savanna fire regime: responses to variation in annual rainfall. *African Journal of Range and Forage Science* **19**: 45–53.

Beckage, B. & Ellingwood, C. (2008) Fire feedbacks with vegetation and alternative stable states. *Complex Systems* **18**: 159–171.

Bond, W. J. & Midgley, G. F. (2012) Carbon dioxide and the uneasy interactions of trees and savannah grasses. *Philosophical Transactions of the Royal Society B – Biological Sciences* **367**: 601–612.

Bond, W., Smythe, K., & Balfour, D. (2001) Acacia species turnover in space and time in an African savanna. *Journal of Biogeography* **28**: 117–128.

Bond, W., Woodward, F., & Midgley, G. (2005) The global distribution of ecosystems in a world without fire. *New Phytologist* **165**: 525–537.

Boutton, T. W. (1996) Stable carbon isotope ratios of soil organic matter and their use as indicators of vegetation and climate change. In: *Mass spectrometry of soils* (eds. T. W. Boutton & S. I. Yamasaki), pp. 47–82. Marcel Dekker, New York.

Brando, P. M., Nepstad, D. C., Balch, J. K., *et al.* (2011) Fire-induced tree mortality in a neotropical forest: the roles of bark traits, tree size, wood density and fire behaviour. *Global Change Biology* **18**: 630–641.

Brooks, P. M. & Macdonald, I. A. W. (1983) An ecological case history of the Hluhluwe–Corridor–iMfolozi Game Reserve Complex, Natal, South Africa. In: *Management of large mammals in African conservation areas* (ed. R. N. Owen-Smith), pp. 51–77. Haum, Pretoria.

Buitenwerf, R., Bond, W. J., Stevens, N., & Trollope, W. S. W. (2011) Increased tree densities in South African savannas: > 50 years of data suggests CO_2 as a driver. *Global Change Biology* **18**: 675–684.

Chesson, P. L. & Warner, R. R. (1981) Environmental variability promotes coexistence in lottery competitive systems. *American Naturalist* **117**: 923–943.

Colinvaux, P. A., De Oliveira, P. E., & Bush, M. B. (2000) Amazonian and neotropical plant communities on glacial time-scales: the failure of the aridity and refuge hypotheses. *Quaternary Science Reviews* **19**: 141–169.

Cowles, R. B. (1959) *Zulu Journal: field notes of a naturalist in South Africa*. University of California Press, Berkeley, CA.

Cramer, M. D., Chimphango, S. B. M., van Cauter, A., Waldram, M. S., & Bond, W. J. (2007) Grass competition induces N-2 fixation in some species of African Acacia. *Journal of Ecology* **95**: 1123–1133.

Deane, N. N. (1966) Ecological changes and their effect on a population of reedbuck. *Lammergeyer* **6**: 2–8.

Delegorgue, A. (1847) *Voyage dans l'Afrique Australe*, reproduced as *Adulphe Delegorgue's travels in Southern Africa*, Vol. 1 (F. Webb. (transl.), S. J. Alexander & C. de B. Webb (eds), 1990). University of Natal Press, Pietermaritzburg.

Dora, C. A. (2004) The influences of habitat structure and landscape heterogeneity on African buffalo (*Syncerus caffer*) group size in Hluhluwe-iMfolozi Game Reserve, South Africa. MSc thesis, Oregon State University, USA.

Downing, B. H. (1979) The biological, research and management significance of bush-clearing during the anti-tsetse *Glossina spp.* campaign, Umfolozi Game Reserve. Unpublished report, Natal Parks Board, Pietermaritzburg.

Downing, B. H. (1980) Changes in the vegetation of the Hluhluwe Game Reserve, Zululand, as regulated by edaphic and biotic factors over 36 years. *Journal of South African Botany* **46**: 225–231.

Eeley, H. A., Lawes, M. J., & Piper, S. E. (1999) The influence of climate change on the distribution of indigenous forest in KwaZulu-Natal, South Africa. *Journal of Biogeography* **26**: 595–617.

Eldridge, D. J., Bowker, M. A., Maestre, F. T., *et al.* (2011) Impacts of shrub encroachment on ecosystem structure and functioning: towards a global synthesis. *Ecology Letters* **14**: 709–722.

Ezemvelo KZN Wildlife (2002) *Hluhluwe iMfolozi Park Management Plan*. Ezemvelo KwaZulu Natal Wildlife, Pietermaritzburg, South Africa.

February, E. C., Higgins, S. I., Bond, W. J., & Swemmer, L. (2013) Influence of competition and rainfall manipulation on the growth responses of savanna trees and grasses. *Ecology* **94**: 1155–1164.

Feely, J. M. (1985) Smelting in the Iron Age of Transkei. *South African Journal of Science* **81**: 10–11.

Foster, W. E. (1955) History of the Umfolozi Game Reserve. Unpublished report, Natal Parks Board, Pietermaritzburg.

Gillson, L. (2006) A 'large infrequent disturbance' in an East African savanna. *African Journal of Ecology* **44**: 458–467.

Gillson, L. (2008) Resilience, thresholds and dynamic landscapes. *South African Journal of Botany* **74**: 357–358.

Gillson, L. (2015) Evidence of a tipping point in a southern African savanna? *Ecological Complexity* **21**: 78–86.

Gillson, L. & Duffin, K. (2007) Thresholds of potential concern as benchmarks in the management of African savannahs. *Philosophical Transactions of the Royal Society B – Biological Sciences* **362**: 309–319.

Good, S. P. & Caylor, K. K. (2011) Climatological determinants of woody cover in Africa. *Proceedings of the National Academy of Sciences* **108**: 4902–4907.

Goucher, C. L. (1981) Iron is iron 'til it is rust: trade and ecology in the decline of West African iron-smelting. *Journal of African History* **22**: 179–189.

Hall, M. (1978) Enkwazini: fourth century Iron Age site on the Zululand coast. *South African Journal of Science* **74**: 70–71.

Hall, M. (1984) Prehistoric farming in the Mfolozi and Hluhluwe valleys of South-east Africa: an archaeo-botanical survey. *Journal of Archaeological Science* **11**: 223–235.

Hall, M. (1987) Archaeology and modes of production in pre-colonial Southern Africa. *Journal of Southern African Studies* **14**: 1–17.

Henkel, J. S. (1937) *Report on the plant and animal ecology of the Hluhluwe Game Reserve with special reference to tsetse flies*. The Natal Witness, Pietermaritzburg, South Africa.

Higgins, S., Bond, W., & Trollope, W. (2000) Fire, resprouting and variability: a recipe for grass–tree coexistence in savanna, *Journal of Ecology* **88**: 213–229.

Hirota, M., Holmgren, M., Van Nes, E. H., & Scheffer, M. (2011) Global resilience of tropical forest and savanna to critical transitions. *Science* **334**: 232–235.

Hoffmann, W., Orthen, B., & Do Nascimento, P. (2003) Comparative fire ecology of tropical savanna and forest trees. *Functional Ecology* **17**: 720–726.

Hoffmann, W., Adasme, R., Haridasan, M., *et al.* (2009) Tree topkill, not mortality, governs the dynamics of savanna–forest boundaries under frequent fire in central Brazil. *Ecology* **90**: 1326–1337.

Holdo, R. M., Sinclair, A. R. E., Dobson, A. P., *et al.* (2009) A disease-mediated trophic cascade in the Serengeti and its implications for ecosystem C. *PLoS Biology* **7**: e1000210.

Holdridge, L. R. (1947) Determination of world plant formations from simple climatic data. *Science* **105**: 367–368.

Hutchinson, G. E. (1961) The paradox of the plankton. *American Naturalist* **95**: 137–145.

Jeffers, E. S., Bonsall, M. B., Brooks, S. J., & Willis, K. J. (2011) Abrupt environmental changes drive shifts in tree–grass interaction outcomes. *Journal of Ecology* **99**: 1063–1070.

Jeltsch, F., Weber, G., & Grimm, V. (2000) Ecological buffering mechanisms in savannas: a unifying theory of long-term tree–grass coexistence. *Plant Ecology* **150**: 161–171.

Kemper, N. (1991) The structure and dynamics of riverine vegetation in the Umfolozi Game Reserve. MSc thesis, University of the Witwatersrand, South Africa.

Kristen, I., Fuhrmann, A., Thorpe, J., *et al.* (2007) Hydrological changes in southern Africa over the last 200 Ka as recorded in lake sediments from the Tswaing impact crater. *South African Journal of Geology* **110**: 311–326.

Lawes, M. J., Eeley, H. A. C., Findlay, N. J., & Forbes, D. (2007) Resilient forest faunal communities in South Africa: a legacy of palaeoclimatic change and extinction filtering? *Journal of Biogeography* **34**: 1246–1264.

Lincoln, D. (1995) Settlement and servitude in Zululand, 1918–1948. *International Journal of African Historical Studies* **28**: 49–67.

Maggs, T. (1980) The Iron Age sequence south of the Vaal and Pongola rivers: some historical implications. *Journal of African History* **21**: 1–15.

Maggs, T. & Whitelaw, G. (1991) A review of recent archaeological research on food-producing communities in southern Africa. *Journal of African History* **32**: 3–24.

Maley, J. (2011) The African rain forest – main characteristics of changes in vegetation and climate from the Upper Cretaceous to the Quaternary. *Proceedings of the Royal Society of Edinburgh Section B Biological Sciences* **104**: 31–73.

Mariotti, A. & Balesdent, J. (1990) 13C natural abundance as a tracer of soil organic matter turnover and paleoenvironment dynamics. *Chemical Geology* **84**: 217–219.

Mayle, F. E. (2000) Millennial-scale dynamics of southern Amazonian rain forests. *Science* **290**: 2291–2294.

Mayle, F. E., Langstroth, R. P., Fisher, R. A., & Meir, P. (2007) Long-term forest–savannah dynamics in the Bolivian Amazon: implications for conservation. *Philosophical Transactions of the Royal Society B – Biological Sciences* **362**: 291–307.

McCracken, D. P. (2008) *Saving the Zululand wilderness: an early struggle for nature conservation*. Jacana Media, Auckland Park.

Mentis, M. T. (1968) An evaluation of the veld conditions in the Umfolozi Game Reserve: an assessment of soil erosion and the amount of available graze and browse at present. Unpublished report, Natal Parks Board, Pietermaritzburg.

Mucina, L. & Rutherford, M. C. (2006) *The vegetation of South Africa, Lesotho and Swaziland*. South African National Biodiversity Institute, Pretoria.

Nepstad, D., Lefebvre, P., Lopes da Silva, U., *et al.* (2004) Amazon drought and its implications for forest flammability and tree growth: a basin-wide analysis. *Global Change Biology* **10**: 704–717.

O'Connor, T. G., Puttick, J. R., & Hoffman, M. T. (2014) Bush encroachment in southern Africa: changes and causes. *African Journal of Range and Forage Science* **31**: 67–88.

Parr, C. L., Gray, E. F., & Bond, W. J. (2012) Cascading biodiversity and functional consequences of a global change-induced biome switch. *Diversity and Distributions* **18**: 493–503.

Pickett, S. T. A., Parker, V. T., & Fielder, P. L. (1992) The new paradigm in ecology: implications for conservation biology about the species level. In: *Conservation biology: the theory and practice of nature conservation, preservation, and management* (eds. P. L. Fielder & S. K. Jain), pp. 65–88. Chapman and Hall, New York.

Plug, I. (2004) Resource exploitation: animal use during the Middle Stone Age at Sibudu Cave, KwaZulu-Natal: Sibudu Cave. *South African Journal of Science* **100**: 151–158.

Prins, H. H. T. & van der Jeugd, H. P. (1993) Herbivore population crashes and woodland structure in East Africa. *Journal of Ecology* **81**: 305–314.

Rees, M. (2001) Long-term studies of vegetation dynamics. *Science* **293**: 650–655.

Riginos, C. (2009) Grass competition suppresses savanna tree growth across multiple demographic stages. *Ecology* **90**: 335–340.

Russell, J. & Ward, D. (2014) Vegetation change in northern KwaZulu-Natal since the Anglo-Zulu War of 1879: local or global drivers? *African Journal of Range and Forage Science* **31**: 89–105.

Sankaran, M., Ratnam, J., & Hanan, N. P. (2004) Tree–grass coexistence in savannas revisited – insights from an examination of assumptions and mechanisms invoked in existing models. *Ecology Letters* **7**: 480–490.

Sankaran, M., Hanan, N., Scholes, R., *et al.* (2005) Determinants of woody cover in African savannas. *Nature* **438**: 846–849.

Scanlon, T., Caylor, K., Manfreda, S., Levin, S., & Rodriguez-Iturbe, I. (2005) Dynamic response of grass cover to rainfall variability: implications for the function and persistence of savanna ecosystems. *Advances in Water Resources* **28**: 291–302.

Scholes, R. & Archer, S. (1997) Tree–grass interactions in savannas. *Annual Review of Ecology and Systematics* **28**: 517–544.

Scholes, R. J. & Walker, B. H. (1993) *An African savanna: synthesis of the Nylsvley study*. Cambridge University Press, Cambridge.

Silva, L. C. R., Sternberg, L., Haridasan, M., *et al.* (2008) Expansion of gallery forests into central Brazilian savannas. *Global Change Biology* **14**: 2108–2118.

Simon, M. F., Grether, R., De Queiroz, L. P., *et al.* (2009) Recent assembly of the Cerrado, a neotropical plant diversity hotspot, by in situ evolution of adaptations to fire. *Proceedings of the National Academy of Sciences* **106**: 20359–20364.

Skowno, A., Midgley, J., Bond, W., & Balfour, D. (1999) Secondary succession in *Acacia nilotica* (L.) savanna in the Hluhluwe Game Reserve, South Africa. *Plant Ecology* **145**: 1–9.

Staver, A. C., Archibald, S., & Levin, S. A. (2011a) Tree cover in sub-Saharan Africa: rainfall and fire constrain forest and savanna as alternative stable states. *Ecology* **92**: 1063–1072.

Staver, A. C., Archibald, S., & Levin, S. A. (2011b) The global extent and determinants of savanna and forest as alternative biome states. *Science* **334**: 230–232.

Staver, A. C., Bond, W. J., & February, E. C. (2011c) History matters: tree establishment variability and species turnover in an African savanna. *Ecosphere* **2**: art49.

Stevens, N. (2014) Exploring the potential impacts of global change on the woody component of South African savannas. PhD thesis, University of Cape Town, South Africa.

te Beest, M., Cromsigt, J. P. G. M., Ngobese, J., & Olff, H. (2012) Managing invasions at the cost of native habitat? An experimental test of the impact of fire on the invasion of *Chromolaena odorata* in a South African savanna. *Biological Invasions* **14**: 607–618.

van der Merwe, N. J. (1980) The advent of iron in Africa. In: *The coming of the age of iron* (eds T. A. Wertime & J. D. Muhly), pp. 463–506. Yale University Press, New Haven, CT.

Wadley, L. (2004) Vegetation changes between 61,500 and 26,000 years ago: the evidence from seeds in Sibudu Cave, KwaZulu-Natal. *South African Journal of Science* **100**: 167–173.

Wadley, L. & Jacobs, Z. (2004) Sibudu Cave, KwaZulu-Natal: background to the excavations of Middle Stone Age and Iron Age occupations. *South African Journal of Science* **100**: 145–151.

Waldram, M. S., Bond, W. J., & Stock, W. D. (2008) Ecological engineering by a mega-grazer: white rhino impacts on a South African savanna. *Ecosystems* **11**: 101–112.

Walker, B., Ludwig, D., Holling, C., & Peterman, R. (1981) Stability of semi-arid savanna grazing systems. *Journal of Ecology* **69**: 473–498.

Walter, H. (1971) *Ecology of tropical and subtropical vegetation*. Oliver and Boyd, Edinburgh.

Watson, H. K. (1995) Management implications of vegetation changes in Hluhluwe-Umfolozi Park. *South African Geographical Journal* **77**: 77–83.

West, A. (1999) Hunting for humans in forest ecosystems: are the traces of Iron-Age people detectable? MSc thesis, University of Cape Town, South Africa.

West, A. G., Bond, W. J., & Midgley, J. J. (2000) Soil carbon isotopes reveal ancient grassland under forest in Hluhluwe, KwaZulu-Natal. *South African Journal of Science* **96**: 252–254.

Whateley, A. & Porter, R. N. (1983) The woody vegetation communities of Hluhluwe–Corridor–Umfolozi Game Reserve Complex. *Bothalia* **14**: 745–758.

Whittaker, R. (1975) *Communities and ecosystems*. Macmillan, New York.

Wigley, B. J., Bond, W. J., & Hoffman, M. T. (2010) Thicket expansion in a South African savanna under divergent land use: local vs. global drivers? *Global Change Biology* **16**: 964–976.

Williams, R., Duff, G., Bowman, D., & Cook, G. (1996) Variation in the composition and structure of tropical savannas as a function of rainfall and soil texture along a large-scale climatic gradient in the Northern Territory, Australia. *Journal of Biogeography* **23**: 747–756.

Willis, K. J. & MacDonald, G. (2011) Long-term ecological records and their relevance to climate change predictions for a warmer world. *Annual Review of Ecology and Systematics* **42**: 267–287.

Willis, K., Kleczkowski, A., & Crowhurst, S. (1999) 124,000-year periodicity in terrestrial vegetation change during the late Pliocene epoch. *Nature* **397**: 685–688.

Willis, K. J., Bailey, R. M., Bhagwat, S. A., & Birks, H. J. B. (2010) Biodiversity baselines, thresholds and resilience: testing predictions and assumptions using palaeoecological data. *Trends in Ecology and Evolution* **25**: 583–591.

Wills, A. J. (1984) The 1984 flood in Umfolozi Game Reserve: ecological effects and management implications. Unpublished report, Natal Parks Board, Pietermaritzburg.

4 · *Temporal Changes in the Large Herbivore Fauna of Hluhluwe-iMfolozi Park*

ELIZABETH LE ROUX, GEOFF CLINNING, DAVE J. DRUCE, NORMAN OWEN-SMITH, JAN A. GRAF, AND JORIS P.G.M. CROMSIGT

4.1 Introduction

This chapter is a reconstruction of the history of the large mammalian herbivore assemblage of Hluhluwe-iMfolozi Park (HiP) and the surrounding area. The region has had a long history of human occupation and human impacts on animal populations were considerable long before the park was proclaimed (Chapter 1). The rich and relatively well-documented history of the park spans ecologically informative time frames and transcends various political and economic agendas that drove past management policies. We focus on changes in large mammalian herbivores because they have formed a central part of the park's monitoring efforts (predators are covered in Chapter 12). Apart from being charismatic, important for ecotourism and highly valued in the wildlife industry, large mammalian herbivores are often central drivers of ecosystem structure and functioning (Chapin *et al.*, 1997; Paine, 2000). We use nearby archaeological deposits that represent mammal remains from the Late Pleistocene up to the Late Iron Age (eleventh to twelfth centuries) to provide information on the prehistoric and early historical faunal composition. Writings of European hunters and explorers and records of exported animal products sketch the situation during the 1800s, when impact intensified due to the arrival of these European hunters. Following park proclamation at the end of the nineteenth century, more regular and formal information on animal population trends becomes available

Conserving Africa's Mega-Diversity in the Anthropocene, ed. Joris P. G. M. Cromsigt, Sally Archibald and Norman Owen-Smith. Published by Cambridge University Press.

in the form of unpublished management reports and some early published accounts on long-term ecological research. Regular surveys since 1986 enabled a rigorous analysis of the population trends of the last three decades. We present the herbivore dynamics within HiP in the context of strong human influence and natural regulation.

4.2 Early Records of Wildlife before the Game Reserves Were Proclaimed

Detailed records of vertebrate remains from Sibudu Cave, a Middle Stone Age human settlement approximately 150 km south of HiP, provide insight into mammalian assemblages since the Late Pleistocene in the vicinity of HiP. The main layers excavated contain faunal remains from the Middle Stone Age that date back to between 61,000 and 38,000 BP (Plug, 2004; Clark and Plug, 2008), while the upper layers extend into Iron Age times as recently as 800–900 BP. Although the species represented in the deposits include all extant species, the excavations suggest much richer large mammal diversity during the Late Pleistocene. The samples from the Middle Stone Age sequence are dominated by medium- to large-sized grazing bovids (Plug, 2004), mostly Burchell's zebra (*Equus quagga*), hartebeest (*Alcelaphus* sp.), and African buffalo (*Syncerus caffer*), although many large bovid bones remain unidentified. Older deposits of the Middle Stone Age sequence include the extinct Cape zebra (*Equus capensis*), long-horned African buffalo (*Pelorovis antiquus*), and a giant alcelaphine (*Megalotragus priscus*), all of which likely preferred predominantly open grassland habitats (Faith, 2014). While the abundance of these large grazers in the Middle Stone Age record may merely reflect a hunting preference of early humans, it does suggest that grassland habitat was widespread during the Late Pleistocene (Plug, 2004). Also present in the excavations are extant species not historically recorded from the region surrounding HiP: a hippotragine (*Hippotragus* sp.), giraffe (*Giraffa camelopardalis*), and impala (*Aepyceros melampus*). *Hippotragus* sp. and giraffe have also been recorded from archaeological sites further inland, in the midlands of KwaZulu-Natal, until around 4000 and 1000 BP, respectively (Plug and Badenhorst, 2001; Plug, 2004). Impala remains, although rare, have been found further to the north but also in the middle Thukela river basin (southward of the current park boundaries) until about 1200 BP (Plug and Badenhorst, 2001).

Early writings of hunters and explorers indicate that a remarkable abundance of wildlife persisted in the greater Zululand region alongside the Nguni people during the early 1800s. Henry Francis Fynn remarked on the 'tremendous abundance of game' seen during his visit to the Zulu king Shaka in 1824 (Fynn, 1950). Particular mention is made of the large numbers of elephant (*Loxodonta africana*), eland (*Taurotragus oryx*), and buffalo (Delegorgue, 1847), and other species regularly mentioned include white rhino (*Ceratotherium simum*), hippo (*Hippopotamus amphibius*), zebra, kudu (*Tragelaphus strepsiceros*), oribi (*Ourebia ourebi*), common and mountain reedbuck (*Redunca arundinum* and *R. fulvorufula*), bushbuck (*Tragelaphus scriptus*), and red duiker (*Cephalophus natalensis*) (Delegorgue, 1847; Drummond, 1875; Ward, 1896; Findlay, 1903; Lydekker, 1908). Delegorgue (1847) describes regular sightings of eland herds within the area later covered by the Umfolozi Game Reserve and downstream along the floodplain of the Mfolozi river (Delegorgue, 1847). Hippos were reported to occur in the Mfolozi and Hluhluwe rivers (Baldwin, 1894; Findlay, 1903) and reedbuck along the Hluhluwe and Manzibomvu rivers (Findlay, 1903).

None of the records of early European hunters and explorers make reference to giraffe in the Zululand area (Goodman and Tomkinson, 1987; McCracken, 2008). For this reason, giraffe are considered to be extralimital to the region (Goodman and Tomkinson, 1987). Historical records place the southern boundary of giraffe distribution north of Swaziland (Skinner and Chimimba, 2005). Similarly, mention of impala in these early records is very scarce and limited to coastal areas to the north-east of HiP (Drummond, 1875; Baldwin, 1894; Selous, 1908). However, the presence of both these species in archaeological remains further to the south and from inland areas is undisputed (Plug and Badenhorst, 2001; Plug, 2004) and the reasons for their disappearance 1000–1200 BP remain unclear (Rowe-Rowe, 1994; Cramer and Mazel, 2007). Whether impala and giraffe can be considered native to the HiP region is thus contentious.

Heavy exploitation of wildlife populations in Zululand by European and later also Zulu hunters continued through the late nineteenth century: 885 metric tonnes of ivory, 19,245 rhino horns, 2,015,246 unspecified animal skins, 22,154 buffalo hides, and 956 hippo hides were exported from Zululand between 1844 and 1904 (McCracken, 2008). Ivory exports dropped precipitously during the 1860s, indicating dwindling elephant numbers (McCracken, 2008). The last elephant within the current boundaries of HiP was shot on the banks of the Black Mfolozi by John Dunn in 1890 (Vincent, 1970). Similarly, the once-abundant eland had become locally extinct by 1880 (Foster, 1955; Vincent, 1970). By the beginning of the twentieth century, very few rhinos were said to remain (Findlay, 1903; Selous, 1908). The first 'official' estimate of 20–30 rhinos came from the first Zululand Game Conservator, Vaughan-Kirby, during the early 1920s, but in hindsight this underestimate was an attempt to foster the protection of the white rhino (Brooks, 2006). The early twentieth century population was likely much larger, because ground counts of white rhino conducted during the late 1920s indicated a minimum of 172 rhino in Umfolozi Game Reserve (Brooks, 2006), and counts in 1932 and 1936 recorded 220 and 226 rhino, respectively (Player and Feely, 1960; see also Chapter 11).

4.3 Early History of the Game Reserves: 1895–1952

The dwindling wildlife populations led to the proclamation of the Hluhluwe and Umfolozi Game Reserves in 1895 to preserve and maintain indigenous wildlife species, particularly white rhino (Brooks and Macdonald, 1983; Chapter 1). The reserves remained unfenced and were separated by the large stretch of land later referred to as the Corridor, which represented a functional corridor for animals to move between both reserves. Most of Umfolozi GR remained unfenced until the early 1950s, while the majority of the Hluhluwe GR boundaries were fenced off by 1941, leaving only a small section linking to the Corridor area to the south of the reserve unfenced (Brooks and Macdonald, 1983; see Chapter 1 for further details).

In 1897 a continent-wide outbreak of rinderpest reached what we now know as KwaZulu-Natal (Ballard, 1983) and decimated wildlife and livestock populations (Brooks and Macdonald, 1983). Cattle losses of almost 70% were reported (Ballard, 1983). According to Foster (1955), the Zululand buffalo populations suffered particularly heavy losses, while kudu, reedbuck, and bushbuck were also affected although 'in no way as serious as buffalo'. By 1905, substantive recovery of wild ungulates following the rinderpest epizootic had occurred (Foster, 1955; Brooks and Macdonald, 1983). This led to a resurgence of tsetse flies (*Glossina* spp.) and hence the incidence of another major livestock disease for which the tsetse fly acts as a vector: nagana or African trypanosomiasis (caused by *Trypanosoma* spp. parasites; Henkel, 1937; Brown, 2008). Farmers attributed increased nagana prevalence to the recovery of wildlife populations (Brown, 2008) and considered the game reserves as sources of infection (Minnaar, 1989). Dissatisfaction from local residents eventually led the provincial administration to organize a concerted campaign from May 1929 to the end of 1930. The aim was to eradicate all ungulate species, except white rhino, from the buffer zones surrounding Umfolozi GR and to drive those that remained into the game reserve. This resulted in the slaughter of >25,000 ungulates in these designated buffer zones, and 377 individuals from inside the boundaries of Umfolozi GR (Mentis, 1970; Minnaar, 1989; Table 4.1). During the same campaign, >5000 animals were killed in buffer zones around Hluhluwe GR, and >2000 inside the current reserve boundaries (Bourquin and Hitchins, 1979). Because Hluhluwe GR was strictly protected during this time, the latter number likely reflects animals shot in the area south of the Hluhluwe river, which was not then part of the proclaimed reserve. Despite this effort, prevailing drought conditions drove wildlife to move beyond the yet unfenced reserves. Authorities thus concluded that alleviating the threat demanded the eradication of ungulates from inside the reserve boundaries. Consequently, between 1932 and 1954, administrative responsibility of Umfolozi GR was officially assigned to the Division of Veterinary Services to undertake this (Vincent, 1970; Brooks and Macdonald, 1983). Between the late 1930s and the early 1950s, more than 70,000 animals were killed in Umfolozi and surrounding buffer zones (Mentis, 1970; Table 4.1). During this second campaign, Hluhluwe GR remained protected, although ~5400 animals were shot in the buffer zone around this reserve (Bourquin and Hitchins, 1979).

Throughout the first and second shooting campaigns, bushbuck and grey duiker (*Sylvicapra grimmia*) represented ~50% of all animals shot (~25,000 individuals of each, Table 4.1). Despite these massive kills, both these species maintained large populations until the early 1950s (Mentis, 1970). Using actual numbers of animals killed and estimates of reproductive rates, Mentis (1970) back-calculated population sizes of various species in 1942 early during the second campaign. The Umfolozi GR bushbuck population was estimated to have been larger than 12,000 individuals in 1942 and described as still 'very numerous' 10 years later (Mentis, 1970), while the Hluhluwe population remained at around 1000 individuals throughout the campaigns (1936 estimate: Henkel, 1937; 1950 estimate: Anon., 1952). Similarly, grey duiker populations in Umfolozi GR were also referred to as still 'very numerous' by 1952 (Mentis, 1970), and the Hluhluwe grey duiker population apparently doubled between 1936 and the early 1950s (Henkel, 1937; Anon., 1952). While these two species seemed to survive

Table 4.1 *Numbers of various large herbivore species shot during the game eradication campaign of 1929–1930 and while the management of Umfolozi Game Reserve (UGR) and surrounding regions was under the control of the veterinary authority during 1942–1954. Few animals were shot within UGR during 1929–1930 because it had protected status and the animals reported shot in Hluhluwe Game Reserve (HGR) then were probably in the region south of the Hluhluwe river that was not part of the proclaimed game reserve at that time. In 1929–1930, UGR and its buffer zones covered 1400 km^2 and in 1942–1950 1150 km^2. The extent of the other sections was not recorded. UGR buffer zones encompassed regions subsequently included in the game reserve, but extended beyond the present-day boundaries in the south and west (extracted from Mentis (1970) and Bourquin and Hitchins (1979))*

	1929–1930					1942–1950			
Species	UGR	UGR buffer zones	HGR	HGR buffer zones	Total	UGR	UGR buffer zones	HGR buffer zone	Total
Buffalo	5	120	7	13	145	13	19	6	38
Bushbuck	22	2173	453	1337	4185	7540	14,692	1138	23,370
Bushpig	0	22	16	39	77	98	299	14	412
Duiker, grey	10	2987	2	383	3382	4728	15,733	1024	21,485
Duiker, red	0	0	0	19	19	0	9	59	68
Impala	0	0	0	0	0	16	3	14	33
Klipspringer	0	5	0	0	5	47	50	2	99
Kudu	2	286	0	15	303	759	679	72	1510
Nyala	0	0	0	0	0	23	16	228	267
Reedbuck, common	0	390	2	449	841	396	2952	198	3546
Reedbuck, mountain	0	12	0	19	31	334	3173	157	3664
Steenbok	1	306	0	17	324	558	1437	50	2045
Warthog	216	3456	524	1109	5305	7661	5173	2225	15,059
Waterbuck	26	504	0	173	703	535	474	157	1166
Wildebeest	26	664	996	1906	3592	721	304	130	1155
Zebra	69	15130	158	666	16023	917	845	78	1840
Grand total	377	26162	2158	5645	34,342	24,346	45,858	5430	75,634

the anti-nagana campaigns well, this could not be said for populations of most of the larger species. By the middle of the century, wildebeest and zebra had been eliminated from Umfolozi GR. Similarly, based on the population estimates reported in Mentis (1970), the Umfolozi GR reedbuck population (including both the common and the mountain reedbuck species) dropped by 96% between 1942 and 1952 (Mentis, 1970) and waterbuck (*Kobus ellipsiprymnus*) dropped to 15% of its former abundance (Mentis, 1970). By 1952, steenbok (*Raphicerus campestris*) numbers were a mere 5% of what they were calculated to have been 10 years prior (Mentis, 1970). Interestingly, only approximately 3% of the estimated population of buffalo was culled in Umfolozi GR and the population was still estimated at approximately 1000 individuals in 1942 (Mentis, 1970). However, following the massive culling of other species during the 1940s, the Umfolozi GR buffalo population was believed to have migrated out of the park to the north (Foster, 1955), leaving only a handful of individuals during the early 1950s (Mentis, 1970; Brooks and Macdonald, 1983). These massive drops in population numbers need to be interpreted with caution as the 1942 estimates were derived from a calculation based on kill numbers and assumed reproductive rates and not from direct counts. As such, it does not take into account surviving animals or migrations to and from the area. Regardless, at the start of the 1950s, populations of most large ungulate species were either entirely eliminated or nearly so in Umfolozi GR.

Hluhluwe GR was less affected, particularly during the second campaign, and by the start of the 1950s the reserve had maintained large populations of ungulates, including many wildebeest and zebra (Bourquin and Hitchins, 1979). Likewise, the buffalo population in Hluhluwe at the end of the nagana campaigns was estimated to be in the region of 700. However, several smaller antelope species experienced severe population collapses in Hluhluwe GR as well. The Hluhluwe populations of both species of reedbuck were halved between 1936 and 1950 (Bourquin and Hitchins, 1979). Foster (1955) refers to 'countless numbers [of reedbuck] on either side of the road' throughout the Corridor in 1906, but they had become a 'vanishing population' by the 1950s. Deane (1966) mentions that reedbuck was the most frequently seen animal in the Corridor during the early twentieth century, but were noticeably rare by 1954. Similarly, steenbok numbers in Hluhluwe estimated at approximately 100 in 1936 had become 'very few' (Bourquin and Hitchins, 1979). Waterbuck numbers in Hluhluwe were low (300) in 1936, but remained unchanged in 1950 (Bourquin and Hitchins, 1979).

Rhinos of both species remained protected throughout the 1929–1952 game eradication campaigns (Minnaar, 1989). By the early 1950s, around 500 white rhinos occurred in Umfolozi GR (Foster, 1955) and they had recolonized the area north of Umfolozi. In 1950, an estimated 20 occurred in Hluhluwe (Bourquin and Hitchins, 1979), while approximately 70 were present in the Corridor in 1958 (Deane, 1966). Black rhino (*Diceros bicornis*) numbers were estimated at 10 individuals in Umfolozi GR and 190 in Hluhluwe during the early 1950s (Foster, 1955; Bourquin and Hitchins, 1979; see Chapter 11).

4.4 Herbivore Population Trends after 1952

4.4.1 A Time of Faunal Restoration – Population Recovery and (Re)Introductions

In 1952 the responsibility of reserve management was assumed by the newly established Natal Parks, Game and Fish Preservation Board (Vincent, 1970). At this time, fencing of Umfolozi GR commenced at a large scale but was only completed by 1965 (Vincent, 1970). As ungulate populations within Hluhluwe GR had been little affected by the anti-nagana campaigns, wildebeest, warthog (*Phacochoerus africanus*), and zebra began dispersing southwards, crossing the Corridor to repopulate Umfolozi GR (Deane, 1966; Vincent, 1970). Dispersal into Umfolozi GR was facilitated following a flood in July 1963 which deposited sand, making the Black Mfolozi river shallower and hence easier to cross for a longer portion of the year (Vincent, 1970). White rhino followed the opposite dispersal route and gradually dispersed northwards into Hluhluwe (Vincent, 1969).

This period also saw several successful and unsuccessful attempts to (re)introduce species. Impala and nyala (*Tragelaphus angasi*) were introduced in Hluhluwe GR in 1936, although a few nyala may have been present in the Hluhluwe river thicket (Hitchins and Vincent, 1972). By the 1940s impala had entered Umfolozi, as 16 of them were shot during the 1942–1950 nagana campaign (Mentis, 1970). Still only about 10 impala were estimated to occur in Umfolozi GR by 1952 (Foster, 1955; Mentis, 1970). Following its initial re-introduction in Hluhluwe, nyala had colonized Umfolozi GR by 1937 (Foster, 1955), and in 1952 a minimum of 50 individuals occurred in Umfolozi GR (Foster, 1955). Foot counts in 1967 came to a minimum of 550 and 850 impala and nyala in Umfolozi GR, respectively (Bourquin, 1968) and both species have increased rapidly since (Figure 4.1). Further introductions included

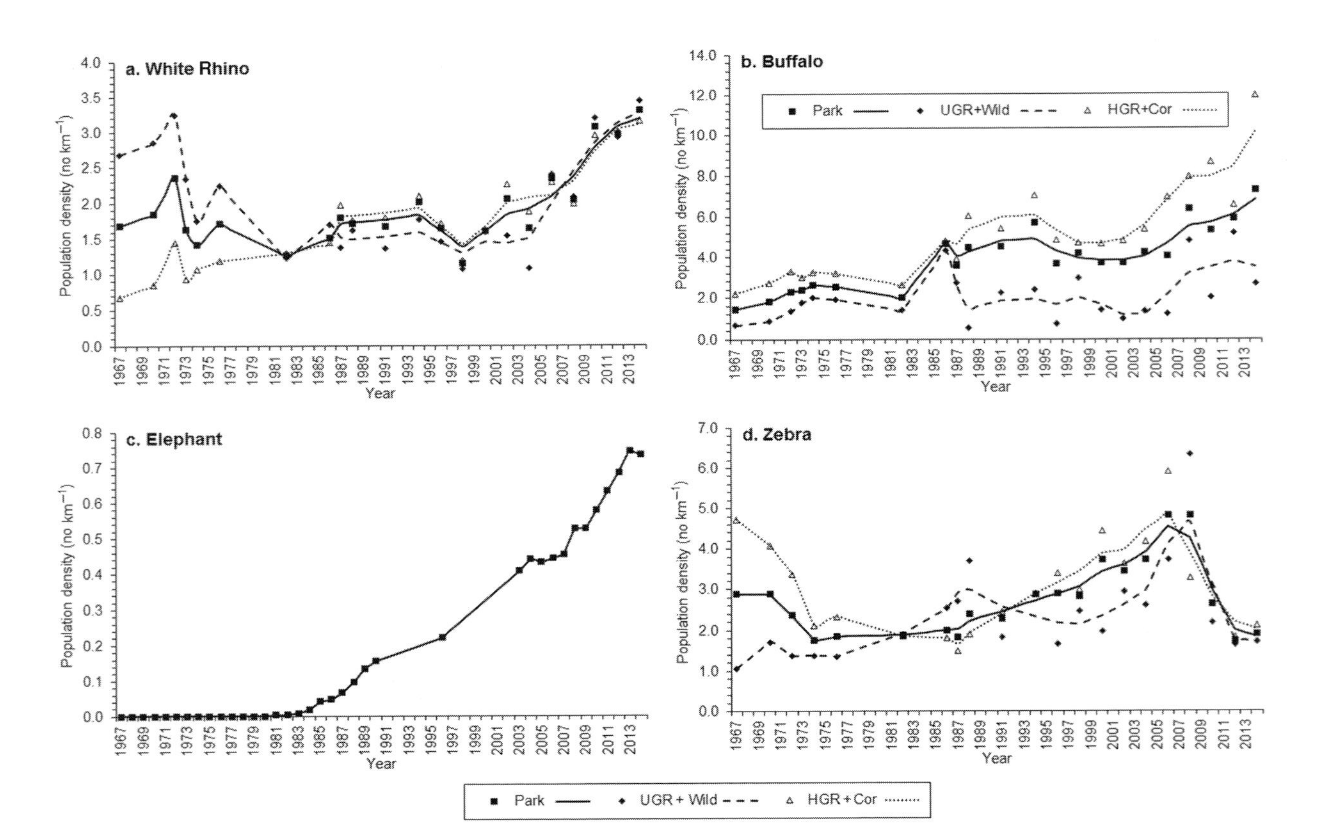
a. White Rhino
b. Buffalo
c. Elephant
d. Zebra
Population density (no km^{-1})
Year
Park
UGR+Wild
HGR+Cor
Park
UGR + Wild
HGR + Cor

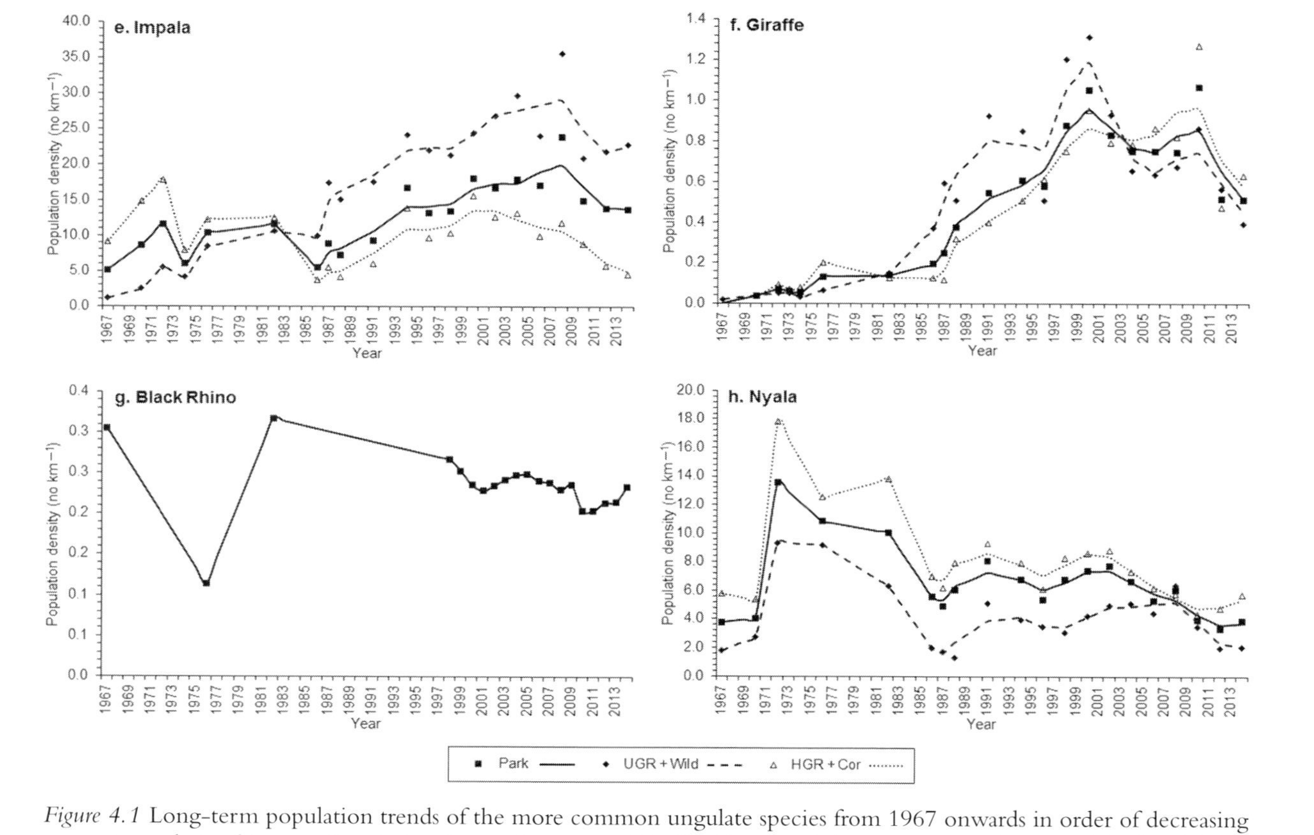

Figure 4.1 Long-term population trends of the more common ungulate species from 1967 onwards in order of decreasing current population biomass. Population density is presented for Hluhluwe and the Corridor sections combined and for Mfolozi and the Wilderness sections combined. No density estimates are available for the Wilderness section between 1986 and 2006 and the density displayed for this period represents only the Mfolozi section. (A) white rhino; (B) buffalo; (C) elephant; (D) zebra; (E) impala; (F) giraffe; (G) black rhino; (h) nyala, (I) wildebeest; (J) kudu; (K) warthog; (L) waterbuck.

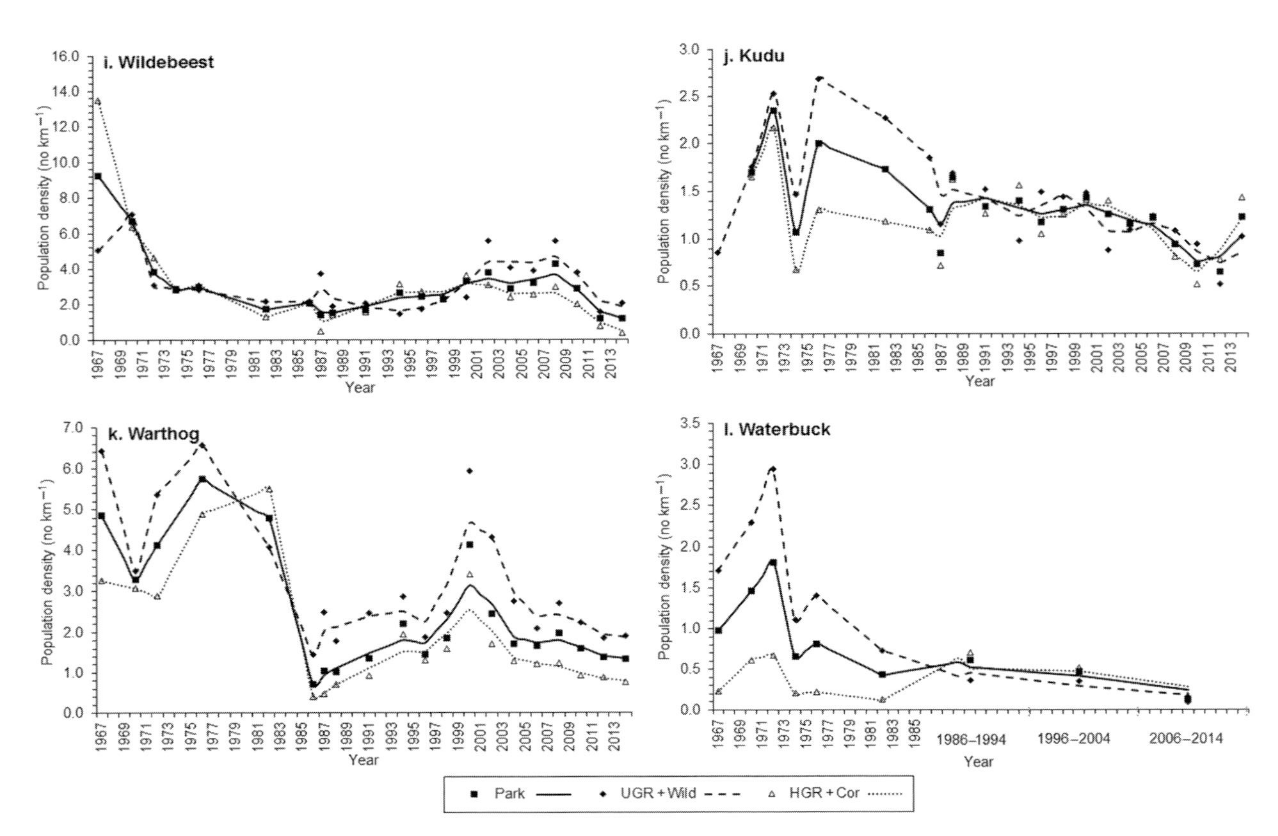

Figure 4.1 (continued)

four surviving giraffe brought into Hluhluwe GR south of the Hluhluwe river between 1949 and 1957 (Bourquin *et al.*, 1971) and nine into Umfolozi GR in 1965/66 (Vincent, 1970; Brooks and Macdonald, 1983). Less successful was the re-introduction of 14 eland from the Drakensberg into Umfolozi in 1964 (Vincent, 1970). All except for a single male died within a few years, apparently due to tick infestations. During the mid-1990s over 20 eland were again released in Mfolozi, but all died of heartwater, a disease spread by ticks, or fell prey to lion (Craig Reid, personal communication). No eland currently remain in HiP. The final herbivore restoration was the re-introduction of elephants in Hluhluwe GR in 1981 (Brooks and Macdonald, 1983) and into Umfolozi GR in 1985 (Wills, 1986; see Chapter 14). Carnivore populations were also re-established during this period with the arrival of the first lion (*Panthera leo*) in 1958 followed by the introduction of more lions in 1965, and re-introductions of cheetahs (*Acinonyx jubatus*) in 1966 and wild dogs (*Lycaon pictus*) in 1980/81 (see Chapter 12).

4.4.2 The Era of Population Management and Concerns about Overgrazing: Late 1950s–1980

The 1960s saw the effective inclusion of the Corridor as part of the game reserve complex through fencing of its eastern and western boundaries. Fencing of the entire reserve complex was completed by the start of the 1970s (Brooks and Macdonald, 1983). The perception at the time was that the restriction on animal movement and lack of predators had led to excessive grazing impacts on the grasslands. Concerns about over-utilization were further fuelled by drought conditions prevailing during the mid-1950s, worsening grass cover losses, soil erosion and shrub encroachment in Hluhluwe GR (Ward, 1958). Consequently, culling was initiated in Hluhluwe GR and the northern Corridor in 1954 directed particularly at wildebeest, zebra, and warthog and extended to Umfolozi GR and the southern Corridor in 1960 (Table 4.2; Brooks and Macdonald, 1983). In 1969, 'Operation Rhino' was instigated to capture white rhinos and distribute the local subspecies more widely through other wildlife areas to reduce the risk that the entire world population, confined at that time mostly within Umfolozi GR, could be wiped out by a disaster such as a disease outbreak. White rhino removals were expanded during the early 1970s because of concern about widening 'overgrazing' during a sequence of low rainfall years, and other species that might

Table 4.2 *Numbers of various large herbivore species removed (live removal as well as shot animals) per decade since 1954. Data from the 1950s to the 1970s were extracted from Brooks and Macdonald (1983) and Bourquin and Hitchins (1979). 1: UGR + corridor south of Hlabisa road, 2: HGR + corridor north of Hlabisa road. Data from the later decades were extracted from the park's animal removal database (EKZNW, unpublished data)*

Species	1954–1959		1960s		1970s		1980s	1990s	2000s	2010–2014	Total
	UGR[1]	HGR[2]	UGR[1]	HGR[2]	UGR[1]	HGR[2]	HiP	HiP	HiP	HiP	
Buffalo	1	0	5	0	193	130	891	2347	1568	172	5307
Bushbuck	2	0	78	0	0	0	0	0	0	0	80
Bushpig	–	0	–	0	0	0	0	0	0	0	0
Duiker, grey	–	0	–	0	0	0	0	0	0	0	0
Duiker, red	–	0	–	0	0	0	0	0	0	0	0
Elephant	0	0	0	0	0	0	0	1	34	7	42
Giraffe	0	0	0	0	0	0	0	99	156	26	281
Impala	1	1599	771	6800	4522	7787	9613	3524	4013	981	39,611
Klipspringer	–	0	–	0	0	0	0	0	0	0	0
Kudu	0	0	7	21	88	4	106	132	60	0	418
Nyala	0	538	78	718	2038	2155	4633	1626	2528	165	14,479
Reedbuck, common	–	0	–	0	0	0	0	0	0	0	0
Reedbuck, mountain	–	0	–	0	0	0	0	0	0	0	0
Rhino, black	0	6	0	0	2	38	44	132	106	37	365
Rhino, white	0	0	74	0	1526	53	1318	896	459	318	4634
Steenbok	–	0	–	0	0	0	0	0	0	0	0
Warthog	302	4670	13,823	4724	3275	2869	997	151	388	32	31,231
Waterbuck	0	0	75	0	48	72	20	87	28	0	330
Wildebeest	25	2537	5038	3806	2027	1896	1366	0	146	115	16,809
Zebra	19	1052	870	684	148	641	651	71	135	58	4329
Grand total	350	10,402	20,819	16,753	13,867	15,645	19,639	8919	9621	1911	117,916

be competing with white rhinos for forage, especially wildebeest and warthog, were heavily culled within Umfolozi GR as well as in Hluhluwe GR (Brooks and Macdonald, 1983). Culling of all of the more common herbivores then became aimed at counteracting the influence of rainfall patterns on veld condition, with more animals being removed in dry years when veld conditions were considered to be vulnerable to overgrazing and fewer removals in wet years when veld conditions improved. During the 1960s, 1970s, and part of the 1980s, very large numbers of warthog, wildebeest, zebra, and nyala were culled in both Umfolozi and Hluhluwe (Table 4.2; Brooks and Macdonald, 1983). Impala were subject to especially heavy culling because they were perceived to be extra-limital (Table 4.2). Few pure browsers, represented by black rhino, kudu, and giraffe, were removed, mainly directed at live sales. The heavy removals resulted in very few herbivores dying of malnutrition during the severe 1982/83 drought, and low wildlife sightings by tourists when conditions improved after the drought (Walker *et al.*, 1987).

4.4.3 Towards Process-Based Management: 1980s

At the start of the 1980s the severity of the culling programme and the appropriateness of the agriculturally based approach to setting carrying capacities were called into question (Walker *et al.*, 1987). A reassessment of management policies led to the implementation of a process-based management approach, curtailing removals and allowing for natural population variation (see Chapter 1 for further details). Herbivore management changed to the manipulation of herbivore landscape use through strategic burning practices (Brooks and Macdonald, 1983). Moreover, culling increasingly included live capture and sale of wildlife to restock other protected areas. Finally, biennially repeated ground surveys using distance sampling (Burnham *et al.*, 1980) were implemented to provide less-biased estimates of total populations than provided by the aerial counts that were used in the late 1960s and 1970s (Knott, 1986; specific details on implementation are provided in Appendix 4.1). The strong reductions in annual animal removals led to population increases in several of the ungulate species from the 1980s until the early 2000s, particularly by wildebeest, zebra, and warthog (Figure 4.1; see Box 4.1 for an overview of the different counting methods and Box 4.2 for a description of population trends). Annual removals were strongly reduced by the mid-1980s for these species, but continued into the 2000s for impala, nyala and buffalo (Table 4.2).

Box 4.1 *Interpreting Results from Different Counting Methods*

Interpreting long-term population trends is particularly problematic due to the changes in estimation techniques. When comparing density estimates between periods, the differing biases in the different methods used must be taken into account. The population trends described in this chapter can be divided into three main periods according to the counting methods used: pre-1960s, 1960s to early 1980s, 1986 to present.

Pre-1960s: Early twentieth-century estimates resulted from foot counts or through expert-based guestimates at the time. For this reason, the early foot count estimates are excluded from the trends represented in the graphs in Figure 4.1.

1960s–1985: The first aerial count covering Hluhluwe GR, Umfolozi GR, and the Corridor was conducted in 1967 by fixed-wing aircraft. Further aerial counts were carried out using a helicopter annually from 1970–1976 (Melton, 1978a). Further aerial counts were done during 1981 (Brooks, 1982) and 1983 (Brooks *et al.*, 1983); a combination of fixed-wing, helicopter, and foot counts in 1981, and by helicopter in 1983. Because aerial counts incorporate an inherent undercount bias depending on the size and behaviour of the species (Melton, 1978b), aerial count totals have been compared with animals counted on the ground in local areas (Melton, 1978b; Bothma *et al.*, 1990; van Hensbergen *et al.*, 1996). This indicated that 70% or more of larger species like wildebeest and zebra are tallied by helicopter but only 33–50% of smaller species like impala and warthog, with more animals missed from a fixed-wing aircraft. The estimates from 1967 onwards presented in this chapter were extracted from Brooks and Macdonald (1983) and have been corrected for this undercount bias using specific correction factors per species reported in Brooks (1978).

1986–present: The dense woody cover over much of HiP, and the continuing uncertainty over the reliability of the aerial counts, led to the decision to move away from aerial counts. Accordingly, after initial trials (Knott and Venter, 1987, 1990) a switch was made starting in 1986 to carry out distance-sampling based ground counts using a set of transects representing most of HiP (see Appendix Figure A4.1 for a map and detailed methods). Since 1994, these ground surveys have been undertaken biennially using volunteers supported by park staff. Because of difficulties in access and restrictions on clearing transects, the wilderness section of Mfolozi was only effectively incorporated from 2006 onwards. These distance sampling estimates are inherently unbiased, but subject to greater sampling error, and therefore have been smoothed in the plots (Figure 4.1).

For consistency, the population trends shown by the more common ungulate species in HiP are plotted as density estimates separately for the Mfolozi section and for Hluhluwe plus the Corridor, as well as for the entire complex combined (Figure 4.1). Furthermore, the animals removed for population control (Table 4.2) must also be considered in interpreting the trends shown.

Box 4.2 *Noteworthy Features of the Population Trends Shown by Various Ungulate Species in Rough Order of their Body Size*

Elephant

Elephant numbers in HiP have increased exponentially following their re-introduction in the early 1980s. There is little sign as yet of any reduction in population growth rate.

White rhino

White rhinos were initially concentrated mostly in Umfolozi GR, but heavy removals substantially reduced the Umfolozi GR subpopulation while numbers were allowed to grow in the Corridor and Hluhluwe GR. Following recent reductions in animals removed, the Umfolozi subpopulation has reached similar density levels to those exhibited around 1970, and very similar densities also now prevail over the rest of HiP.

Black rhino

Pre-1960, the Hluhluwe black rhino population approached the highest density for the species recorded anywhere in Africa. After 1960, this population crashed following deaths from unknown causes. There was some compensation from increasing numbers in Umfolozi, but overall fewer black rhinos currently persist within HiP than the park had originally supported.

Giraffe

Giraffe grew rapidly in numbers following introductions and currently fluctuate around an overall density of 0.7 animals/km^2.

Buffalo

The buffalo population has grown progressively despite quite heavy removals during the late 1990s when culling was aimed at eliminating animals infected with bovine tuberculosis (see Chapter 13).

Zebra

Following the cessation of heavy culling after 1982, zebra initially increased towards the density levels that they had attained in the 1960s. However, since 2008 there has been a downturn in their numbers over the whole park.

Wildebeest

Despite removals being greatly reduced after 1982, wildebeest did not regain the numbers that they had attained in the late 1960s. Wildebeest numbers went into steep decline after 2008 and hardly any wildebeest remained in Hluhluwe by 2014.

Waterbuck

Waterbuck have been a source of concern because of the progressive decline in their numbers (currently estimated at roughly 120 animals in total). Very few waterbuck were removed.

Kudu

As for waterbuck, kudu have also incurred little culling. Nevertheless, their population density has also declined progressively since the late 1970s.

Nyala

Heavy culling during the 1970s and early 1980s reduced nyala densities considerably below their peak levels attained around the early 1970s. From the mid-1980s their numbers were stabilized by continuing removals. However, their density declined after 2008, particularly in Mfolozi, despite the curtailment of culling since the late 2000s.

Impala

Heavy culling kept impala density at or below 10 animals/km^2 during the 1970s and 1980s. With reduced removals after 1982, the impala population grew towards a density approaching 30 animals/km^2 within Mfolozi. A decline in impala density became evident in Hluhluwe after 2002 and in Umfolozi after 2008.

Warthog

The high density of warthogs that occurred prior to the 1980s was greatly reduced by culling. The population recovered until 2001, but has declined strongly during the 2000s without massive culling.

Less-Common Ungulate Species

Several ungulate species are currently too uncommon or localized to be counted reliably, so that only broad decadal trends can be described (Table 4.3). Among these species, grey duiker, red duiker, and bushpig (*Potamochoerus larvatus*) seem to be maintaining their abundance during recent decades. However, grey duiker had been far more numerous prior to the 1970s when their local density in western Umfolozi GR was estimated to be four to five animals/km^2 (Owen-Smith, 1973). Common reedbuck, mountain reedbuck, bushbuck, blue duiker, and steenbok have declined to critically low numbers.

4.4.4 Population Declines by Several Small- to Medium-Sized Antelope Species

Despite an increase in populations of many species since the early 1980s, several small- to medium-sized antelopes continued their declines to numbers that can hardly be regarded as viable. While more than 1700 waterbuck were reported in 1972 (Brooks and Macdonald, 1983), the current estimates show a mere 120 remaining park-wide (Figure 4.1). Similarly, common and mountain reedbuck populations that were estimated to still number 623 and 152 individuals, respectively, during 1972 (Brooks and Macdonald, 1983) have now almost disappeared

Table 4.3 *Population estimates for species that currently occur in low numbers within Hluhluwe-iMfofozi Park. The 1982 estimate is from Brooks and Macdonald (1983). From 1986 onwards estimates are based on distance sampling records aggregated over 10 years to ensure a robust approximation*

Period	Bushpig	Grey duiker	Red duiker	Blue duiker	Bushbuck	Common reedbuck	Mountain reedbuck	Steenbok
1982	346	1410	455	?	479	165	136	235
1986–1994	106	794	343	6	73	56	40	93
1995–2004	100	398	232	6	58	13	16	9
2005–2014	97	442	329	5	25	10	21	3

(Table 4.3). Other smaller antelope species that have shown drastic declines are bushbuck, steenbok, and blue duiker (*Philantomba monticola*), all with population sizes of < 50. While bushbuck were still numerous at the end of the anti-nagana campaigns in 1952 (1000s), the current population estimate is 25 for the entire park.

Several factors could have been responsible for the population declines by these species. For reedbuck (Deane, 1966; Anderson, 1979), and steenbok (Anderson, 1979; Brooks and Macdonald, 1983), deterioration of grasslands due to ongoing encroachment by woody plants since the early 1950s has been invoked (Ward, 1958; see also Chapter 3). Competition with other grazers has been suggested for reedbuck (Deane, 1966) and waterbuck (Melton, 1987). Declines in bushbuck, and perhaps blue duiker, have been attributed to competition with the increasing nyala population (Anderson, 1979; Brooks and Macdonald, 1983). On a local scale within Hluhluwe, wetland drainage brought about by the construction of roads may have altered grassland quality, possibly further isolating subpopulations of specialist feeders such as reedbuck and waterbuck and contributing to their population crashes (Tony Whateley, personal comments). All in all, there is no conclusive evidence for any of the mentioned drivers of population decline, and a combination of drivers likely played a role.

4.4.5 Effects of Increasing Predator Populations: 2000s

During the past few decades, predator populations have recovered following successful re-introductions (see Chapter 12). Since 2004 the lion population has increased about twofold whereas the wild dog population increased three- or fourfold between 2002 and 2010 (although it has shrunk again since). This increase in predators coincides with drops in the numbers of zebra, wildebeest, nyala, and impala (specifically since 2008; Figure 4.1). All of these species have been shown to be the preferred prey species of predators (Chapter 12). In contrast, buffalo showed a slight increase over the past several decades (Figure 4.1B) despite their vulnerability to lion predation. However, over the past 15 years the buffalo population has been affected by a bovine tuberculosis (*Mycobacterium bovis*) control programme (Chapter 13). This programme has dramatically reduced the prevalence of this disease in the HiP buffalo population, which may have boosted the population growth rate through lowered mortality rates and enhanced fecundity. Such positive effects on growth rate may be counteracting any effects of possible increases in predation pressure.

Interestingly, the decline in numbers for species such as impala and wildebeest started earlier and has been much steeper in Hluhluwe than in other regions of the park (Figure 4.1E,I), which may indicate an exacerbated effect of predation due to the woody plant encroachment within Hluhluwe (Chapter 3).

4.4.6 Changes in Functional Composition

HiP has experienced some consistent trends in the functional composition of its ungulate assemblage during the last 40–50 years. Following the 1960s' and 1970s' culling removals, overall herbivore biomass increased substantially (Figure 4.2). However, while mixed feeders and, to a lesser extent, browsers have increased park-wide, increase among grazers has been restricted to the Corridor. White rhino and buffalo currently contribute equally to more than 90% of the grazer biomass and growth in grazer biomass can be mostly attributed to these two species. Perhaps even more striking is the mounting influence of the largest herbivores (body mass > 400 kg) during the last 40–50 years, specifically elephant, giraffe, black and white rhino, and buffalo (Figure 4.3). Hippos also still occur in the park, but numbers are very low and fluctuate quite strongly driven by variation in water levels in perennial water sources in the park. Species in this size range contribute more than three times as much to total herbivore biomass than they did in 1980, whereas species the size of zebra and smaller have decreased in aggregate biomass by a third in the same period. The strong increase in large-bodied species is mostly due to the expansion of numbers of white rhino within the Corridor and Hluhluwe plus the recovery of their numbers within Mfolozi towards earlier peak densities, combined with the growth in the elephant, and to some extent buffalo, populations. This has led to a situation where the total biomass of large herbivores within HiP currently exceeds 10,000 kg/km^2, rivalling that in the Serengeti ecosystem and double that in Kruger National Park (see Figure 5.1 in Chapter 5). Another striking pattern is the drastic decline of the small body-size category (dominated by impala and nyala) during the early 1980s, particularly in Hluhluwe GR. The early 1980s were characterized by a severe drought, a final phase of heavy culling of impala and nyala and the introduction of a main predator of impala and nyala – wild dogs – into Hluhluwe GR. The combination of these three factors could have driven the drastic decline of small-bodied ungulates in Hluhluwe GR during the early 1980s.

Changes in functional composition of herbivores could have far-reaching consequences for the structure and composition of HiP's savanna

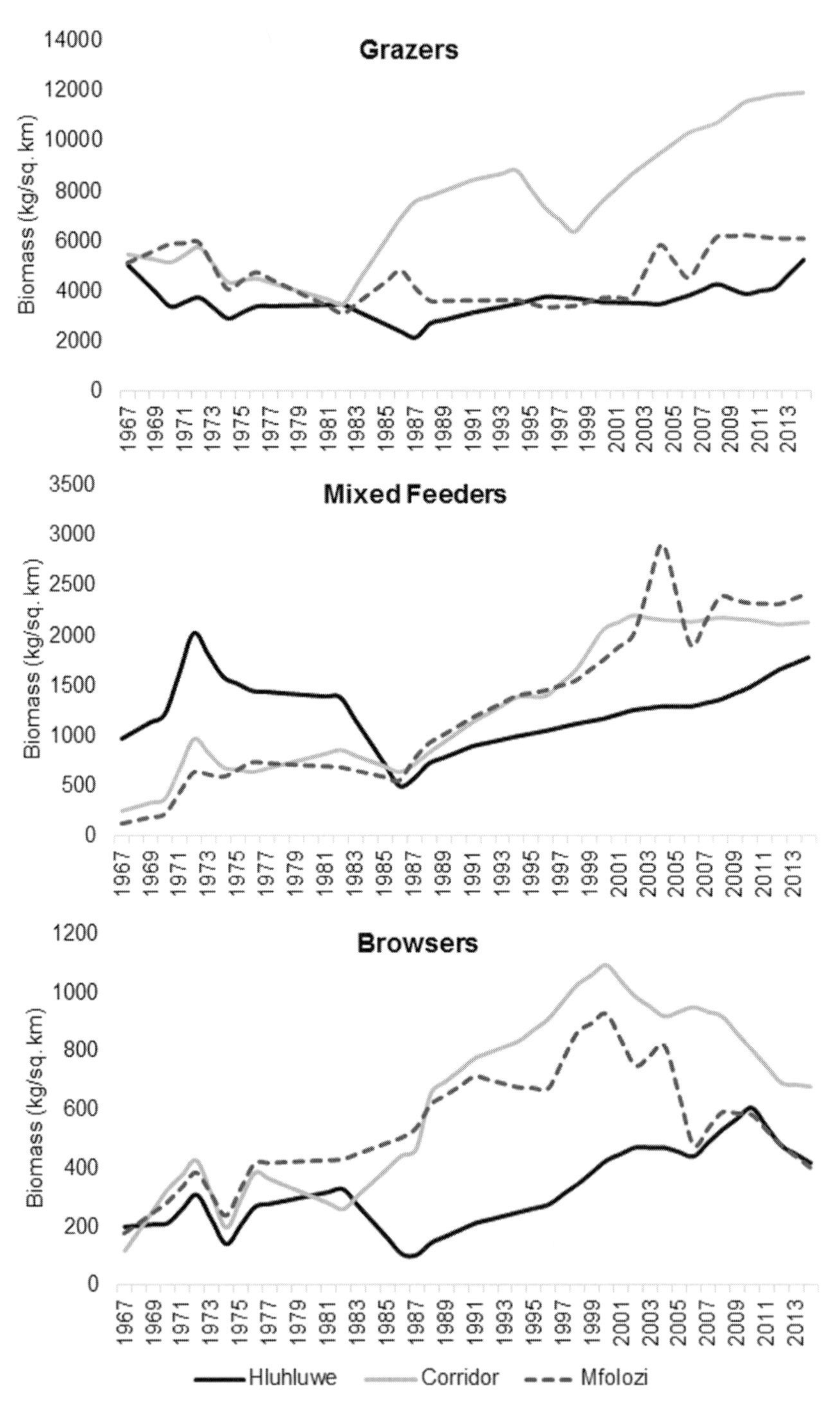

Figure 4.2 Changes in biomass density for herbivores of different functional groups between the years 1967 and 2014. Functional groups are categorized as follows:

grasslands and woodlands (refer to Chapters 5 and 6 for white rhino, and Chapter 14 for elephant). Competitive and facilitative relationships between the expanding white rhino population and other grazers are considered in Chapter 5.

4.5 Concluding Remarks

HiP's ungulate populations showed wide fluctuations and varying trends prior to 1982, driven mainly by human influences. Following the curtailment of animal removals, densities of the main grazers increased, until recent downturns especially in wildebeest, zebra, and warthog. Even the two browsers least affected by culling, i.e. giraffe and kudu, showed little indication of a persistent 'carrying capacity'. While studies elsewhere have demonstrated how rainfall variation controls population trends of savanna herbivores (Ogutu and Owen-Smith, 2003), such a relationship is not evident in the HiP data. Rainfall in HiP tended to be above average through the 1990s and below average after 2000 (Chapter 2). Density estimates from distance sampling are subject to quite wide confidence limits, and are obtained only biennially, possibly obscuring the short-term effects of annual or seasonal rainfall on population growth.

The most recent trends in population patterns, particularly of grazers such as zebra and wildebeest, raise intriguing questions about predator–prey dynamics in small fenced reserves. Theory suggests that the growing abundance of predators will lead to coupled oscillations in predator–prey abundance. Will this happen in HiP, or will populations of herbivores stabilize at some lower abundance by becoming confined to habitats where they are relatively more secure from predation? Although we currently lack the answer to this question, it seems clear that we are

Figure 4.2 (caption continued) grazers (white rhino, buffalo, zebra, wildebeest, waterbuck, warthog, common reedbuck, and mountain reedbuck); browsers (black rhino, giraffe, kudu, bushbuck, grey duiker, red duiker, steenbok, and blue duiker); and mixed feeders (elephant, nyala, and impala). Separate estimates are presented for Hluhluwe, the Corridor, and Mfolozi. Mfolozi biomass estimates prior to 1986 and after 2006 are calculated for an area including the Wilderness. Elephant numbers were not available per section as they roam across the entire park. For the purpose of this figure we assumed that elephants were distributed evenly across the park and therefore added the total biomass contribution of elephants to each section weighted by section area. Three-quarters average female body-weight estimates, obtained from Owen-Smith (1988), were used in biomass calculations.

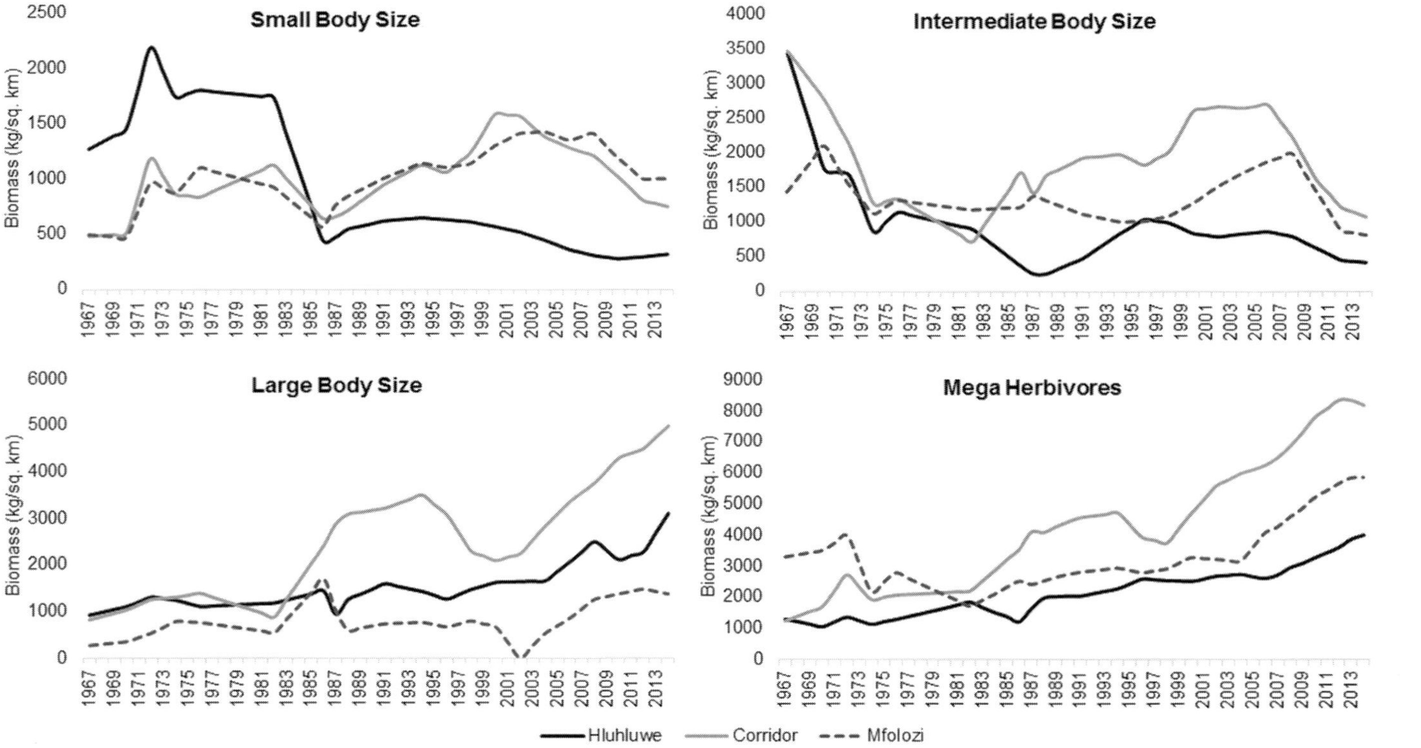

Figure 4.3 Changes in biomass density for herbivores of different size classes between the years 1967 and 2014, derived as in Figure 4.2. For the rare species for which small sample size only allowed park-wide estimates to be calculated, we assumed that the proportional distribution across sections remained the same as it was in 1982. Size classes were distinguished as small (adult male weight less than 150 kg, including blue duiker, steenbok, red duiker, grey duiker, mountain reedbuck, bushbuck, impala, common reedbuck, warthog, and nyala); intermediate (adult male weight 150–400 kg, including kudu, waterbuck, wildebeest, and zebra); large (adult male weight 400–1000 kg, including buffalo only); and megaherbivores (adult male weight > 1000 kg, including giraffe, black rhino, white rhino, and elephant).

currently experiencing a re-establishment of the predator–prey dynamics in HiP. One of the key current concerns and questions for the conservation management of HiP is what trajectory the carnivore–prey dynamics will follow (see Chapter 12).

The herbivore species that have declined to numbers threatening their survival in the park are certain of the smaller ungulate species. They appear unable to cope with habitat changes, increased competition with expanding populations of other ungulate species or higher predator numbers. This is contrary to general assumptions about the kinds of species most vulnerable to extinction. It is widely assumed that very large species are most in danger because of their slow population growth rates (see e.g. Cardillo *et al.*, 2005). On the contrary, observations in HiP show that it is some of the smaller species that are more vulnerable, because of their narrower habitat requirements and vulnerability to predation. This situation can be drastically altered when humans become predators on the very largest species, albeit merely for parts such as tusks and horns (Chapter 11). Regrettably, the current rhino and elephant poaching crisis spreading across Africa could soon alter the upwards trends of even these megaherbivores.

Appendix 4.1. Details on Methods for Estimating Population Abundances from the Distance-Sampling Transects

From 1986 onwards, the estimation technique was a ground census technique using line transects with visibility bias corrected using distance sampling protocols developed by Burnham *et al.* (1980) and refined by Buckland *et al.* (1993). Since 1994, these surveys have been undertaken biennially.

The transect network consists of 34 transects (Figure A4.1) ranging in length from 3.88 to 12.94 km and totalling approximately 281 km. Transects are evenly distributed across the park and cover roughly 30% of the area. All except the eight transects situated in the Wilderness section are cut prior to each census. In an attempt to adhere to the management policy of minimal interference in the wilderness, population surveys in this section were initially performed using point transects. However, by 2006, the point transect method was abandoned as unsuccessful and the line transect method was extended to the wilderness, although the transect lines remained uncut. For this reason only the Wilderness estimates from 2006 onwards are incorporated in the trends discussed in this chapter.

Figure A4.1 Map of Hluhluwe-iMfolozi Park showing line transect positions in each of the four sections (shaded).

Each transect is walked approximately 14 times on a three-day rotation. Surveys are completed in the late dry season between July and October.

Census-derived density estimates were calculated using distance sampling methods and the software DISTANCE (version 6.2, Thomas *et al.*, 2010). By using information on the position of each animal relative to the position of the observer, this technique takes into account the decreasing possibility of detecting animals that are further from the observer (Buckland *et al.*, 1993). Depending on sighting frequency, a survey-specific detection function was calculated either separately per section or a park-wide detection function was used to calculate section-specific densities.

For the purposes of this chapter, census derived density estimates were calculated separately per section. For a few rare species, notably blue duiker, red duiker, waterbuck, bushbuck, common reedbuck, mountain reedbuck, steenbok, and bushpig, observations were amalgamated over a 10-year period to ensure a statistically robust density estimate.

While suitable for most large mammalian species, ground counts are inappropriate for species with shy habits such as black rhino or dangerous species such as elephant. In this instance, these species were surveyed through a combination of aerial counts aimed specifically at quantifying these populations and mark–recapture techniques using trap cameras and ad hoc staff sightings.

4.6 References

Anderson, J. L. (1979) The effects of past vegetation changes on the mammals of the complex. Paper presented at the workshop on vegetation dynamics of the central complex, Hluhluwe Game Reserve, 1–2 August 1979. Unpublished report, Natal Parks Board, Pietermaritzburg.

Anon. (1952) Nature Reserves of Natal and Zululand. *Oryx* **1**: 235–242.

Baldwin, W. C. (1894) *African hunting and adventure from Natal to the Zambezi, including Lake Ngami, the Kalahari Desert, etc., from 1852 to 1860*. Richard Bentley/ Macmillan & Co., London.

Ballard, C. (1983) The great rinderpest epidemic in Natal and Zululand: a case study of ecological break-down and economic collapse. Seminar paper, African Studies Institute, University of the Witwatersrand.

Bothma, J. P., Peel, M. J. S., Pettit, S., & Grossman, D. (1990) Evaluating the accuracy of some commonly used game-counting methods. *South African Journal of Wildlife Research* **20**: 26–32.

Bourquin, O. (1968) Summary of results of game counts carried out in the Umfolozi/Corridor/Hluhluwe complex, 1967. Unpublished report, Natal Parks Board, Pietermaritzburg.

Bourquin, O. & Hitchins, P. M. (1979) Faunal and vegetation changes in the Hluhluwe Game Reserve 1937–1970. Unpublished Report, Natal Parks Board, Pietermaritzburg.

Bourquin, O., Vincent, J., & Hitchins, P. M. (1971) The vertebrates of the Hluhluwe Game Reserve–Corridor (State-land)–Umfolozi Game Reserve Complex. *The Lammergeyer* **14**: 1–58.

Brooks, P. M. (1978). Ungulate population estimates for H.G.R. and the N. Corridor for July 1978, based on foot and helicopter counts. Unpublished report, Natal Parks Board, Pietermaritzburg.

Brooks, P. M. (1982) Population estimates and stocking rates of ungulates in Hluhluwe Game Reserve and northern Corridor calculated for April 1982 for use in the 1982–1984 removal cycle. Unpublished report, Natal Parks Board, Pietermaritzburg.

Brooks, P. M. & Macdonald, I. A. W. (1983) The Hluhluwe-Umfolozi Game Reserve: an ecological case history. In: *Management of large mammals in African conservation areas* (ed. R. N. Owen-Smith), pp. 51–77. Haum, Pretoria.

Brooks, P. M., Knott, A. P., & Whateley, A. (1983) Helicopter count results for Hluhluwe and Umfolozi Game Reserves for August 1983, and their implications for the 1982–1985 game control programme. Unpublished report, Natal Parks Board, Pietermaritzburg.

Brooks, S. (2006) Human discourses, animal geographies: imagining Umfolozi's white rhinos. *Current Writing* **18**: 6–27.

Brown, K. (2008) From Ubombo to Mkuzi: disease, colonial science, and the control of nagana (livestock trypanosomosis) in Zululand, South Africa, c. 1894–1953. *Journal of the History of Medicine and Allied Sciences* **63**: 285–322.

Buckland, S. T., Anderson, D. R., Burnham, K. P., *et al.* (1993) *Distance sampling: estimating abundance of biological populations*. Chapman & Hall, London.

Burnham, K. P., Anderson, D. R., & Laake, J. L. (1980) Estimation of density from line transect sampling of biological populations. *Wildlife Monographs* **72**: 3–202.

Cardillo, M., Mace, G. M., Jones, K. E., *et al.* (2005) Multiple causes of high extinction risk in large mammal species. *Science* **19**: 123–241.

Chapin, F. S., Walker, B. H., Hobbs, R. J., *et al.* (1997) Biotic control over the functioning of ecosystems. *Science* **277**: 500–504.

Clark, J. L. & Plug, I. (2008) Animal exploitation strategies during the South African Middle Stone Age: Howiesons Poort and post-Howiesons Poort fauna from Sibudu Cave. *Journal of Human Evolution* **54**: 886–898.

Cramer, M. D. & Mazel, A. D. (2007) The past distribution of giraffe in KwaZulu-Natal. *South African Journal of Wildlife Research* **37**: 197–201.

Deane, N. N. (1966) Ecological changes and their effect on a population of reedbuck (*Redunca arundinum* (Boddaert)). *Lammergeyer* **6**: 2–8.

Delegorgue, A. (1847) *Voyage dans l'Afrique Australe*, reproduced as *Adulphe Delegorgue's travels in Southern Africa*, Vol. 1 (F. Webb. (transl.), S. J. Alexander & C. de B. Webb (eds), 1990). University of Natal Press, Pietermaritzburg.

Drummond, W. H. (1875) *The large game and natural history of South and South-East Africa, from the journals of the Hon. W. H. Drummond*. Edmonston and Douglas, Edinburgh.

Faith, J. T. (2014) Late Pleistocene and Holocene extinctions on continental Africa. *Earth-Science Reviews* **128**: 105–121.

Findlay, F. R. N. (1903) *Big game shooting and travels in South-East Africa: an account of shooting trips in the Cheringoma and Gorongoza divisions of Portuguese South-East Africa and in Zululand*. T. Fisher Unwin, London.

Foster, W. (1955) A history of Umfolozi Game Reserve. Unpublished report, Natal Parks Board, Pietermaritzburg.

Fynn, H. F. (1950) *The diary of Henry Francis Fynn*, compiled from original sources and edited by J. Stuart and D. McK. Malcolm. Shuter & Shooter, Pietermaritzburg.

Goodman, P. S. & Tomkinson, A. J. (1987) The past distribution of giraffe in Zululand and its implications for reserve management. *South African Journal of Wildlife Research* **17**: 28–32.

Henkel, J. S. (1937) *Report on the plant and animal ecology of the Hluhluwe Game Reserve with special reference to tsetse flies*. The Natal Witness, Pietermaritzburg, South Africa.

Hitchins, P. M. & Vincent, J. (1972) Observations on range extension and dispersal of impala (*Aepyceros melampus Lichtenstein*) in Zululand. *Journal of the South African Wildlife Management Association* **2**: 3–8.

Knott, A. P. (1986) Minutes of Workshop on Management Strategies in Hluhluwe and Umfolozi Game Reserves held at Masinda Lodge Lounge, U.G.R. 2nd–4th December 1985. Unpublished report, Natal Parks Board, Pietermaritzburg.

Knott, A. P. & Venter, J. (1987) An observation on the relative accuracy of two transect census techniques. *Lammergeyer* **38**: 2–1.

Knott, A. P. & Venter, J. (1990) A field test on the accuracy and repeatability of a line transect method. *Lammergeyer* **41**: 1–2.

Lydekker, R. (1908) *The game animals of Africa*. Rowland Ward, London.

McCracken, D. P. (2008) *Saving the Zululand wilderness. An early struggle for nature conservation*. Jacana Media, Johannesburg.

Melton, D. A. (1978a) The validity of helicopter counts as indices of trend. *Lammergeyer* **26**: 3–3.

Melton, D. A. (1978b) Undercounting bias of helicopter censuses in Umfolozi Game Reserve. *Lammergeyer* **26**: 1–6.

Melton, D. A. (1987) Waterbuck (*Kobus ellipsipyrmnus*) population dynamics: the testing of an hypothesis. *African Journal of Ecology* **25**: 13–45.

Mentis, M. T. (1970) Estimates of natural biomasses of large herbivores in the Umfolozi Game Reserve area. *Mammalia* **34**: 363–393.

Minnaar, A. de V. (1989) Nagana, big-game drives and the Zululand Game Reserves (1890s to 1950s). *CONTREE* **25**: 12–21.

Ogutu, J. O. & Owen-Smith, N. (2003) ENSO, rainfall and temperature influences on extreme population declines among African savanna ungulates. *Ecology Letters* **6**: 41–19.

Owen-Smith, R. N. (1973) The behavioural ecology of the white rhinoceros. PhD thesis, University of Wisconsin.

Owen-Smith, R. N. (1988) *Megaherbivores. The influence of very large body size on ecology*. Cambridge University Press, Cambridge.

Paine, R. T. (2000) Phycology for the mammologist: marine rocky shores and mammal dominated communities: how different are the structuring processes? *Journal of Mammalogy* **81**: 637–648.

Player, I. C. & Feely, J. M. (1960). A preliminary report on the square-lipped rhinoceros *Ceratotherium simum simum*. *The Lammergeyer* **1**: 2.

Plug, I. (2004) Resource exploitation: animal use during the Middle Stone Age at Sibudu Cave, KwaZulu-Natal. *South African Journal of Science* **100**: 151–158.

Plug, I. & Badenhorst, S. (2001) The distribution of macromammals in southern Africa over the past 30 000 years. *Transvaal Museum Monograph* **12**, Transvaal Museum, Pretoria.

Rowe-Rowe, D. T. (1994) *The ungulates of Natal*. Natal Parks Board, Pietermaritzburg.

Selous, F. C. (1908) *African nature notes and reminiscences*. Macmillan and Co., Ltd, London.

Skinner, J. D. & Chimimba, C. T. (2005) *The mammals of the southern African subregion*, 3rd edn. Cambridge University Press, Cambridge.

Thomas, L., Buckland, S. T., Rexstad, E. A., *et al.* (2010) Distance software: design and analysis of distance sampling surveys for estimating population size. *Journal of Applied Ecology* **47**: 5–14.

Van Hensbergen, H. J., Berry, M. P. S., & Junitz, J. (1996) Helicopter-based line transect estimates of some southern African game populations. *South African Journal of Wildlife Research* **26**: 8–7.

Vincent, J. (1969) The status of the square-lipped rhinoceros, *Ceratotherium simum simum* (Burchell), in Zululand. *Lammergeyer* **10**: 12–21.

Vincent, J. (1970) The history of Umfolozi Game Reserve, Zululand, as it relates to management. *Lammergeyer* **11**: 7–49.

Walker, B. H., Emslie, R. H., Owen-Smith, R. N., *et al.* (1987) To cull or not to cull: lessons from a southern African drought. *Journal of Applied Ecology* **24**: 381–401.

Ward, C. J. (1958) Report on the Proceedings on the Occasion of the Visit of the Ecologist Mr. C. J. Ward to a Meeting of the Board on 8 August 1958. Unpublished report, Natal Parks Board, Pietermaritzburg.

Ward, R. (1896) *Records of big game containing an account of their distribution, descriptions of species, lengths and weights, measurements of horns and field notes – for the use of sportsmen and naturalists*. Rowland Ward, London.

Wills, A. J. (1986) Re-establishment of elephant in the Hluhluwe and Umfolozi Game Reserves, Natal, South Africa. *Pachyderm* **7**: 1–3.

Part II

Theoretical Advances in Savanna Ecology

5 · *Megaherbivores, Competition and Coexistence within the Large Herbivore Guild*

NORMAN OWEN-SMITH, JORIS P. G. M. CROMSIGT, AND RANDAL ARSENAULT

5.1 Introduction

A special feature of the Hluhluwe-iMfolozi Park (HiP) is the presence of an abundant 'mega-grazer' in the form of the white rhinoceros (*Ceratotherium simum*). The study of this species in HiP (Owen-Smith, 1974) led to the identification of common features shared by white rhinos with other extremely large herbivores and hence to the recognition that these 'megaherbivores' constitute a distinct life form (Owen-Smith, 1988, 2013a). Defined strictly, the label 'megaherbivore' encompasses terrestrial mammals exceeding one metric tonne (i.e. a mega-gram) in adult body mass. The distinguishing ecological and life-history features of these megaherbivores include (1) invulnerability to non-human predation in the adult stage; (2) birth interval exceeding 1 year; (3) maximum rate of population growth typically less than 10% per year; (4) dominance of large herbivore biomass; (5) dietary tolerance for plant structural fibre; and (6) capacity to transform vegetation structure. Extant species manifesting this syndrome include two species of elephant, four rhino species, the hippopotamus, and, marginally, the giraffe (Table 5.1).

While palaeontologists have applied the label 'megafauna' to encompass species weighing more than 45 kg (100 pounds), no functional transition is associated with the latter size threshold. Prior to the end of the Pleistocene, megaherbivores were widely represented on all continents (Owen-Smith, 2013a,b). Those formerly present in Europe and northern Asia included the woolly mammoth (*Mammuthus primigenius*)

Conserving Africa's Mega-Diversity in the Anthropocene, ed. Joris P. G. M. Cromsigt, Sally Archibald and Norman Owen-Smith. Published by Cambridge University Press.

Table 5.1. *Extant megaherbivores, their maximum body mass and historical distributions*

Common name	Scientific name	Maximum body mass (kg)		Historical distribution
		Male	Female	
African elephant	*Loxodonta africana*	6300	3500	Africa-wide
Asian elephant	*Elephas maximus*	5400	3000	Tropical Asia
White rhinoceros	*Ceratotherium simum*	2300	1700	Southern and north-eastern Africa
Black rhinoceros	*Diceros bicornis*	1300	1300	Africa-wide
Indian rhinoceros	*Rhinoceros unicornis*	2100	1600	Northern India and adjoining countries
Javan rhinoceros	*R. sondaicus*	1300	1300	Java
Hippopotamus	*Hippopotamus amphibius*	2000	1850	Africa-wide
Giraffe	*Giraffa camelopardalis*	1500	1050	Africa-wide

and woolly rhino (*Coelodonta antiquitatus*), both largely grazers, plus two browsing rhinos. North and South America were inhabited by additional species of mammoth, the mastodont, gomphotheres, giant ground sloths, and a hippo-like notoungulate. Australia formerly housed a giant marsupial in the family Diprotodontidae. In Africa, a grazing elephant (*Elephas recki*) and second hippo (*Hippopotamus gorgops*) survived until the mid–late Pleistocene. Cascading extinctions of all megaherbivores and numerous other large mammals followed shortly after the entry of modern humans on all continents outside Africa and tropical Asia during the late Pleistocene (Barnovsky *et al*., 2004). HiP retains a full suite of extant African megaherbivores, including grazing white rhinos and hippos, mixed-feeding elephants, and browsing black rhinos and giraffes, alongside many less-large grazers and browsers (see Chapter 4).

White rhinos were abundant through much of southern Africa prior to the arrival of Europeans with firearms. Harris (1838) reported seeing 80 white rhinos while hunting near the Magaliesberg in what is now the North-West Province of South Africa, while Smith (1849) encountered over 100 white rhinos during a day's journey with ox-wagons through this region. White rhinos were especially common in Botswana

(Andersson, 1856), and Selous (1899) encountered them throughout Zimbabwe while hunting during the 1870s. The Zambezi river formed the northern limit of their distribution in southern Africa, while the area that became HiP lay at their southern limit. However, white rhinos were absent from the Highveld grassland region. A northern subspecies was present to the west of the Nile river in north-central Africa, but is now extinct in the wild. The absence of white rhinos from apparently suitable habitat in the region between the Zambezi and Nile rivers is evidently of quite recent origin. White rhino remains are abundant in fossil deposits at Olduvai Gorge in Tanzania dated at early to mid Pleistocene. Teeth found on the surface in Tanzania and Kenya plus cave paintings indicate that white rhinos persisted in eastern Africa into the Holocene (Hooijer and Patterson, 1972). It seems likely that the distribution gap is a legacy of past human predation, paralleling the extinctions of megaherbivores elsewhere in the world after the arrival of human hunters with effective spears (Barnovsky and Lindsey, 2010). The fortuitous survival of white rhinos south of the Zambezi river may have been due to the livestock dependency of the Iron Age pastoralists and cultivators who displaced the earlier inhabitants with Stone Age technology (Chapter 1).

White rhinos currently contribute nearly half of the grazing biomass and consume over one-third of all grass eaten by large herbivores in HiP, allowing for effects of body size on metabolic requirements (Owen-Smith, 1988; Waldram *et al.*, 2008). Consequently, the biomass of large herbivores that HiP supports matches that in the Serengeti ecosystem and is twice as great as that in the southern half of Kruger Park (Figure 5.1). Moreover, populations of the largest herbivores in HiP were still growing following the cessation of most culling and the re-introduction of elephants (Chapter 4). Hence it is only in HiP that the full community and ecosystem impacts of a widespread mega-grazer at regional densities approaching three animals/km^2 and local densities exceeding five animals/km^2 can be observed. The closest approach to these conditions is in southern Kruger Park where re-introduced white rhinos have reached local densities of up to two animals/km^2 (Cromsigt and te Beest, 2014).

Coexisting alongside white rhinos in HiP are all of the other large mammalian herbivores that were historically present in the region (Chapter 4), apart from eland (*Taurotragus oryx*; Table 5.2). The continuing growth in the white rhino population, despite annual offtakes under the sink management strategy (Chapter 11), potentially has both beneficial and detrimental consequences for other large herbivores and for the structure and composition of the vegetation. The issues that we will

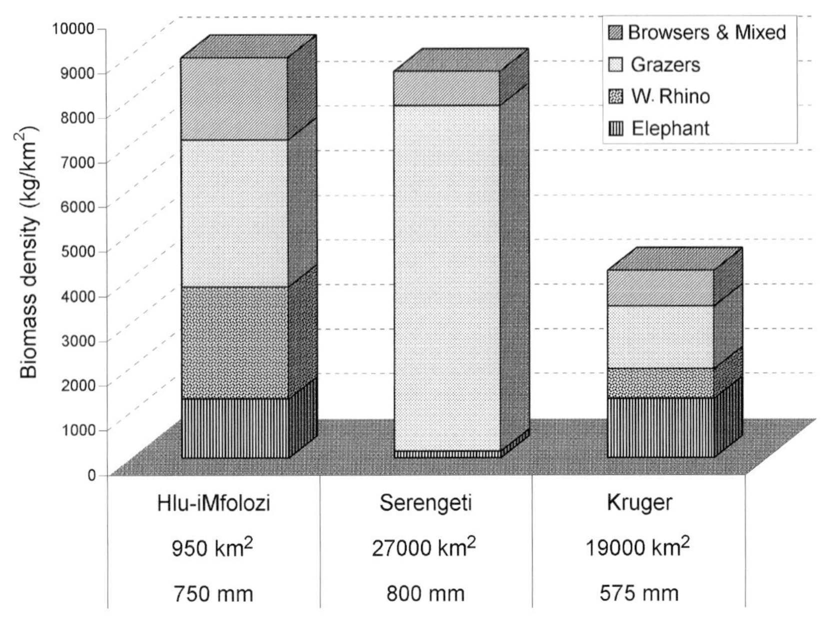

Figure 5.1 Comparative herbivore biomass densities subdivided by dietary categories for Hluhluwe-iMfolozi Park (from mean population totals for 1986–2008 shown in Table 5.2), Serengeti ecosystem (from Mduma and Hopcraft, 1995), and the southern half of Kruger National Park (mean of aerial counts 1980–1993 corrected for undercounting bias except for elephants and rhinos for which most recent estimates were used). Mean body mass is taken to be three-quarters of adult female body mass, from Owen-Smith (1988).

address in this chapter are as follows. (1) How might the growth of the white rhino population ultimately be regulated through its trophic interaction with vegetation, in the absence of predation on adult animals? (2) How do the grazing impacts of white rhinos alter the structure and composition of the herbaceous vegetation layer? (3) What are the consequences of the vegetation transformations wrought by white rhinos for competitive and facilitative interactions within the grazer guild, and hence for the coexistence of other large herbivores?

5.2 Population Regulation in the Absence of Predation

White rhinos do not feature among kills by lions and other large carnivores recorded in HiP (Chapter 12), and there are no records of adult white rhinos being killed by lions or other predators in Kruger National

Table 5.2. *Herbivore populations within the Hluhluwe-iMfolozi Park related to feeding style and body mass (Owen-Smith, 1988; Kingdon and Hoffmann, 2013). Population totals represent the mean over 1986–2008 derived from the density estimates obtained from distance sampling projected over the 950 km² total extent of the park (Chapter 4) except for elephants (2014 total from helicopter counts), and hippo (approximate estimate)*

Common name	Scientific name	Diet category	Maximum body mass (kg)		Population total
			Male	Female	
African elephant	*Loxodonta africana*	Mixed	8000	4000	698
White rhino	*Ceratotherium simum*	Grazer	2300	1700	1678
Hippo	*Hippopotamus amphibius*	Grazer	2065	1850	30
Black rhino	*Diceros bicornis*	Browser	1300	1200	230
Giraffe	*Giraffa camelopardalis*	Browser	1500	1050	600
African buffalo	*Syncerus caffer*	Grazer	860	640	4193
Plain's zebra	*Equus quagga*	Grazer	375	385	2979
Greater kudu	*Tragelaphus strepsiceros*	Browser	345	210	1186
Waterbuck	*Kobus ellipsiprymnus*	Grazer	290	210	510
Blue wildebeest	*Connochaetes taurinus*	Grazer	280	230	2476
Nyala	*Tragelaphus angasi*	Mixed	140	80	6086
Warthog	*Phacochoerus africanus*	Grazer	105	75	2462
Common reedbuck	*Redunca arundinum*	Grazer	105	65	35
Bushpig	*Potamochoerus porcus*	Omnivore	80	55	102
Impala	*Aepyceros melampus*	Mixed	75	55	13,288
Bushbuck	*Tragelaphus scriptus*	Browser	55	40	66
Mountain reedbuck	*Redunca fulvorufula*	Grazer	40	35	29
Grey duiker	*Sylvicapra grimmia*	Browser	21	25	645
Red duiker	*Cephalophus natalensis*	Browser	17	18	289
Steenbok	*Raphicerus campestris*	Browser	14	15	51
Blue duiker	*Cephalophus monticola*	Browser	4	5	6

Park, although immature white rhinos sometimes fall victim (Pienaar, 1969; Owen-Smith and Mills, 2008). When disturbed, subadult white rhinos adopt a defensive formation standing with rumps pressed together, facing outwards. This response seems designed to ward off potential predators. Furthermore, white rhino mothers with small calves stand protectively over the infant, rather than running off. White rhinos flee only in response to the threat posed by human intruders. While young white rhinos are seldom attacked by lions or spotted hyenas in HiP, there are records of black rhino calves being killed by spotted hyenas in HiP (Hitchins and Anderson, 1983) and by lions elsewhere (Brain *et al.*, 1999). The hippos killed by lions in Kruger Park are mostly young animals. In Botswana and Zimbabwe, young elephants as large as half-grown may be killed by lions hunting in large prides (Joubert, 2006; Loveridge *et al.*, 2006).

With little impact from predation, white rhino populations must ultimately be regulated either through effects of their feeding on food resources or via social mechanisms. While territoriality could be an effective regulating mechanism for some species (e.g. for most carnivores), territorial exclusion affects only the distribution of male white rhinos, and hence cannot control overall population growth (Owen-Smith, 1975). Accordingly, population regulation must eventually come about through the effects of malnutrition on life-history features, specifically the birth interval, age at first reproduction, and postnatal survival of offspring, as documented for elephants in East Africa (Laws *et al.*, 1975) and black rhinos in HiP (Hitchins and Anderson, 1983).

By 1970, the total white rhino population within the 950 km^2 extent of HiP had exceeded 2000 animals, with local densities of over five rhinos/km^2 attained in sections of western Mfolozi. Despite the grassland transformation that was occurring, the population was still increasing at over 9% per year, close to its maximum potential rate (Owen-Smith, 1988). This growth rate was generated by a mean birth interval of around 2.5 years, age at first parturition of 6–7 years, and offspring survival rate > 90% over the first year. Adult mortality rates were estimated to be only 1.5% per year among females and 3.5% among males, from accidents or injuries incurred in fights. Few adults died of old age, because such animals would have been born 40 years earlier when the population was very small. There were no signs that the physical well-being of white rhinos was being affected by the density levels attained, even in low rainfall years. Nevertheless, erosion gullies were expanding and soils bared of much grass cover were washing downslope, threatening the sustainability of the resource base. The concern of park managers was that a rhino 'slum' would develop, with starving white rhinos existing within a degraded habitat. Such a situation could be disastrous for other species, and hence for the wider aims of biodiversity conservation.

Somehow the growing white rhino population needed to be transformed into an effectively stable one. How might this come about solely through the interaction with food resources? With inter-birth intervals spanning multiple years, the critical reproductive stages (gestation, birth, and weaning) are less responsive to annual variation in food availability than for annually breeding ungulates. The effects of malnutrition on reproductive rates become expressed in the population growth rate only a generation later, due to the continuing recruitment of animals already born into the adult segment. Because of this delay, habitat deterioration could pass critical thresholds before the growth of the white rhino population was halted. The only mechanism capable of counteracting the growth of megaherbivore populations sufficiently promptly is dispersal, i.e. animals moving from where they were born to settle elsewhere. This implies a source–sink structure, with animals moving out from the crowded core region into less-favourable localities where the population might not persist in the absence of immigration.

There was evidence that dispersal was indeed taking place before the boundary fence enclosing the game reserve was completed, undertaken mainly by subadult white rhinos of both sexes plus some adult males (Owen-Smith, 1988). Compared with the overall population structure, the high-density core showed an excess of adult females with calves, and peripheral regions a preponderance of adult males plus subadults. The rate of local density increase in the most densely populated region was only half of the overall rate. Dispersal rates estimated from changes in the population composition in core and border regions between successive censuses, taking into account animals removed, indicated that about 7.5% of subadults moved out of the core region per year. A simulation model showed that this rate of dispersal could potentially prevent the population from exceeding the threshold density leading to progressively diminishing food resources. However, with HiP becoming completely fenced, how could such dispersal take place?

The proposed resolution of this management dilemma was the establishment of dispersal sinks within the boundaries of HiP (Owen-Smith, 1974, 1981, 1983). To maintain these low-density regions, or 'vacuum zones', rhinos settling within them would need to be captured and relocated elsewhere. Rather than imposing some arbitrary ceiling on the white rhino population through widespread removals, the animals themselves would indicate when resources became effectively limiting by moving from the core area into sink zones. Furthermore, the sink zones would provide habitat refuges for plant and animal species adversely affected by the grassland transformations brought about by white rhinos in the core region. This management strategy pioneering the application of concepts of source–sink population dynamics was eventually adopted, and its implementation and outcomes are described in Chapter 11.

The key finding here was the importance of dispersal for the regulation of megaherbivore populations in response to diminishing food availability, because demographic changes would be too slow-acting to avoid potential over-shoot of the ultimate carrying capacity (Laws, 1969; Caughley, 1976). Comprehensive population surveys covering HiP enabled dispersal rates by white rhinos to be measured and incorporated into models of coupled herbivore–plant dynamics (Owen-Smith, 1988). However, the uncertainty was what level of grazing would lead to progressive vegetation deterioration and hence to the irruptions and crashes to which herbivore populations are prone in the absence of predation and when opportunities for dispersal are precluded (Caughley, 1976; Gross *et al.*, 2010). Nevertheless, the implementation of dispersal sink concepts has alleviated concerns about overgrazing up to the present time (Chapter 11).

5.3 Diet Selection and Grassland Impacts of a Mega-Grazer

Field observations of the food consumed by white rhinos in HiP confirmed that they are strictly grazers. Forbs (non-grassy herbs) contributed no more than 1% to the material ingested (Owen-Smith, 1988; Shrader *et al.*, 2006), and no browsing on woody plants was observed. Using their broad mouths (20 cm in diameter) and lip-plucking technique, white rhinos are able to crop grass as low as 25 mm above soil level. Their cropping action promotes grass species that are low-growing and spread via stolons or rhizomes, forming grazing lawns (McNaughton, 1984; Waldram *et al.*, 2008; Hempson *et al.*, 2015; Figures 5.2 and 5.3). Due to their low stature, lawn grasses have less structural fibre than taller grasses, and hence constitute the most nutritious forage on offer to grazers. A broad mouth enables white rhinos to obtain an adequate rate of intake from grass swards that would otherwise be too short to meet their quantitative food intake requirements (Owen-Smith, 1988; Shrader *et al.*, 2006). Hippos with even wider mouths than white rhinos

Figure 5.2 White rhino male grazing short grasses (photo: Norman Owen-Smith).

Figure 5.3 Extensive grazing lawn promoted by white rhino grazing in western Mfolozi during 1970 (photo: Norman Owen-Smith).

Figure 5.4 White rhinos grazing in tall grassland dominated by red grass during the dry season, but seeking local patches of short grass associated with termite mounds (photo: Norman Owen-Smith).

similarly cultivate the formation of grazing lawns (Olivier and Laurie, 1974). However, the grazing impacts of hippos are confined to the vicinity of the rivers and lakes where they seek refuge during the day, whereas those of white rhinos are spread more broadly across regional landscapes.

White rhinos concentrate their grazing on the lawn grasslands as long as these retain sufficient forage (Shrader *et al.*, 2006). Once lawn grasses have become reduced to stubble, white rhinos shift their grazing to stands of taller grass (Owen-Smith, 1988). Initially they seek guinea grass (*Panicum maximum*), the most nutritious of the bunch grasses, which typically grows under tree canopies. At a later stage in the dry season, they mow down the tall grasslands dominated primarily by red grass (*Themeda triandra*; Figure 5.4). After stands of red grass become closely cropped, white rhinos move onto hill slopes where tall grass remains available (Figure 5.5; Owen-Smith, 1988). Accordingly, different grassland components support white rhinos through different stages of the seasonal cycle (Owen-Smith, 1988). The functional distinctions are between (1)

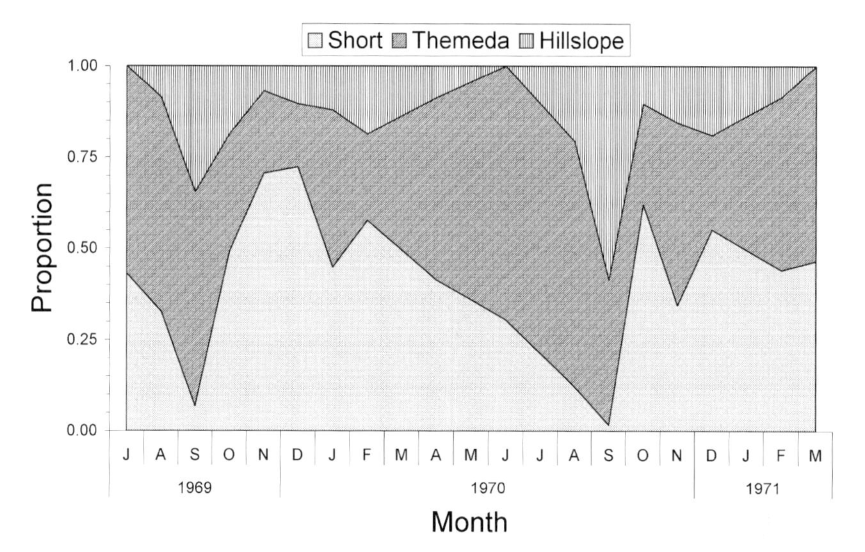

Figure 5.5 Proportional use of grassland types by white rhinos in western Mfolozi during 1969–1971. Short grassland is represented by grazing lawns, Themeda grassland is the medium–tall grassland dominated by *Themeda triandra* (red grass) on relatively flat terrain, and hillslope grassland is constituted by predominantly *T. triandra* growing on steeper slopes (adapted from figure 3.7 in Owen-Smith, 1988).

short-grass lawns providing staple high-quality forage; (2) relatively nutritious bunch grasses grazed during the course of the dry season, providing reserve forage; and (3) less-nutritious or less-accessible grasslands, serving to buffer starvation rates after other sources of forage have mostly been consumed.

It has been inferred that greater size and consequently prolonged digestive retention enables larger herbivores to digest forage more completely than smaller herbivores, and that ruminants lose their advantage in digestive efficiency over hindgut fermenters once a body mass of 1000 kg is surpassed (Demment and Van Soest, 1985). However, there is a limit to how effectively grass tissues can be digested, dependent on the fraction constituted by structural cellulose (Muller *et al.*, 2013). White rhinos and other megaherbivores with hindgut fermentation can tolerate a greater dilution of digestible material by structural fibre than can medium–large ruminants because of their very large size coupled with hindgut fermentation (Clauss *et al.*, 2003). The downside of faster digestive passage is greater sodium losses through the digestive tract, increasing dietary requirements for sodium (demonstrated for black rhinos by Clauss *et al.*, 2007). Lawn grasses favoured by white rhinos have higher concentrations of sodium as well as other nutrients than surrounding bunch grasses (Stock *et al.*, 2010). The high sodium requirement could help explain why white rhinos were historically absent from cooler Highveld grasslands where there is less evaporation and hence little concentration of sodium in the topsoil.

Through their grazing, white rhinos promote a mosaic of lawn grasslands amidst the prevalent bunch grasslands (Owen-Smith, 1988; Waldram *et al.*, 2008). During the early 2000s, grazing lawns covered about 13% of HiP overall, but extended over as much as a quarter of the landscape in western Mfolozi where the highest white rhino densities occurred (Cromsigt, 2006; Arsenault and Owen-Smith, 2011). The grazing down of bunch grasslands and establishment of grazing lawns restricts the spread of fires (Owen-Smith, 1988; Waldram *et al.*, 2008). Grazing by white rhinos also counteracts the accumulation of dead or moribund grass tissues that builds up in the absence of fire or grazing. It might be expected that woody plants would invade following the suppression of fires in grazing lawns, but this seems not to happen. Conditions might be too dry for seedling establishment, and woody seedlings that emerge are exposed to browsers like impalas. White rhinos also damage woody plants by horning small bushes and then dragging their feet over them during their urine marking ceremonies (Owen-Smith, 1975).

Hence, while elephants radically transform the tree layer (Laws, 1970), white rhinos alter the structure, composition, and functioning of the herbaceous layer, along with other major drivers of savanna vegetation dynamics (Waldram *et al.*, 2008; see Chapter 6). The grassland mosaic that white rhinos promote modifies the landscape heterogeneity underlain by geology and soils (Chapter 2). This in turn influences food availability for other grazers, with possibly both positive (facilitative) and negative (competitive) outcomes for these species.

5.4 Resource Partitioning, Competition, and Facilitation within the Grazer Guild

Over evolutionary time frames, competitive relationships among species sharing the same basic food resource should theoretically lead to niche partitioning. For mammalian herbivores, a major influence on resource partitioning comes from physiological mechanisms dependent on body size (Prins and Olff, 1998). Metabolic requirements for energy and nutrients increase allometrically with body mass with a power coefficient of approximately 0.75, meaning that larger animals have lower specific requirements per unit of body mass. On the other hand, the volume capacity of the gut to accommodate food varies in direct relation to body mass. This means that larger animals can either eat relatively less food per day per kg of body mass, or consume a similar amount of food but of a lower nutritional content than smaller animals (Bell, 1971; Jarman, 1974). For mammalian herbivores, the nutritional value of the forage consumed is diluted by the structural fibre content of plant tissues that is chemically ligno-cellulose. Larger herbivores can thus tolerate a greater dietary fibre content than can smaller herbivores. This does not mean that they should preferentially seek a high-fibre diet, but rather that during stressful times they can survive on fibrous plant tissues that provide inadequate nutrition for smaller herbivores. Accordingly, smaller species should specialize on the best quality food resources, while larger species should exploit a wider range in quality (Jarman, 1974). This implies that the smallest grazers should selectively graze nutritious short grasses or especially nutritious grass parts, while larger species spread their grazing over taller more fibrous grasses. Differences in digestive system between ruminants and non-ruminants modify relative efficiency in exploiting food quality, because hindgut fermenters have a faster digestive turnover and hence can tolerate higher fibre contents than can ruminants (Janis, 1976).

Observations made in HiP showed that distinctions in muzzle width relative to body mass modify the effects of body size differences (Arsenault and Owen-Smith, 2008). Despite being the largest grazer, white rhinos concentrate on the shortest grass, while considerably smaller impalas graze grass heights intermediate between those cropped by wildebeest and zebra. This is because the wide mouth of white rhinos, coupled with their lip-plucking technique, enables these animals to exploit short grass very effectively (Owen-Smith, 1988). On the other hand, impalas with relatively narrow muzzles can pluck the most nutritious leaves from within both short and comparatively tall-grass swards.

The expected gradient in grass height and quality in the forage consumed by herbivores of different body size was found in HiP, over the size range from warthog through impala and zebra to buffalo (Kleynhans *et al.*, 2011; Figure 5.6). However, wildebeest exploited shorter grass than impala, enabled by their relatively broad muzzles (Arsenault and Owen-Smith, 2008), and during the wet season white rhinos selected grass as short as grazed by warthogs (Cromsigt, 2006). During the dry season

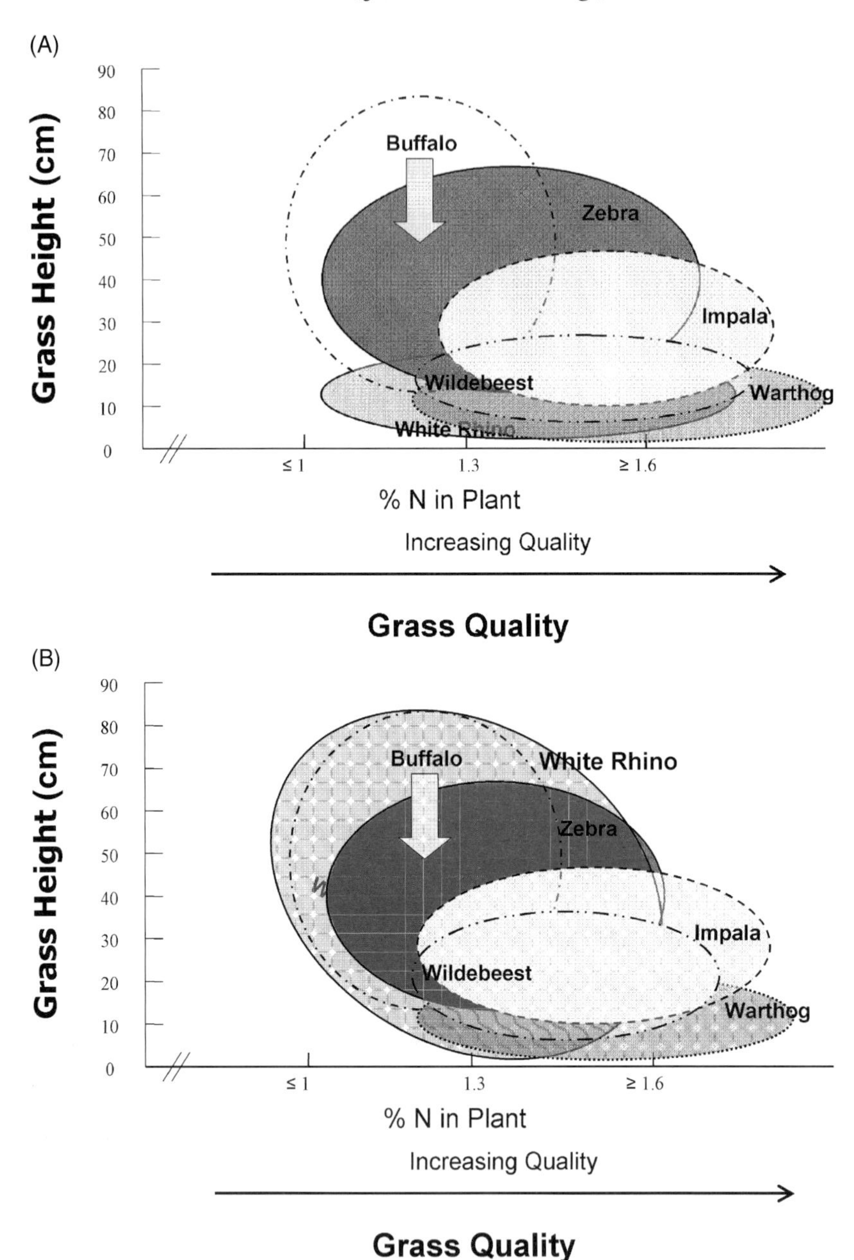

Figure 5.6 Resource partitioning among large grazers in Hluhluwe-iMfolozi expressed in axes of forage quality and quantity (adapted from Kleynhans *et al.*, 2011). (A) Wet season pattern, indicating how white rhinos select short grass of higher quality than expected for their size. (B) Dry season pattern, showing how white rhinos extend their food selection to encompass taller grass of lower quality.

when white rhinos turned their attention towards taller grass, they overlapped with other grazers like buffalo and zebra in the grass height and quality that they utilized. There was also much overlap between white rhinos and other grazers in grass species consumed (Arsenault and Owen-Smith, 2011; Kleynhans *et al.*, 2011). White rhinos exploited mainly lawn grasses, plus guinea grass growing under tree canopies during the wet season, shifting to red grass during the course of the dry season. Warthogs concentrated mostly on lawn grasses throughout the year, but dug up underground grass parts during the dry season. Wildebeests and impalas utilized mainly lawn grasses during the wet season, but reduced their use of these short grasses during the dry season. Wildebeests then exploited a variety of grass species while impalas shifted towards browse. Zebras concentrated especially on guinea grass throughout the year, while buffalos favoured red grass particularly during the dry season.

Resource partitioning could also be effective spatially, with smaller herbivores able to exploit more localized patches of favourable vegetation than larger species (Ritchie and Olff, 1999). This was investigated experimentally within HiP by creating short-grass patches at different size scales within plots that were either fertilized to increase forage quality or left unfertilized (Cromsigt and Olff, 2006). The set-up mimicked the tall–short grass mosaics generated by rhino foraging. As expected, the smallest species, warthog and impala, strongly selected for fertilized plots. However, contrary to expectations, warthog and impala avoided plots with the finest grain. The larger species, white rhino and zebra, appeared unselective and also used the small short-grass plots.

When white rhinos extend their habitat use into predominantly tall-grass areas during the dry season, they initially concentrate their grazing on termite mounds only a few square metres in extent (Owen-Smith, 1988). These local patches may be avoided by small grazers because of the perceived predation risk associated with the surrounding tall grass and shrub cover. Habitat heterogeneity at larger spatial scales could also facilitate spatial separation among grazers. Dung accumulations indicated that larger ruminants were more evenly distributed than smaller ones through occupying a wider diversity of habitats, including those of lower quality (Cromsigt *et al.*, 2009). Furthermore, non-ruminants were more evenly distributed than ruminants of similar size. Hence, size-dependent habitat partitioning among these grazers seemed consistent with habitat relationships among browsers in Kruger Park described by du Toit and Owen-Smith (1989).

The high degree of overlap among grazers in grass height, quality, species, and habitats utilized observed in HiP suggests that there should be considerable potential for competitive displacements. Because of their efficiency in exploiting very short grass, white rhinos should outcompete smaller grazers, like wildebeest and zebra, through depleting stands of the best-quality short grasses. On the other hand, white rhinos promote the spatial spread of nutritious short grasses for lesser herbivores by expanding the grazing lawns and through mowing down fibrous tall grasses. It thus seems surprising that the growing population of white rhinos in HiP (Chapter 11) has apparently had such little influence on the abundance of most other grazers (Chapter 4).

To explain this enigma, Arsenault and Owen-Smith (2002, 2011) suggested that the competitive and facilitative effects of grazing by white rhinos on food availability for other herbivores compensate seasonally. Evidence of competitive displacement is provided by the shift by wildebeest from lawn grasslands towards taller bunch grasslands earlier in the dry season than shown by white rhino, and by observations that zebra made much use of lawn grasslands only in the wetter of 2 years when grasses remained greener into the dry season. However, the extent of the short grass available to other grazers is increased by the grassland impacts of white rhinos, not only through the expansion of grazing lawns, but also by the reduction in height of bunch grasslands during the dry season. The latter impact also restricts the build up of dead grass tissues that otherwise occurs in the absence of fire. As a result, other grazers benefit through having access to better-quality forage during the wet season and early dry season, at the cost of less forage remaining to support them later in the dry season. Hence the overall outcome for the abundance of other grazers could be negligible.

5.5 Resource Partitioning among Browsers

Comparisons have been made between the effects of mega-grazers like white rhinos in cultivating grazing lawns, and those of mega-browsers, like elephants, black rhinos, and giraffe, in promoting the development of browsing 'lawns' (Fornara and du Toit, 2007; Cromsigt and Kuijper, 2011) or 'hedges' (du Toit and Olff, 2014). Pruning of the twigs and branches of tree saplings can maintain these plants within the height range where they are readily accessible to large browsers, while also retaining foliage within the height range of smaller browsers (Makhabu *et al.*, 2006). In this way, mega-browsers like elephants can facilitate food access by other browsers. This depends on the extent to which the plant species favoured by these herbivores overlap, and also on habitat choices. Resource partitioning may also occur through distinctions in height ranges of trees and shrubs browsed, dependent on the size of the browser (du Toit, 1990).

However, the widened use of habitat types with increasing body size found by du Toit and Owen-Smith (1989) in Kruger Park was not evident among the five browsers compared in HiP by O'Kane *et al.* (2011, 2013). The two smallest species, impala and nyala, overlapped most broadly with elephants in habitat use. Partial separation in tree heights exploited was evident between the two tallest browsers, represented by elephant and giraffe, and the two smallest browsers, with kudu intermediate. At plant species level, eight woody species were common to the core diets of all five browsers, with the dietary species range being narrowest for giraffe and widest for elephant.

Investigations on how food availability for smaller browsers is affected by the browsing impacts of the three mega-browsers have yet to be undertaken. An important issue still unresolved is whether increasing numbers of elephants (Chapter 14) will reduce or enhance food availability for black rhinos.

5.6 Consequences of the Biomass Dominance by White Rhinos

The high biomass density attained by white rhinos within HiP contributes largely to an overall large herbivore biomass matching that in the Serengeti region of Tanzania, where migratory wildebeest predominate, and that in parts of Uganda where elephant and hippo are the major contributors (Field and Laws, 1970). Moreover, the white rhino population in HiP has been restricted by ongoing harvests (Chapter 11), while the elephant population there is still growing (Chapter 14). The densities that might ultimately be attained by these megaherbivores in HiP remain uncertain. Elephants have attained regional densities exceeding two animals/km^2 in parts of Zimbabwe (Chamaillé-Jammes *et al.*, 2008) and Uganda (Laws *et al.*, 1975) under similar rainfall regimes to HiP, while hippos have reached effective grazing densities exceeding 10 animals per km^2 in the lakeshore or riparian grasslands that they exploit in Uganda (Field, 1970) and Zambia (Marshall and Sayer, 1976). Other herbivores have continued to coexist at these densities. However, the grazing impacts of white rhinos are more widely distributed away from surface water than those of hippos, and the grazing lawn grasslands promoted by white rhinos in upland regions would be somewhat less productive than the lawns generated by hippos in riparian margins where both soil moisture and nutrients are concentrated. The only sign of density feedbacks slowing the population increase of white rhinos is a reduction in the overall growth rate from over 9% per year prior to 1970 to around 7% per year currently (Chapter 11). This continuing upward trend highlights the importance of dispersal, enabling the incoming generation of white rhinos to move from regions that are heavily exploited towards places retaining more food. However, dispersal does not ultimately avoid the regulating effects of increased mortality; it only shifts it elsewhere. In the case of HiP, such mortality is pre-empted by live removals within the context of the sink management strategy (Chapter 11).

The high densities attained by white rhinos do not necessarily have a negative effect on effective food availability and hence population growth, because by expanding the extent of grazing lawns white rhinos increase the availability of high-quality forage to the benefit of their reproductive performance. The white rhino population is likely to become limited eventually by the extent of the tall-grass reserves available to support animals through the dry season. However, the fragmentation of tall grasslands by the lawn mosaic restricts the spread of fires and hence the loss of forage for grazers that occurs through incineration (Waldram *et al.*, 2008). The consequences may be beneficial for other grazers, not only for those similarly favouring short grass (wildebeest, warthog) but also for species dependent more on tall grass (buffalo). For the short-grass grazers, white rhinos expand nutritious grazing lawns, but consume much of this grass themselves. For species requiring taller grass, white rhinos pre-empt the loss of this forage to fire while restricting the build up of moribund tissues by fostering annual regrowth.

While conventional wisdom has it that smaller herbivores can outcompete larger ones through being able to survive on sparser forage (Prins and Olff, 1998), white rhinos are not threatened by any smaller grazer because of their capability to crop the shortest grass (Arsenault and Owen-Smith, 2008). Both white rhinos and hippos gain an advantage over smaller species when food is short by being able to survive for longer on their fat reserves, as a consequence of the lower mass-specific metabolic rates (Shrader *et al.*, 2006). Hence, white rhinos neither threaten the coexistence of other grazers, nor are threatened by the latter, at least under the density levels they have attained thus far. This has implications for whether the high biomass densities reached by elephants and hippos elsewhere would negatively affect the coexistence of other browsers and grazers, especially if wider dispersal is precluded.

In the past, mobile grazers like wildebeest and zebra probably migrated beyond the current boundaries of HiP during the wet season, either towards fertile basaltic soils adjoining the Lebombo hills (see Chapter 2) or towards higher-elevation grasslands where fires lit by humans during the dry season had promoted regrowth. The region near the Mfolozi and Hluhluwe rivers with perennial surface water constituting HiP would have formed a dry season concentration area. This seasonal exodus of mobile herbivores beyond the vicinity of the rivers would have enabled some alleviation of grazing pressure on grasslands during the wet season when grasses are most sensitive to overexploitation. Currently, the year-round grazing by these herbivores coupled with that by sedentary white rhinos plus introduced impala could have negative consequences for the sustainability of grasslands and hence for the herbivore populations dependent on them within the confines of HiP. Thus, the possibility of genuine overgrazing in the form of a reduction in productive capacity of the forage resource cannot be discounted (see Chapter 6).

5.7 Concluding Remarks

Studies on the ecology of white rhinos within HiP, including their interactions with vegetation and other herbivores, have revealed the central role of this mega-grazer in community and ecosystem dynamics. Features shared by white rhinos with other megaherbivores include their dominant contribution to overall herbivore biomass, delayed demographic response to resource depletion and hence dependence on dispersal, capacity to extensively transform structural features of the vegetation, and both competitive and facilitative relationships with other large herbivores (Owen-Smith, 1974, 1988). They function both as keystone species (Owen-Smith, 1987) and as ecosystem engineers (Jones *et al.*, 1997).

Megaherbivores, both grazers and browsers, were formerly present in species assemblages in all continents before the arrival of human hunters with effective weapons (Owen-Smith, 2013b). Observers familiar only with extant ecosystems in northern continents have not fully appreciated the radical changes in both woody and herbaceous vegetation that must have occurred following the extirpation of megaherbivores by human hunting towards the end of the Pleistocene, in association with the effects of climate change. Evidence is progressively revealing the transformation in vegetation structure and fire regimes that have taken place outside

Africa since late Pleistocene times (Owen-Smith, 1987, 1989; Vera *et al.*, 2006; Gill *et al.*, 2009; Johnson, 2009; Rule *et al.*, 2012). The effects of herbivory as well as fire must be adequately taken into account for explaining the global distribution of savannas and other grassy biomes (Bond, 2005). Recognition of the grazing and trampling effects of the extinct herbivore fauna on the herbaceous cover, coupled with nutrient enhancements from their dung, is guiding attempts to restore the grassy steppe that formerly extended from Siberia into Alaska in place of the current shrub tundra (Zimov *et al.*, 1995; Olofsson *et al.*, 2001; van der Wal *et al.*, 2004; Blinnikov *et al.*, 2011). If white rhinos had not been preserved so effectively within HiP, these attempts at ecological restoration would have lacked the observational support documented in this review.

5.8 References

Andersson, C. J. (1856) *Lake Ngami*. Facsimile reprint in 1967 by C. S. Struik, Cape Town.

Arsenault, R. & Owen-Smith, N. (2002) Facilitation versus competition in grazing herbivore assemblages. *Oikos* **97**: 313–318.

Arsenault, R. & Owen-Smith, N. (2008) Resource partitioning by grass height among grazing ungulates does not follow body size relation. *Oikos* **117**: 1711–1717.

Arsenault, R. & Owen-Smith, N. (2011) Competition and coexistence among short grass grazers in the Hluhluwe-iMfolozi Park, South Africa. *Canadian Journal of Zoology* **89**: 900–907.

Barnovsky, A. D. & Lindsey, E. L. (2010) Timing of Quaternary megafaunal extinctions in South America in relation to human arrival and climate change. *Quaternary International* **217**: 10–29.

Barnovsky, A. D., Kock, P. L., Feranec, R. S., Wing, S. L., & Shabel, A. B. (2004) Assessing the causes of late Pleistocene extinctions on the continents. *Science* **306**: 70–75.

Bell, R. H. V. (1971) A grazing ecosystem in the Serengeti. *Scientific American* **225**: 86–93.

Blinnikov, M. S., Gaglioti, B. V., Walker, D. A., Wooller, M. J., & Zazula, G. D. (2011) Pleistocene graminoid-dominated ecosystems in the Arctic. *Quaternary Science Review* **30**: 2906–2929.

Bond, W. J. (2005) Large parts of the world are brown or black: a different view on the 'Green World' hypothesis. *Journal of Vegetation Science* **16**: 261–266.

Brain, C., Forge, O., & Erb, P. (1999) Lion predation on black rhinoceros in Etosha National Park. *African Journal of Ecology* **37**: 107–109.

Caughley, G. (1976) Plant–herbivore systems. In: *Theoretical ecology* (ed. R. M. May), pp. 94–113. Blackwell, Oxford.

Chamaillé-Jammes, S., Fritz, H., Valeix, M., Murindagoma, F., & Clobert, J. (2008) Resource availability, aggregation and direct density dependence in an open context: the local regulation of an African elephant population. *Journal of Animal Ecology* **77**: 135–144.

Clauss, M., Frey, R., Kieffer, B., *et al.* (2003) The maximum attainable body size of herbivorous mammals: morphological constraints on foregut, and adaptations of hindgut fermenters. *Oecologia* **136**: 14–27.

Clauss, M., Castell, J. C., Kienzle, E., *et al.* (2007) Mineral absorption in the black rhinoceros as compared with the domestic horse. *Journal of Animal Physiology and Animal Nutrition* **91**: 193–204.

Cromsigt, J. P. G. M. (2006) Large herbivores in space: resource partitioning among savanna grazers in a heterogeneous environment. PhD thesis, University of Groningen, the Netherlands.

Cromsigt, J. P. G. M. & Kuijper, D. P. J. (2011) Revisiting the browsing lawn concept: evolutionary interaction of pruning herbivores? *Perspectives in Plant Ecology, Evolution and Systematics* **13**: 207–215.

Cromsigt, J. P. G. M. & Olff, H. (2006) Resource partitioning among savanna grazers mediated by local heterogeneity: an experimental approach. *Ecology* **87**: 1532–1541.

Cromsigt, J. P. G. M. & te Beest, M. (2014) Restoration of a megaherbivore: landscape-level impacts of white rhinoceros in Kruger National Park, South Africa. *Journal of Ecology* **102**: 566–575.

Cromsigt, J. P. G. M., Prins, H. H. T., & Olff, H. (2009) Habitat heterogeneity as a driver of ungulate diversity and distribution patterns: interaction of body mass and digestive strategy. *Diversity and Distributions* **15**: 513–522.

Demment, M. W. & Van Soest, P. J. (1985) A nutritional explanation for body size patterns of ruminant and nonruminant herbivores. *American Naturalist* **125**: 641–672.

du Toit, J. T. (1990) Feeding-height stratification among African browsing ruminants. *African Journal of Ecology* **28**: 55–61.

du Toit, J. T. & Olff, H. (2014) Generalities in grazing and browsing ecology: using cross-guild comparisons to control contingencies. *Oecologia* **174**: 1075–1083.

du Toit, J. T. & Owen-Smith, N. (1989) Body size, population metabolism and habitat specialization among African large herbivores. *The American Naturalist* **133**: 736–740.

Field, C. R. (1970) A study of the feeding habits of the hippopotamus in the Queen Elizabeth National Park, Uganda, with some management implications. *Zoologica Africana* **5**: 71–86.

Field, C. R. & Laws, R. M. (1970) The distribution of the larger herbivores in the Queen Elizabeth National Park, Uganda. *Journal of Applied Ecology* **7**: 273–294.

Fornara, D. A. & du Toit, J. T. (2007) Browsing lawns? Responses of *Acacia nigrescens* to ungulate browsing in an African savanna. *Ecology* **88**: 200–209.

Gill, J. L., Williams, J. W., Jackson, S. T., Lininger, K. B., & Robinson, G. S. (2009) Pleistocene megafaunal collapse, novel plant communities, and enhanced fire regimes in North America. *Science* **326**: 1100–1103.

Gross, J. E., Gordon, I. J., & Owen-Smith, N. (2010) Irruptive dynamics and vegetation interactions. In: *Dynamics of large herbivore populations in changing environments* (ed. N. Owen-Smith), pp. 117–140. Blackwell, Oxford.

Harris, W. C. (1838) *Narrative of an expedition into southern Africa during the years 1836 and 1837*. John Murray, London.

Hempson, G., Archibald, S., Bond, W., *et al.* (2015) Ecology of grazing lawns in Africa. *Biological Reviews* **90**: 979–994.

Hitchins, P. M. & Anderson, J. L. (1983) Reproduction, population characteristics and management of the black rhinoceros in the Hluhluwe/Corridor/Umfolozi Game Reserve. *South African Journal of Wildlife Research* **13**: 78–85.

Hooijer, D. A. & Patterson, B. (1972) Rhinoceroses from the Pliocene of north-western Kenya. *Bulletin of the Museum of Comparative Zoology (Harvard University)* **144**: 1–26.

Janis, C. (1976) The evolutionary strategy of the Equidae and the origins of rumen and cecal digestion. *Evolution* **30**: 757–776.

Jarman, P. J. (1974) The social organization of antelope in relation to their ecology. *Behaviour* **48**: 215–267.

Johnson, C. N. (2009) Ecological consequences of late Quaternary extinctions of megafauna. *Proceedings of the Royal Society Series B* **276**: 2509–2519.

Jones, C. G., Lawton, J. H., & Shachak, M. (1997) Positive and negative effects of organisms as physical ecosystem engineers. *Ecology* **78**: 1946–1957.

Joubert, D. (2006) Hunting behaviour of lions (*Panthera leo*) on elephants (*Loxodonta africana*) in the Chobe National Park, Botswana. *African Journal of Ecology* **44**: 279–281.

Kingdon, J. & Hoffmann, M. (2013) *Mammals of Africa Volume VI*. Bloomsbury, London.

Kleynhans, E. J., Jolles, A. E., Bos, M. R. E., & Olff, H. (2011) Resource partitioning along multiple niche dimensions in differently sized African savanna grazers. *Oikos* **120**: 591–600.

Laws, R. M. (1969) The Tsavo Research Project. *Journal of Reproduction and Fertility, Supplement* **6**: 495–531.

Laws, R. M. (1970) Elephants as agents of habitat and landscape change in East Africa. *Oikos* **21**: 1–15.

Laws, R. M., Parker, I. S. C., & Johnstone, R. C. B. (1975) *Elephants and their habitats. The ecology of elephants in North Bunyoro, Uganda*. Clarendon Press, Oxford.

Loveridge, A. J., Hunt, J. E., Murindagomo, F., & Macdonald, D. (2006) Influence of drought on predation of elephant calves by lions in an African savanna woodland. *Journal of Zoology* **270**: 523–530.

Makhabu, S. W., Skarpe, C., & Hytteborn, H. (2006) Elephant impact on shoot distribution on trees and on rebrowsing by smaller browsers. *Acta Oecologia* **30**: 136–146.

Marshall, P. J. and Sayer, J. A. (1976) Population ecology and response to cropping of a hippopotamus population in eastern Zambia. *Journal of Applied Ecology* **13**: 391–404.

McNaughton, S. J. (1984) Grazing lawns: animals in herds, plant form and coevolution. *American Naturalist* **124**: 863–886.

Mduma, S. A. R. & Hopcraft, J. G. C. (1995) The main herbivorous mammals and crocodiles in the Greater Serengeti Ecosystem. In: *Serengeti III* (eds A. R. E. Sinclair, C. Packer, S. A. R. Mduma, & J. M. Fryxell), pp. 497–499. University of Chicago Press, Chicago.

Muller, D. W. H., Codron, D., Meloro, C., *et al.* (2013) Assessing the Jarman–Bell Principle: scaling of intake, digestibility, retention time and gut fill with body mass in mammalian herbivores. *Comparative Biochemistry and Physiology Part A* **164**: 129–149.

O'Kane, C. A. J., Duffey, K. J., Page, B. R., & Macdonald, D. W. (2011) Overlap and seasonal shifts in use of woody plant species amongst a guild of savanna browsers. *Journal of Tropical Ecology* **27**: 249–258.

O'Kane, C. A. J., Duffey, K. J., Page, B. R., & Macdonald, D. W. (2013) Effects of resource limitation on habitat usage by the browser guild in Hluhluwe-iMfolozi Park, South Africa. *Journal of Tropical Ecology* **29**: 39–47.

Olivier, R. C. D. & Laurie, W. A. (1974) Habitat utilization by hippopotamus in the Mara river. *East African Wildlife Journal* **12**: 249–272.

Olofsson, J., Kitti, H., Rautiainen, P., Stark, S., & Oksanen, L. (2001) Effects of summer grazing by reindeer on composition of vegetation, productivity and nitrogen cycling. *Ecography* **24**: 13–24.

Owen-Smith, N. (1974) The social system of the white rhinoceros. In: *The behaviour of ungulates and its relation to management* (eds V. Geist & F. Walther), pp. 341–351. IUCN Publication new series no. 24, IUCN, Morges.

Owen-Smith, N. (1975) The social ethology of the white rhinoceros. *Zeitschrift fur Tierpsychologie* **38**: 337–384.

Owen-Smith, N. (1981) The white rhinoceros overpopulation problem, and a proposed solution. In: *Problems in management of locally abundant wild mammals* (eds P. A. Jewell, S. Holt, & D. Hart), pp. 129–150. Academic Press, New York.

Owen-Smith, N. (1983) Dispersal and the dynamics of large herbivore populations in enclosed areas. In: *Management of large mammals in African conservation areas* (ed. R. N. Owen-Smith), pp. 127–143. Haum, Pretoria.

Owen-Smith, N. (1987) Pleistocene extinctions: the pivotal role of megaherbivores. *Paleobiology* **13**: 351–362.

Owen-Smith, N. (1988) *Megaherbivores. The influence of very large body size on ecology.* Cambridge University Press, Cambridge.

Owen-Smith, N. (1989) Megafaunal extinctions: the conservation message from 11 000 years BP. *Conservation Biology* **3**: 405–412.

Owen-Smith, N. (2013a) Megaherbivores. In: *Encyclopedia of biodiversity* (ed. S. A. Levin), pp. 223–239. Academic Press, Waltham, MA.

Owen-Smith, N. (2013b) Contrasts in the large herbivore faunas of the southern continents in the late Pleistocene and the ecological implications for human origins. *Journal of Biogeography* **40**: 1215–1224.

Owen-Smith, N. & Mills, M. G. L. (2008) Shifting prey selection generates contrasting herbivore dynamics within a large-mammal predator–prey web. *Ecology* **89**: 1120–1133.

Pienaar, U. de V. (1969) Predator–prey relationships among the large mammals of the Kruger National Park. *Koedoe* **12**: 108–176.

Prins, H. H. T. & Olff, H. (1998) Species richness of African grazer assemblages: towards a functional explanation. In: *Dynamics of tropical ecosystems* (eds D. N. Newberry, H. H. T. Prins, & N. Brown), pp. 449–490. Blackwell, Oxford.

Ritchie, M. E. & Olff, H. (1999) Spatial scaling laws yield a synthetic theory of biodiversity. *Nature* **400**: 557–560.

Rule, S., Brook, B. W., Haberle, S. G., *et al.* (2012) The aftermath of megafaunal extinction: ecosystem transformation in Pleistocene Australia. *Science* **335**: 1483–1486.

Selous, F. C. (1899) The white or square-mouthed rhinoceros. In: *Great and small game of Africa* (ed. H. A. Bryden), pp. 52–67. Rowland Ward, London.

Shrader, A. M., Owen-Smith, N. & Ogutu, J. O. (2006) Food intake rate and nutrient gains of a mega-grazer, the white rhinoceros, through the dry season. *Functional Ecology* **20**: 376–384.

Smith, A. (1849) *Illustrations of the zoology of South Africa. Mammals.* London.

Stock, W. D., Bond, W. J., & van de Vijver, C. A. D. M. (2010) Herbivore and nutrient control of lawn and bunch grass distributions in a southern African savanna. *Plant Ecology* **206**: 15–27.

van der Wal, R., Bardgett, R. D., Harrison, K. A., & Stien, A. (2004) Vertebrate herbivores and ecosystem control: cascading effects of faeces on tundra ecosystems. *Ecography* **134**: 242–252.

Vera, F. W. M., Bakker, E. S. & Olff, H. (2006) Large herbivores: missing partners of western European light-demanding tree and shrub species? In: *Large herbivore ecology, ecosystem dynamics and conservation* (eds K. Danell, R. Bergstrom, P. Duncan, & J. Pastor), pp. 203–231. Cambridge University Press, Cambridge.

Waldram, M. S., Bond, W. J., & Stock, W. D. (2008) Ecological engineering by a mega-grazer: white rhino impacts on a South African savanna. *Ecosystems* **11**: 101–112.

Zimov, S. A., Chuprynin, V. I., Oreshko, A. P., *et al.* (1995) Steppe–tundra transition: a herbivore-driven biome shift at the end of the Pleistocene. *American Naturalist* **146**: 765–794.

6 · *The Functional Ecology of Grazing Lawns: How Grazers, Termites, People, and Fire Shape HiP's Savanna Grassland Mosaic*

JORIS P. G. M. CROMSIGT, MICHIEL P. VELDHUIS, WILLIAM D. STOCK, ELIZABETH LE ROUX, CLEO M. GOSLING, AND SALLY ARCHIBALD

6.1 Introduction

The grasslands of Hluhluwe-iMfolozi Park (HiP) are highly heterogeneous, both in terms of their species composition and their structure. Grazing lawns are a particularly striking feature of this heterogeneity (Figure 6.1). These grazing lawns are short, repeatedly grazed grassland patches ranging in size from several square metres to a few hectares. These patches consist of grass species that grow in a prostrate form with a high leaf to stem ratio. Back in 1960, Vesey-Fitzgerald (1960) had already noted how the effects of grazing and trampling by various-sized ungulates maintained floodplain grasses in a short and actively growing condition and how bohor reedbuck *Redunca redunca* grazed on these 'short-grass lawns' in the Rukwa Valley in Tanzania. Olivier and Laurie (1974) used the term 'grazing lawn' for grasslands impacted by hippo grazing along the Mara river in the Maasai Mara National Park. Around the same time, similar short grasslands promoted by heavy grazing and preferred by short-grass grazers such as blue wildebeest *Connochaetes taurinus*, hippopotamus *Hippopotamus amphibius*, and warthog *Phacochoerus africanus* were described by others (Bell, 1971; Lock, 1972; Eltringham, 1974).

Conserving Africa's Mega-Diversity in the Anthropocene, ed. Joris P. G. M. Cromsigt, Sally Archibald and Norman Owen-Smith. Published by Cambridge University Press.

Figure 6.1 Heterogeneous grassland in Hluhluwe, showing grazing lawn interspersed with patches of taller bunch grass (photo: Joris Cromsigt, around 2006).

From the mid-1950s, conservation management in HiP was increasingly concerned about the impact of expanding herbivore populations on grassland conditions in an increasingly fenced reserve (see Chapters 1 and 4). During this time, herbivore carrying capacities and assessments of grazing impacts were heavily influenced by agricultural viewpoints and knowledge. As a result, short grasslands were generally perceived as overgrazed. This perspective prompted the culling of short-grass grazers such as warthog and blue wildebeest from the 1950s to early 1980s (see Chapter 4; Brooks and Macdonald, 1983). It was also an important driving force behind the start of live white rhino *Ceratotherium simum* removals in the early 1960s (see Chapter 11). It was during these early years that work in HiP by Norman Owen-Smith, and later Richard Emslie, indicated that these short 'degraded grass patches' were in fact quite resilient to repeated grazing and represented important resource areas for grazers such as white rhino, wildebeest, and warthog (Owen-Smith, 1973, 1988). Their work in HiP has been the basis of a paradigm shift in the conservation management of HiP and beyond: we now recognize that grazing lawns are not degraded grasslands but in fact an important functional component of savannas. The grazing lawn concept was formalized by McNaughton (1984) through his studies on the short grasslands in the

Serengeti. Lawns have since been described for terrestrial grazing systems around the globe: East Africa (McNaughton, 1984); Southern Africa (Owen-Smith, 1988); West Africa (Verwij *et al.*, 2006); North America (Knapp *et al.*, 1999); the (sub)Arctic (Person *et al.*, 2003); Asia (Karki *et al.*, 2000); and Australasia (Roberts, 2009). Here we review the processes driving the creation and maintenance of these systems in HiP and elsewhere, as well as their impacts on savanna heterogeneity and functioning.

6.2 Drivers of Lawn Creation and Maintenance: Nutrient versus Drought-Tolerance Loops

6.2.1 The Nutrient Loop

Central to McNaughton's (1984) grazing lawn concept is that grazing increases grass productivity and/or leaf nutrient concentrations, which attracts further grazing, and thus leads to a positive consumer–resource interaction. The increase in grass productivity in response to grazing has been attributed to multiple mechanisms, including: (1) grazing reduces within-plant light competition and thus enhances photosynthesis of remaining leaves, and (2) frequent grazing prevents leaf senescence (see McNaughton, 1983; Craig, 2010). In addition, the input of dung and urine by grazing herbivores stimulates soil microbial productivity and increases N and P cycling and availability, facilitating grass productivity (McNaughton *et al.*, 1997; Frank *et al.*, 2000). So what evidence do we have from HiP's grazing lawns for enhanced productivity and leaf nutrient concentrations?

6.2.1.1 Productivity of Lawns versus Bunch Grasslands in HiP

Veldhuis *et al.* (2016) estimated productivity of grazing lawns in five different sites (three in Hluhluwe, two in Mfolozi) using a similar temporary exclosure approach as McNaughton (1985). They found that during the nine-month growing season (September 2013 to May 2014), above-ground productivity in lawns increased with rainfall from ~150 g/m^2 at 200 mm to ~350 g/m^2 at 500 mm rainfall. This productivity is slightly lower than that found for the short and mid-grass grasslands in the Serengeti (equation in Table 5 in McNaughton 1985: productivity = 0.96 × rainfall + 68), possibly due to the high-fertility volcanic soils in East Africa (Scholes, 2003). Veldhuis *et al.* (2016) also compared the above-ground productivity of grazing lawns with that of recently burnt bunch grassland. Productivity of bunch grasslands was two to three times higher than that of grazing lawns during the growing season following the burns. However, they did not measure the productivity of bunch grassland beyond this first post-burn growing season. Because the productivity of bunch grassland declines with time since burn (Morgan and Lunt, 1999), it remains unclear how the productivity of an average grazing lawn compares with that of average bunch grassland.

6.2.1.2 Lawns as Nutrient Hot Spots in HiP

In the Serengeti, grazers are attracted to lawn grass sites because they are areas of high macro- and micronutrient concentrations, particularly Mg, Na, and P (McNaughton, 1988). In HiP, the difference in sodium between individual green leaves of lawn and bunch grass species is particularly striking (Table 6.1). From Table 6.1 there are no clear leaf-level differences in the content of other nutrients between lawn and bunch grass species. However, clipping studies in HiP which measured nutrient contents of the grass sward suggest that lawns in HiP do act as hot spots for a diversity of nutrients. Leaf concentrations of N, P, K, Mg, and Na, and also the micronutrients Cu and Zn, were higher in mixed-species grass samples from grazing lawns than from nearby bunch grassland (Cromsigt, 2006; Hagenah *et al.*, 2009; Stock *et al.*, 2010; Veldhuis *et al.*, 2016). The fact that the leaf-level results (Table 6.1) only showed a difference in sodium while the sward-level studies showed a difference in multiple nutrients may be due to the fact that lawns do not only consist of typical 'lawn grass species', but also include short-grazed, high-quality bunch grass species such as *Eragrostis superba* (see Section 6.4 for more discussion on the species composition of lawns). Moreover, the sward samples were taken from short, intensely grazed lawns and contrasted with much less intensely grazed bunch grassland, while the leaf-level samples did not necessarily represent this grazing contrast. In fact, some of the leaf-level samples of 'lawn grass species' in Table 6.1 may have been from mildly grazed swards, while some of the samples from 'bunch grass species' may have been from more intensely grazed swards. This may also have reduced the difference in nutrient content between lawn and bunch grass species shown in Table 6.1, because intense grazing increases the nutrient content in leaves for grasses in general (Van der Vijver *et al.*, 1999).

Whether herbivores concentrate on lawns for protein or minerals such as sodium remains an open question. Because sodium is a limiting nutrient for most animals, high sodium levels in grazing lawn grasses may be an important reason for herbivores to visit lawns (McNaughton, 1988; Tracy and McNaughton, 1995). Based on feeding trials in zoos,

Table 6.1. *Leaf nutrient concentrations (mean ± standard deviation) for four common grazing lawn grass species and eight species that commonly dominate bunch grassland. Only green leaves were collected, and # gives the number of individual grass plants that were sampled. Reference 2 (Hagenah* et al.*, 2009) sampled grass in 10 different sites distributed across HiP (five sites in Hluhluwe and five in Mfolozi) during the late wet season in 2003 (March). Reference 1 (Cromsigt, unpublished data) sampled grass from five different sites distributed across Hluhluwe during the early wet season 2003 (September).*

Species	Type	#	N (%)	P (%)	Ca (%)	Mg (%)	Na (mg/kg)	Reference
Dactyloctenium australe	Lawn	5	1.81 (0.08)	–	0.70 (0.07)	–	9517 (691)	1
Urochloa mosambicensis	Lawn	13	2.70 (0.25)	0.4 (0.03)	0.7 (0.06)	0.4 (0.03)	9471 (1078)	2
Sporobolus nitens	Lawn	9	2.60 (0.27)	0.3 (0.03)	0.4 (0.27)	0.2 (0.01)	5407 (627)	2
Digitaria longiflora	Lawn	10	1.70 (0.16)	0.3 (0.03)	0.4 (0.06)	0.2 (0.03)	6313 (384)	2
Panicum maximum	Bunch	16	2.50 (0.13)	0.3 (0.03)	0.5 (0.05)	0.2 (0.03)	2077 (242)	2
Botriochloa insculpta	Bunch	11	2.20 (0.09)	0.3 (0.01)	0.4 (0.03)	0.2 (0.03)	446 (146)	2
Eragrostis superba	Bunch	13	1.90 (0.18)	0.2 (0.04)	0.6 (0.04)	0.2 (0.04)	971 (156)	2
Aristida congesta	Bunch	5	1.90 (0.11)	0.2 (0.03)	0.2 (0.06)	0.1 (0.03)	604 (107)	2
Heteropogon contortus	Bunch	7	1.70 (0.11)	0.2 (0.02)	0.3 (0.04)	0.2 (0.04)	334 (61)	2
Themeda triandra	Bunch	8	1.50 (0.07)	0.2 (0.01)	0.3 (0.04)	0.2 (0.04)	372 (74)	2
Eragrostis curvula	Bunch	10	1.40 (0.09)	0.2 (0.03)	0.3 (0.03)	0.1 (0.01)	997 (98)	2
Sporobolus africanus	Bunch	10	1.30 (0.09)	0.2 (0.03)	0.3 (0.01)	0.1 (0.01)	383 (55)	2

Clauss *et al.* (2007) suggest that sodium (and potassium) may be particularly limiting for rhino species, because they have relatively high faecal and urinal losses of these nutrients.

6.2.1.3 The Role of Soil Nutrients – Stocks versus Fluxes

There is an ongoing debate as to what maintains high leaf nutrient concentrations and productivity in grazing lawns: pre-existing high nutrient stocks in the underlying soil or increased nutrient fluxes induced by continuous grazing and deposition of dung and urine (Stock *et al.*, 2010). The location of grazing lawns on volcanic soils in the Serengeti (McNaughton, 1984), sodic sites in KNP (Grant and Scholes, 2006), termitaria (Davies *et al.*, 2014), and other nutrient-enriched areas such as abandoned kraal sites (Young *et al.*, 1995; Valls-Fox *et al.*, 2015) argues for the importance of stocks in attracting grazers and facilitating the creation and maintenance of grazing lawns. There is also ample evidence that the cycling and availability of nutrients such as N and P, but also Na, may be enhanced by grazing (e.g. McNaughton *et al.*, 1997; Frank *et al.*, 2000). In HiP there is conflicting evidence regarding the importance of high nutrient stocks. Stock *et al.* (2010) quantified soil fertility in 10 sites distributed across HiP situated along a grazing gradient, from heavily grazed lawns to lightly grazed bunch grass sites, and showed that there was no clear relationship between grazing intensity and total soil nutrients. However, another study that directly compared grazing lawns with paired bunch grasslands at five sites distributed across Hluhluwe found that levels of P, K Zn, and pH were higher in grazing lawn than bunch grassland soils (the effect for Zn was only present in the top soil) (Figure 6.2), whereas Ca, Mg, N, and organic C levels did not differ between grazing lawn and bunch grassland soils. The apparently conflicting results of these two studies may be due to the fact that some grazing lawns in HiP originate from abandoned kraal sites or termite mounds (both associated with elevated levels of soil nutrient stocks), while other lawns may result from patch-selective grazing and trampling without underlying differences in soil nutrient stocks (see sections below). Note that neither of these studies provided evidence for increased levels of N stocks in grazing lawn soils. However, there is some evidence for increased N fluxes in lawn soils. Based on isotope analyses, Coetsee *et al.* (2010) suggested that repeated grazing enhances soil N cycling and availability in lawns in HiP. This implies that lawns can become nutrient hot spots for grazers through feedbacks with grazing without underlying differences in the soils (see also Section 6.3.5).

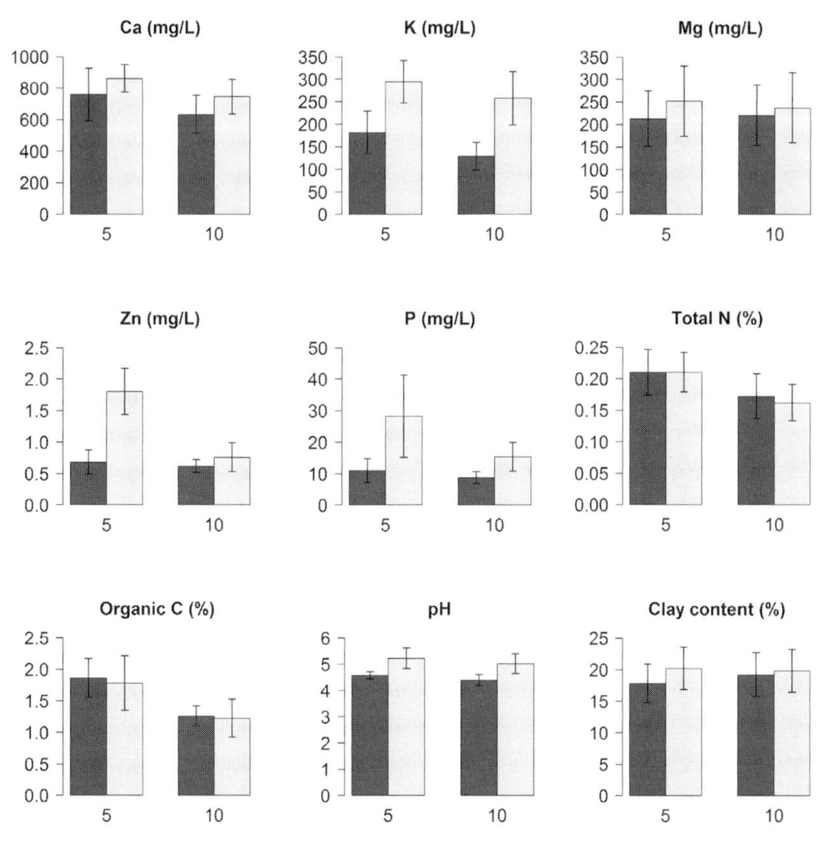

Figure 6.2 Nutrient content (mean ± standard deviation) of soil from bunch grasslands (dark grey) versus paired grazing lawns (light grey) from five sites across Hluhluwe (Joris Cromsigt, unpublished data). Distance between sites was at least 1 km. Five soil samples were taken per site, divided into top (0–5 cm) and deep (5–10 cm) soil, and then pooled per site for nutrient analysis. P ($P = 0.045$), K ($P = 0.01$), and pH ($P = 0.029$) levels were higher in grazing lawns than in bunch grassland for both soil layers. Zn concentrations were higher in grazing lawn soils ($P = 0.019$), but only in the topsoil (grassland type × soil depth: $P = 0.06$). Levels of N, Mg, Ca, organic C, and clay content did not differ significantly between lawn and bunch soils ($P \geq 0.1$).

6.2.2 The Drought Tolerance Pathway

Besides nutrients, studies from HiP and elsewhere highlight the importance of another pathway for grazing lawn formation and maintenance; a drought tolerance pathway. Recent work by Anderson *et al.* (2013), using grasses from HiP and the Serengeti, showed that in a greenhouse experiment lawn and bunch grasses increased biomass production and

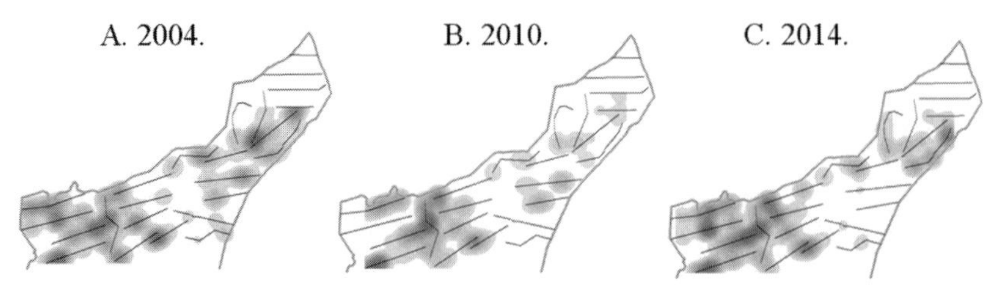

Figure 6.3 Grazing lawn distribution as recorded on line transects (black lines) during 3 years. Transects are on average 8 km long (range 4–11 km), and for every 5 m plot on the transect lawn cover was recorded as yes or no (Cromsigt and Le Roux, unpublished data). We defined grazing lawn as present when lawn grass species dominated (> 75%) a 5 m plot and extended for several metres away on both transect sides (see Cromsigt *et al.*, 2009). The maps represent point density interpolations using Spatial Analyst in ArcGIS including all the 5 m plots that were covered by lawns as points.

photosynthetic rates in similar ways following clipping. However, following defoliation, bunch grasses had higher transpiration rates and lower water-use efficiency than lawn grasses. Thus, grazing may make lawn grasses competitively superior over bunch grasses under drier conditions. This reflects earlier work proposing that herbivory and water availability acted as convergent selection pressures on lawn grass traits such as prostrate growth and rapid shoot regrowth (Coughenour, 1985; Milchunas *et al.*, 1988; Augustine and McNaughton, 1998). This interaction between water availability and grazing is clear at a landscape level in HiP. Grazing lawn cover increases towards the southern sections, with as much as 30% lawn cover across the south-western parts of Mfolozi (Figure 6.3), reflecting an increase in lawn cover with declining annual rainfall (Figure 6.4). Where rainfall may determine the extent to which grazers create lawns across larger scales, at smaller scales large herbivores may actively create the dryer conditions that facilitate grazing lawn grasses through trampling and defoliation (Veldhuis *et al.*, 2014). Herbivore trampling may compact the soil (Kim *et al.*, 2010), especially on intensely used patches such as grazing lawns. Grazing lawns in HiP indeed have higher soil bulk densities than nearby bunch grass areas (Veldhuis *et al.*, 2014) and also lower water infiltration rates (Van der Plas *et al.*, 2013). Grazing herbivores can also induce dry conditions by reducing grass cover and increasing the exposure of bare soil which increases evaporation rates in grazing lawn soils when compared with bunch grass areas (Van der Plas *et al.*, 2013). Altogether, soil compaction and defoliation by large herbivores thus induces the dry conditions in grazing lawn soils that facilitate lawn

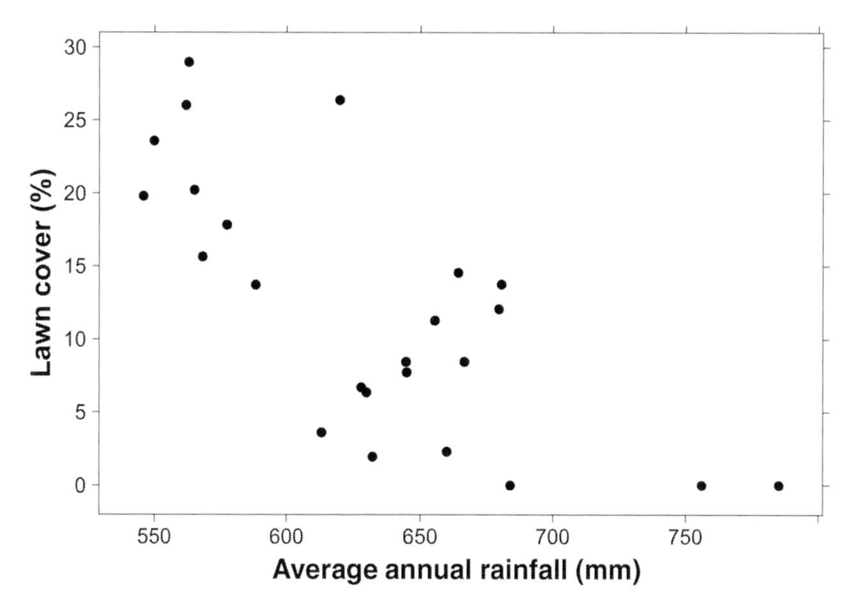

Figure 6.4 The percentage of grazing lawn cover for each of the 24 transects walked in 2004 (see Figure 6.3) versus the average annual rainfall per transect (unpublished data, Hluhluwe Research Centre, Ezemvelo KZN Wildlife) over the period 2000–2006 (see Figure 2.12 for rainfall patterns for similar time period).

grass species over bunch grasses. There is another interesting dimension to this drought-tolerance loop. Drought tolerance can be physiologically achieved through accumulation of osmolytes (e.g. Na and K) in the plant vacuoles to increase plant water potential and turgor (Girma and Krieg, 1992; Bartlett *et al*., 2012). High sodium levels in lawn grasses thus could represent a drought-tolerance adaptation. Hence, the drought-tolerance loop provides an explanation for high leaf nutrient concentrations in lawn grasses.

In conclusion, herbivores play a central role in both pathways that drive grazing lawn development and maintenance: the consumer–nutrient interaction loop and the drought-tolerance loop. As soon as herbivores repeatedly return to a patch, their grazing, trampling, and nutrient deposition induces these interacting loops. Although the consumer–resource interaction and drought-tolerance loops are important reinforcing proximate mechanisms of lawn development and maintenance, they do not explain what attracts and concentrates grazers in a patch in the first place and thus provides the initial conditions for lawn development. In the next section we will discuss a diverse suite of factors that concentrate or disperse herbivores and thus represent the ultimate mechanisms driving grazing lawn development and persistence in HiP.

6.3 Factors that Provide Initial Conditions for Lawn Development by Concentrating Herbivores

6.3.1 Former Human Land Use

HiP has been extensively used by humans over a long period of time (Feely, 1980; Chapter 1). Homesteads were present throughout many parts of the reserve until the end of the twentieth century, especially in western Mfolozi and the Corridor section. In the Corridor area, between the Black Mfolozi river in the south and Hluhluwe river in the north, homesteads prevailed until the 1940s (Brooks, 2001; Chapter 1). Within these homesteads, woody vegetation was cleared and livestock were kept overnight in a kraal (boma). This resulted in removal of the tall-grass layer and continuous trampling and input of nutrients through dung and urine for at least decades. Moreover, the area of the homestead around the huts and kraal was swept bare daily, creating a concrete-like surface that remained after abandonment (Jim Feely, personal communication). Following abandonment of homesteads, the compacted soils with enhanced nutrient levels are colonized by palatable, creeping grasses, attracting wild grazers, and initiating the consumer–resource and drought-tolerance loops that lead to grazing lawn development. Similar processes have been described for abandoned homesteads elsewhere in southern and eastern Africa (Blackmore *et al.*, 1990; Young *et al.*, 1995; Augustine, 2003; Valls-Fox *et al.*, 2015). Such abandoned kraal or boma sites may persist as nutrient rich glades over many decades, and even centuries (Blackmore *et al.*, 1990), maintained by wild grazers, and only degrading slowly from the edges inwards (Augustine and McNaughton, 2006). We have several lines of evidence that former homesteads are an important driver of grazing lawn extent in HiP. Early observations from park staff based in HiP confirm the association of lawns with former homesteads. For example, Jim Feely, game ranger during the 1950s and 1960s, comments: 'My observations of grazing lawns in Mfolozi were that many were situated on the sites of Zulu homesteads abandoned at various times during the nineteenth or early twentieth centuries, i.e. for periods of 70 or more years. They were dominated by short grasses (*Cynodon*, *Urochloa*) and favoured by some grazers such as white rhino and warthog' (personal communication with J. Feely). During Feely's time in HiP, signs of former human use such as grinding stones were still very abundant on grazing lawns throughout the park. Moreover, in an archaeo-botanical study, Hall (1984) surveyed 28 known former homestead sites in the Mfolozi section of HiP and found that almost all of these sites (24) were characterized by a specific grass community. Whereas the dominant grass surrounding the sites was the tall grass *Themeda triandra*, the abandoned homestead sites were covered by stoloniferous, prostrate species such as *Urochloa mosambicensis* and *Digitaria longiflora*, and were heavily grazed in contrast to the surrounding grassland. Interestingly, the 50–60 m diameter of the archeological sites indicated by Hall (1984) is very similar to the average diameter of present-day lawns in HiP (79.3 m ± 121.7 m stdev, $N = 353$: Joris Cromsigt, unpublished data). The role of Late Iron Age farming practices also explains why many of the grazing lawns in HiP are found on crests of hills and ridges: in contrast to the Early Iron Age, farmers in the Late Iron Age located their homesteads away from the rivers on crests to avoid parasites such as mosquitoes and tsetse flies (see Chapter 1; Feely, 2004). All in all, this suggests that the extensive and relatively recent (between 60 and 150 years ago) agro-pastoral land-use practices in HiP have left strong legacy effects. It is likely that many of the grazing lawns in HiP may reflect abandoned kraal sites that have been maintained and perhaps further extended by wild grazers such as white rhino.

6.3.2 Termite Activity

Studies in other savannas have previously shown that termite mounds may be characterized by a ring of lawn grass species that are not found elsewhere in the savanna matrix (Arshad, 1982; Davies *et al.*, 2014). In HiP, *Macrotermes natalensis* build the largest mounds, covering an estimated 10% of the surface area of the park, and show a strong association with lawn grass cover (Gosling *et al.*, 2011; Gosling, 2014; Chapter 9). In addition, *Odontotermes* spp. mounds are characterized by a ring of bare ground followed by a ring of lawn grasses (Arshad, 1982; Gosling, 2014). Termites influence lawn development and maintenance through their foraging and working of the soil. Termitaria in HiP, and throughout Africa, are areas of high macro- and micronutrient availability (Jouquet *et al.*, 2004; Mills *et al.*, 2009; Gosling *et al.*, 2011), because termites bring clay-rich soils from deep layers up to the surface. In addition, termites may remove large amounts of grass biomass around their mounds (Wood and Sands, 1978; Gandar *et al.*, 1982). Termites may thus create the initial conditions for lawn grass species to colonize a patch by removing tall-grass cover and increasing soil nutrient availability (Jouquet *et al.*, 2004; Cromsigt and Olff, 2008). These initial conditions attract grazing by large herbivores on mounds, which then may start the consumer–resource feedback and drought-tolerance loops that lead to lawns (Gosling *et al.*, 2011, Cromsigt and te Beest 2014). Cromsigt and Olff (2008) suggest that the spatial scale of initial termite disturbance may be crucial for this interaction between termites and large mammal grazers. In a field experiment they simulated differently sized disturbances of the tall-grass layer by mowing and fertilizing plots. Following these treatments, grazing by white rhino and warthog kept bunch grass short in the large fertilized plots of 8 × 8 m, but not in the smaller and unfertilized plots. Moreover, 2 years after the initial treatments lawn grasses strongly increased in cover in the fertilized 8 × 8 m plots, but not in the other plots. This work confirms that nutrient inputs are important for lawn development and that initial disturbances of the tall-grass layer have to be big enough to attract grazers in order to develop and maintain grazing lawns.

6.3.3 Role of Fire

Fire may influence the persistence and distribution of grazing lawns in the landscape by shaping the spatial–temporal patterns of grazing (Archibald *et al.*, 2005; Archibald, 2008; Chapter 10). Recently burnt patches attract grazers away from grazing lawns, because the regrowth is palatable (Van de Vijver *et al.*, 1999; Archibald and Bond, 2004). This temporary reduction in grazing pressure in the grazing lawns may allow bunch grasses to recover and prevent spread of lawn species (Archibald *et al.*, 2005). Hence, if fire frequency is high and the extent of burnt areas large, grazing pressure becomes too diffuse to maintain lawns (Archibald *et al.*, 2005). Alternatively, very small fires could have the opposite impact of creating grazing lawns by concentrating animals and initiating the positive-feedback mechanisms above. Whether or not fire facilitates or limits grazing lawn development and persistence may thus depend on a delicate balance of several fire regime characteristics, including the frequency of burns, size and intensity of burn and timing of burning (see Chapter 10 for more details).

6.3.4 Role of Predation Risk

A prevailing hypothesis is that reduced predation risk is an important driver behind the creation and maintenance of grazing hot spots (Young *et al.*, 1995; Van der Waal *et al.*, 2011). By concentrating herbivores in 'safe' areas, predation risk potentially influences the pathways described in Section 6.2, the consumer–resource interaction pathway, but also possibly the drought-tolerance pathway because the concentration leads to increased trampling and soil compaction. Augustine *et al.* (2003) show that during the wet season impala both rested and foraged on grazing lawns in open glades in their Kenyan study site. However, during the dry season, impala used the glades to rest and defecate, while foraging elsewhere in the landscape. Hence, they suggested that perceived predation risk is crucial for the persistence of the lawns in these glades as nutrient hot spots; during the dry season, large impala herds return the nutrients that are removed during the wet season. Although it remains to be tested, this role of predation risk is relevant only for lawns maintained by grazers besides white rhinos; in HiP lawns commonly occur in closed woodland with a well-developed shrub layer and hence low horizontal visibility (Gosling, 2014). Data from camera traps confirm that these 'risky' lawns are visited proportionally more by white rhino than by mesoherbivores (Le Roux and Cromsigt, unpublished data).

6.3.5 Animal Behavioural Mechanisms

Apart from the factors described above, there are certain behavioural mechanisms that may concentrate animals in a patch and thus induce lawn development. Prime examples are the territorial males of certain antelope species, such as blesbok *Damaliscus pygargus*, springbok *Antidorcas marsupialis*, and black wildebeest *Connochaetes gnou* that occupy a small patch of a few square metres, also referred to as a 'stamping ground', for several months or even the whole year (Du Plessis, 1972) to display themselves to females. During this time they graze, defecate, and trample on this patch, which leads to small grazing lawns of a few square metres (Novellie, 1990; Novellie and Gaillard, 2013). Males of blue wildebeest, and perhaps impala, display similar behaviour and may be responsible for some of the smaller lawns in HiP.

Another possible behavioural mechanism that concentrates herbivory and may initiate lawn development is the role of territorial male white rhinos. Due to their very large body size, white rhino are relatively invulnerable to predation, allowing them to be very predictable in their spatial movements (Owen-Smith, 1988; Chapter 5). By returning to the same patches again and again, rhinos may turn a bunch grass patch into a grazing lawn. It remains unclear, however, whether rhino would have this effect anywhere in the landscape without any of the 'concentrating' factors described above. A study in Kruger National Park found that rhino increased the number of lawns in the landscape, but the majority of these lawns were associated with large termite mounds (Cromsigt and te Beest, 2014). Owen-Smith (1988) also describes how rhino in Mfolozi could create lawns by grazing outwards from termite mounds. Waldram *et al.* (2008) showed that the proportion of short-grass ('lawn') patches was much lower around wallows in areas after white rhino had been removed when compared with control areas where rhinos remained, particularly in mesic areas of the park ($\geq$ 750 mm rainfall per year).

6.4 Grazing Lawn Species Composition and Lawn Grass Traits

McNaughton's (1984) original formalization of the grazing lawn concept suggested that grazing lawns consist of particular grass species with specific traits that increase the grazing- and drought-tolerance of these species. This has led to suggestions that the definition of a grazing lawn includes a species composition shift following intense grazing (e.g. Cromsigt and Kuijper, 2011). This definition is currently disputed: in some of the examples of grazing lawns cited above there was no floristic difference, and the maintenance of a community in a short, palatable state was considered sufficient – as long as the positive feedbacks between grazing and forage quality described above prevail (Arnold *et al.*, 2014). Nevertheless, species that persist in grazing lawn grass communities need to show traits that allow tolerance of heavy defoliation (Hempson *et al.*, 2015), and grazing lawns in HiP are strikingly different in grass species composition from adjacent tall grassland (Figure 6.5). Average annual rainfall influences which grass species dominate lawns in HiP, with *Digitaria longiflora* dominating under wetter conditions in Hluhluwe and *Urochloa mosambicensis*, *Panicum coloratum* and *Sporobolus nitens* under drier conditions in Mfolozi (Downing, 1972; Owen-Smith, 1973, 1988; Stock *et al.*, 2010; Vos, 2010). Other common species in the lawns include *Digitaria argyrograpta*, *Cynodon dactylon* (on more sandy soils), and *Eragrostis superba* (Cromsigt, 2006; Arsenault and Owen-Smith, 2011). *Dactyloctenium australe* forms shade-tolerant grazing lawns that are mostly restricted to *Spirostachys africana* and *Acacia burkei* woodlands, and *Euclea* woodlands or thickets, often close to the main rivers in Hluhluwe as well as Mfolozi (Swemmer, 1998; Vos, 2010). In contrast, surrounding tall grassland is dominated by caespitose species such as *Themeda triandra*,

Figure 6.5 Grazing exclusion plot in western Mfolozi showing a grazing lawn dominated by *Panicum coloratum* and *Urochloa mosambicensis* in front of the fence and tall bunch grassland dominated by *Themeda triandra* in the exclosure (photo: Norman Owen-Smith, 1969).

Eragrostis spp., *Sporobolus* spp., and, in more shaded areas, *Panicum maximum* (Downing, 1972; Owen-Smith, 1973; Stock *et al.*, 2010). Most of the lawn species are strongly stoloniferous – although some have the ability to switch to upright growth forms when not heavily grazed (defined as facultative lawn species by Hempson *et al.*, 2015; e.g. *Panicum coloratum*) and some lawns consist of species that never become stoloniferous, but have prostrate leaves, such as *Sporobolus nitens* and *Digitaria argyrograpta*. An unexplored aspect of grazing lawns is the presence of forbs (Hempson *et al.*, 2015). Forbs often cover a substantial proportion of grazing lawns (Cromsigt and te Beest, 2014), but the identity of those forbs and their role in lawn functioning is unclear.

6.5 Utilization of Grazing Lawns by Different-Sized Grazers

Several short-grass grazers in HiP use grazing lawns for a large part of the year, particularly wildebeest, white rhino, warthog, impala, and to some extent plains zebra *Equus quagga* (Cromsigt *et al.*, 2009; Arsenault and Owen-Smith, 2011; Kleynhans *et al.*, 2011). In a detailed observational study, Bonnet *et al.* (2010) showed that these grazing herbivores closely adjusted their grazing activity to the productivity of grazing lawns. Following rainfall-driven regrowth pulses, grazers consumed the newly produced biomass, keeping standing biomass in grazing lawns at relatively low levels. Like McNaughton, these authors concluded that use of grazing lawns is driven by short-term production, and not standing biomass. Other studies have shown that white rhino strongly favour grazing lawns as long as the lawns retain sufficient biomass (Owen-Smith, 1973, 1988), and continue doing so into the dry season (Shrader and Perrin, 2006; Arsenault and Owen-Smith, 2011). Below-average rainfall, or high rhino densities, may restrict their use of lawns to the wet season (Owen-Smith 1973, 1988). Arsenault and Owen-Smith (2011) documented the use of lawns by three grazers during both relatively dry and relatively wet years. Rhino and wildebeest spent 50% or more of their grazing time on lawns throughout both years, but wildebeest used lawns somewhat less during the dry season of the drier year. Zebra use only reached similar levels during the wet year. Arsenault and Owen-Smith (2008) found that white rhino, and to a lesser extent wildebeest and impala, concentrated on the shortest parts of the lawn (≤ 5 cm), while zebra also used the taller grass tufts (6–20 cm) within the lawn. The fact that rhino use lawns more uniformly through the year, and focus on the shortest parts, suggests that they play the strongest role in maintaining lawns, followed by wildebeest, impala, and probably warthog. Zebra, by focusing on the taller grass within the lawns, may be important for preventing bunch grasses from invading lawns and may perhaps contribute to lawn expansion by grazing on tall grasses bordering the lawn.

Grazing lawns provide high-quality wet season resources (for nutrients such as P and K) that support females during lactation and different stages of their reproductive cycle (Illius and O'Connor, 2000; Owen-Smith, 2004), but they do not provide sufficient forage during the dry season. A mix of grazing lawns and dry season tall grassland reserves would lead to the highest long-term densities of short-grass grazers (see Chapter 5 for more details; Owen-Smith, 2004).

6.6 Cascading Impacts of Lawns on the Savanna Ecosystem

6.6.1 Role of Lawns for Savanna Biodiversity

As well as promoting certain grass and forb species, and particular suites of grazing mammals, grazing lawns represent a unique habitat for a range of other fauna. Three of the 12 specialist grassland bird species in HiP are strongly associated with lawns (Krook *et al.*, 2007): African pipit *Anthus cinnamomeus*, crowned lapwing *Vanellus coronatus*, and the sabota lark *Calendulauda sabota*. Two more species seemed to depend on a lawn–bunch mosaic (yellow-throated longclaw *Macronyx croceus*, rufous-naped lark *Mirafra africana*), while seven species were tall grassland specialists. An interesting suggestion by Krook *et al.* that remains to be tested is that post-burn bunch grasslands could provide an ephemeral habitat for the grazing lawn specialists. Similar studies from across the globe confirm the association between short-grass habitat and certain bird species (Agnew *et al.*, 1986; Desmond, 2004; Askins *et al.*, 2007; Augustine and Skagen, 2014). Mgobozi (2008) found that spider abundance, diversity, and species richness was higher in grazing lawns than in bunch grassland patches. At both a species (Currie, 2003) and functional (Van der Plas *et al.*, 2012) level, grasshopper communities on grazing lawns are distinct, although grasshopper richness and density is higher in bunch grassland.

6.6.2 Lawns and Savanna Trees

Walters *et al.* (2004) found reduced germination of several acacia species in grazing lawn sites, likely due to relatively high evaporation rates and, therefore, low water availability in lawn soils (Van der Plas *et al.*, 2013). This, in combination with browsing from mixed feeders like impala, probably limits tree recruitment on lawns. Observations elsewhere confirm that tree recruitment, and resulting woody cover, is restricted on grazing lawns relative to surrounding tall grassland (e.g. Van der Waal *et al.*, 2011). Grazing lawns may thus not be subject to the increasing woody encroachment that is described for many savannas, despite the fact that lawns generally do not burn (see Chapter 7). Bond *et al.* (2001) suggest that a reduction in grazing lawn cover in Hluhluwe from the 1960s to the 1990s has resulted in a shift from high recruitment of browse-tolerant *Acacia nilotica* in the 1960s to high recruitment of the more fire-tolerant *Acacia karroo* in more recent years, because a reduction in lawns led to more frequent, and more intense, fires. Grazing lawns may thus affect tree dynamics in savannas beyond the scale of the lawn patches through their impacts on the fire regime (Archibald *et al.*, 2005).

6.6.3 Between Grazing Lawns and Overgrazing

Overgrazing, with respect to grazing lawns, can be defined in two ways: (1) an excessive proportion of grazing lawns in the landscape, and (2) reduced grass cover in the grazing lawns. Large increases in the extent of grazing lawns could lead to declines in grazer species that are not adapted to exploit short-grass lawns. For example, the declines in numbers of common reedbuck and waterbuck in HiP could have been partly related to the effect of increasing numbers of short-grass grazers such as wildebeest and white rhino (Deane, 1966; Chapters 4 and 5). Similarly, birds or insects that are tall-grass specialists would be negatively affected by strong increases in lawn cover. As described in Section 6.5, even populations of short-grass grazers would be negatively affected if the proportion of lawns in the landscape is too large relative to the proportion of dry season taller-grass reserves (Owen-Smith, 2004). From a savanna conservation and management point of view, heterogeneous grassland with a mix of short- and tall-grass patches will thus be most desirable in terms of maintaining biodiversity and ecosystem functioning, reflecting the *rangeland heterogeneity paradigm* that was introduced for North American grassland systems (Fuhlendorf and Engle, 2001) and the concept of *functional heterogeneity* as presented by Owen-Smith (2004).

Another issue of overgrazing is the degradation of grazing lawns themselves. Heavily utilized lawns are characterized by a lack of aerial plant cover especially in the dry season, which may lead to serious soil erosion. Such concerns about soil erosion were in fact central to the heavy culling of warthog and wildebeest, and the removals of white rhino, during the 1960s and 1970s (Brooks and Macdonald, 1983). Besides soil erosion, lawns may also degrade through their invasion by unpalatable plant species, including grasses such as *Aristida* spp. and certain forb species, making them unsuitable foraging areas for grazers (Hempson *et al.*, 2015). We do not fully understand under what conditions locally intense grazing leads to this degraded state of lawns (Hempson *et al.*, 2015).

6.7 Recent Declines in Grazing Lawn Cover

The extent of grazing lawns appears to be declining in HiP. From a maximum cover in the western Mfolozi around 50% during the early 1970s, the overall lawn extent has declined quite strongly recently (Table 6.2). The reasons for this are not clear. There have been reductions in numbers of some of the short-grass grazers in HiP, particularly impala and wildebeest, during the last 5–10 years (Chapter 4), possibly due to increased predation pressure from growing numbers of wild dog and lion during the 2000s (Chapter 12). This may have contributed to the decline in lawns. However, white rhino numbers increased during the same period (Chapter 4), which may suggest that rhino alone do not maintain the

Table 6.2. *Estimates of grazing lawn cover (average % of area surveyed, standard deviation (SD) and maximum % recorded on a single transect) in Hluhluwe-iMfolozi Park throughout the last 45 years. Estimates for 2004 (Cromsigt* et al., *2009), 2010 (Vos, 2010) and 2014 (Le Roux and Cromsigt, unpublished data) reflect visual lawn cover measurements using the exact same methodology while walking transects across the park (see caption of Figure 6.3 and Cromsigt* et al., *2009). Estimates for 1999 (Arsenault and Owen-Smith, 2011) and 2006 (Gosling, 2014) were also collected on foot, recording lawn cover visually. The 1999 estimates from Archibald* et al. *(2005) were based on satellite images covering the whole park and were likely underestimates due to difficulties with assessing lawn cover underneath closed-canopy woodlands. The same may be true for the estimate by Downing (1972).*

Area	Year	Average	SD	Maximum	Methods	Reference
Mfolozi	1969	~50	–	–	Aerial photographs	Downing 1972
Mfolozi	1999	28	–	–	Line transects	Arsenault and Owen-Smith, 2011
HiP	1999	8	–	24	Satellite images	Archibald *et al.*, 2005
HiP	2004	12.2	8.7	29.0	Line transects	Cromsigt *et al.*, 2009
HiP	2006	15.0	14.7	28.8	Transects	Gosling, 2014
HiP	2010	4.2	4.7	16.3	Line transects	Vos, 2010
HiP	2014	6.9	6.5	21.5	Line transects	Le Roux and Cromsigt, unpublished data

historically high grazing lawn cover in HiP. Changes in the fire regime within HiP may also have contributed to the decline in lawn cover. Fire frequency was low throughout HiP during the 1960s, with a typical area burning only once per decade. This frequency increased during the 1970s and 1980s up to two to three burns per decade and was maintained at this level during the 1990s and 2000s in the area north of the Black Mfolozi river (Chapter 10, Figure 10.3). This two- to three-fold increase in fire frequency may have reduced lawn persistence by dispersing herbivores away from lawns (see Section 6.3.3). However, recommending a reduction in fire frequency to allow grazing lawn cover to increase is problematic because fires are used to manage other aspects of ecosystem change, particularly bush encroachment (see Chapter 10 for a detailed discussion). At a more local scale, white rhino removals, white rhino poaching, and depletion of the legacy effects of abandoned homesteads may also have led to the disappearance of lawns. All in all, however, the reasons behind the drastic decline in lawn cover that HiP has experienced since 2004 remain speculative.

6.8 Concluding Remarks

We have described how former agro-pastoral land use, termites, fire, and predation influence where grazing herbivores may create and maintain the grazing lawns that form such a key part of HiP's grassland heterogeneity. HiP has displayed a strikingly high abundance of grazing lawns (at least 10–15% of extent of the whole park) for many decades. In other areas where cover has been estimated, lawns form only a few percent of the local landscape (e.g. in Nylsvley (Blackmore *et al.*, 1990), Ithala Game Reserve (Valls-Fox *et al.*, 2015), Mpala Research Centre (Young *et al.*, 1995), and Kruger National Park (Cromsigt and te Beest, 2014)). So why has lawn cover been so high in HiP? One aspect that sets HiP apart from these areas is that white rhino never went locally extinct and have been present at high densities for the last 50–60 years at least. Moreover, they used a landscape with clear legacy effects of former human land use. This combined effect of former human land use and persistently high densities of white rhino may distinguish HiP from most if not all current savanna systems. This may reflect a situation that used to be common across Africa where people and this mega-grazer coexisted, but now represents a relict mechanism promoting African savanna diversity.

6.9 References

Agnew, W., Uresk, D. W., & Hansen, R. M. (1986) Flora and fauna associated with prairie dog colonies and adjacent ungrazed mixed-grass prairie in western South Dakota. *Journal of Range Management* **39**: 135–139.

Anderson, T. M., Fokkema, W., Valls-Fox, H., & Olff, H. (2013) Distinct physiological responses underlie defoliation tolerance in African lawn and bunch grasses. *International Journal of Plant Sciences* **174**: 769–778.

Archibald, S. (2008) African grazing lawns – how fire, rainfall, and grazer numbers interact to affect grass community states. *The Journal of Wildlife Management* **72**: 492–501.

Archibald, S. & Bond, W. J. (2004) Grazer movements: spatial and temporal responses to burning in a tall-grass African savanna. *International Journal of Wildland Fire* **13**: 1–9.

Archibald, S., Bond, W. J., Stock, W. D., & Fairbanks, D. H. K. (2005) Shaping the landscape: fire–grazer interactions in an African savanna. *Ecological Applications* **15**: 96–109.

Arnold, S. G., Anderson, T. M., & Holdo, R. M. (2014) Edaphic, nutritive, and species assemblage differences between hotspots and matrix vegetation: two African case studies. *Biotropica* **46**: 387–394.

Arsenault, R. & Owen-Smith, N. (2008) Resource partitioning by grass height among grazing ungulates does not follow body size relation. *Oikos* **117**: 1711–1717.

Arsenault, R. & Owen-Smith, N. (2011) Competition and coexistence among short-grass grazers in the Hluhluwe-iMfolozi Park, South Africa. *Canadian Journal of Zoology* **89**: 900–907.

Arshad, M. A. (1982) Influence of the termite *Macrotermes michaelseni* (Sjost) on soil fertility and vegetation in a semi-arid savannah ecosystem. *Agro-Ecosystems* **8**: 47–58.

Askins, R. A., Chávez-Ramírez, F., Dale, B. C., *et al.* (2007) Conservation of grassland birds in North America: understanding ecological processes in different regions. *Ornithological Monographs* **64**: 1–46.

Augustine, D. J. (2003) Long-term, livestock-mediated redistribution of nitrogen and phosphorus in an East African savanna. *Journal of Applied Ecology* **40**: 137–149.

Augustine, D. J. & McNaughton, S. J. (1998) Ungulate effects on the functional species composition of plant communities: herbivore selectivity and plant tolerance. *Journal of Wildlife Management* **62**: 1165–1183.

Augustine, D. J. & McNaughton, S. J. (2006) Interactive effects of ungulate herbivores, soil fertility, and variable rainfall on ecosystem processes in a semi-arid savanna. *Ecosystems* **9**: 1242–1256.

Augustine, D. J. & Skagen, S. K. (2014) Mountain plover nest survival in relation to prairie dog and fire dynamics in shortgrass steppe. *Journal of Wildlife Management* **78**: 595–602.

Augustine, D. J., McNaughton, S. J., & Frank, D. A. (2003) Feedbacks between soil nutrients and large herbivores in a managed savanna ecosystem. *Ecological Applications* **13**: 1325–1337.

Bartlett, M. K., Scoffoni, C., & Sack, L. (2012) The determinants of leaf turgor loss point and prediction of drought tolerance of species and biomes: a global meta-analysis. *Ecology Letters* **15**: 393–405.

Bell, R. H. V. (1971) A grazing ecosystem in the Serengeti. *Scientific American* **225**: 86–94.

Blackmore, A. C., Mentis, M. T., & Scholes, R. J. (1990) The origin and extent of nutrient-enriched patches within a nutrient-poor savanna in South Africa. *Journal of Biogeography* **17**: 463–470.

Bond, W. J., Smythe, K., & Balfour, D. A. (2001) Acacia species turnover in space and time in an African savanna. *Journal of Biogeography* **28**: 117–128.

Bonnet, O., Fritz, H., Gignoux, J., & Meuret, M. (2010) Challenges of foraging on a high-quality but unpredictable food source: the dynamics of grass production and consumption in savanna grazing lawns. *Journal of Ecology* **98**: 908–916.

Brooks, P. M. & Macdonald, I. A. W. (1983) The Hluhluwe-Umfolozi Reserve: an ecological case history. In: *Management of large mammals in African conservation areas* (ed. N. Owen-Smith), pp. 51–77. Haum Educational Publishers, Pretoria.

Brooks, S. J. (2001) Changing nature: a critical historical geography of the Umfolozi and Hluhluwe Game Reserves, Zululand, 1887 to 1947. PhD thesis, Queen's University, Kingston, Canada.

Clauss, M., Castell, J. C., Kienzle, E., *et al.* (2007) Mineral absorption in the black rhinoceros (*Diceros bicornis*) as compared with the domestic horse. *Journal of Animal Physiology and Animal Nutrition* **91**: 193–204.

Coetsee, C., Stock, W. D., & Craine, J. M. (2010) Do grazers alter nitrogen dynamics on grazing lawns in a South African savannah? *African Journal of Ecology* **49**: 62–69.

Coughenour, M. B. (1985) Graminoid responses to grazing by large herbivores – adaptations, exaptations, and interacting processes. *Annals of the Missouri Botanical Garden* **72**: 852–863.

Craig, T. P. (2010) The resource regulation hypothesis and positive feedback loops in plant–herbivore interactions. *Population Ecology* **52**: 461–473.

Cromsigt, J. P. G. M. (2006) Large herbivores in space: resource partitioning among savanna grazers in a heterogeneous environment. PhD thesis, University of Groningen, Groningen.

Cromsigt, J. P. G. M. & Kuijper, D. P. J. (2011) Revisiting the browsing lawn concept: evolutionary interactions or pruning herbivores? *Perspectives in Plant Ecology, Evolution or Systematics* **13**: 207–215.

Cromsigt, J. P. G. M. & Olff, H. (2008) Dynamics of grazing lawn formation: an experimental test of the role of scale-dependent processes. *Oikos* **117**: 1444–1452.

Cromsigt, J. P. G. M. & te Beest, M. (2014) Restoration of a megaherbivore: landscape-level impacts of white rhinoceros in Kruger National Park, South Africa. *Journal of Ecology* **102**: 566–575.

Cromsigt, J. P. G. M., Prins, H. H. T. & Olff, H. (2009) Habitat heterogeneity as a driver of ungulate diversity and distribution patterns: interaction of body mass and digestive strategy. *Diversity and Distributions* **15**: 513–522.

Currie, G. (2003) The impact of megaherbivore grazers on grasshopper communities via grassland conversion in a savannah ecosystem. Honours thesis, University of Cape Town, Cape Town.

Davies, A. B., Robertson, M. P., Levick, S. R., *et al.* (2014) Variable effects of termite mounds on African savanna grass communities across a rainfall gradient. *Journal of Vegetation Science* **25**: 1405–1416.

Deane, N. N. (1966) Ecological changes and their effect on a population of reedbuck (*Redunca arundinum* (Boddaert)). *Lammergeyer* **6**: 2–8.

Desmond, M. J. (2004) Effects of grazing practices and fossorial rodents on a winter avian community in Chihuahua, Mexico. *Biological Conservation* **116**: 235–242.

Downing, B. H. (1972) A plant ecological survey of the Imfolozi Game Reserve, Zululand. PhD thesis, University of Natal, Durban.

Du Plessis, S. S. (1972) Ecology of blesbok with special reference to productivity. *Wildlife Monographs* **30**: 1–70.

Eltringham, S. K. (1974) Changes in the large mammal community of Mweya Peninsula, Rwenzori National Park, Uganda, following removal of hippopotamus. *Journal of Applied Ecology* **11**: 855–865.

Feely, J. M. (1980) Did Iron Age man have a role in the history of Zululand's wilderness landscapes? *South African Journal of Science* **76**: 150–152.

Feely, J. M. (2004) Prehistoric use of woodland and forest by farming peoples in South Africa. In: *Indigenous forests and woodlands: policy, people and practice* (eds M. J. Lawes, H. A. C. Eeley, C. M. Shackleton, & B. G. S. Geach), pp. 284–286. University of KwaZulu-Natal Press, Pietermaritzburg.

Frank, D. A., Groffman, P. M., Evans, R. D., & Tracy, B. F. (2000) Ungulate stimulation of nitrogen cycling and retention in Yellowstone Park grasslands. *Oecologia* **123**: 116–121.

Fuhlendorf, S. D. & Engle, D. M. (2001) Restoring heterogeneity on rangelands: ecosystem management based on evolutionary grazing patterns. *BioScience* **51**: 625–632.

Gandar, M. V., Huntley, B. J., & Walker, B. H. (1982) Trophic ecology and plant/herbivore energetics. In: *Ecology of tropical savannas* (eds B. J. Huntley & B. H. Walker), pp. 514–534. Springer Verlag, Berlin.

Girma, F. S. & Krieg, D. R. (1992) Osmotic adjustment in sorghum. 1. Mechanisms of diurnal osmotic potential changes. *Plant Physiology* **99**: 577–582.

Gosling, C. M. (2014) Biotic determinants of heterogeneity in a South African savanna. PhD thesis, University of Groningen, Groningen.

Gosling, C. M., Cromsigt, J. P. G. M., & Olff, H. (2011) Effects of erosion from mounds of different termite genera on distinct functional grassland types in an African savanna. *Ecosystems* **15**: 128–139.

Grant, C. C. & Scholes, M. C. (2006) The importance of nutrient hot-spots in the conservation and management of large wild mammalian herbivores in semi-arid savannas. *Biological Conservation* **130**: 426–437.

Hagenah, N., Prins, H. H. T., & Olff, H. (2009) Effects of large herbivores on murid rodents in a South African savanna. *Journal of Tropical Ecology* **25**: 483–492.

Hall, M. (1984) Prehistoric farming in the Mfolozi and Hluhluwe valleys of southeast Africa: an archaeo-botanical survey. *Journal of Archaeological Science* **11**: 223–235.

Hempson, G. P., Archibald, S., Bond, W. J., *et al.* (2015) Ecology of grazing lawns in Africa. *Biological Reviews* **90**: 979–994.

Illius, A. W. & O'Connor, T. G. (2000) Resource heterogeneity and ungulate population dynamics. *Oikos* **89**: 283–294.

Jouquet, P., Boulain, N., Gignoux, J. & Lepage, M. (2004) Association between subterranean termites and grasses in a West African savanna: spatial pattern analysis shows a significant role for *Odontotermes* n. *pauperans*. *Applied Soil Ecology* **27**: 99–107.

Karki, J. B., Jhala, Y. V., & Khanna, P. P. (2000) Grazing lawns in Terai Grasslands, Royal Bardia National Park, Nepal. *Biotropica* **32**: 423–429.

Kim, H., Anderson, S. H., Motavalli, P. P., & Gantzer, C. J. (2010) Compaction effects on soil macropore geometry and related parameters for an arable field. *Geoderma* **160**: 244–251.

Kleynhans, E. J., Jolles, A. E., Bos, M. R. E., & Olff, H. (2011) Resource partitioning along multiple niche dimensions in differently sized African savanna grazers. *Oikos* **120**: 591–600.

Knapp, A. K., Blair, J. M., Briggs, J. M., *et al.* (1999) The keystone role of bison in North American tallgrass prairie. *BioScience* **49**: 39–50.

Krook, K., Bond, W. J., & Hockey, P. A. R. (2007) The effect of grassland shifts on the avifauna of a South African savanna. *Ostrich – Journal of African Ornithology* **78**: 271–279.

Lock, J. M. (1972) The effects of hippopotamus grazing on grassland. *Journal of Ecology* **60**: 445–467.

McNaughton, S. J. (1983) Compensatory growth as a response to herbivory. *Oikos* **40**: 329–336.

McNaughton, S. J. (1984) Grazing lawns: animals in herds, plant form, and coevolution. *American Naturalist* **124**: 863–886.

McNaughton, S. J. (1985) Ecology of a grazing ecosystem: the Serengeti. *Ecological Monographs* **55**: 259–294.

McNaughton, S. J. (1988) Mineral-nutrition and spatial concentrations of African ungulates. *Nature* **334**: 343–345.

McNaughton, S. J., Banyikwa, F. F., & McNaughton, M. M. (1997) Promotion of the cycling of diet-enhancing nutrients by African grazers. *Science* **278**(5344): 1798–1800.

Mgobozi, M. P. (2008) Spider community responses to *Chromolaena odorata* invasion, grassland type and grazing intensities. MSc thesis, University of Pretoria, Pretoria.

Milchunas, D. G., Sala, O. E., & Lauenroth, W. K. (1988) A generalized-model of the effects of grazing by large herbivores on grassland community structure. *American Naturalist* **132**: 87–106.

Mills, A. J., Milewski, A., Fey, M. V., Groengroeft, A., & Petersen, A. (2009) Fungus culturing, nutrient mining and geophagy: a geochemical investigation of *Macrotermes* and *Trinervitermes* mounds in southern Africa. *Journal of Zoology* **278**: 24–35.

Morgan, J. W. & Lunt, I. D. (1999) Effects of time-since-fire on the tussock dynamics of a dominant grass (*Themeda triandra*) in a temperate Australian grassland. *Biological Conservation* **88**: 379–386.

Novellie, P. (1990) Habitat use by indigenous grazing ungulates in relation to sward structure and veld condition. *Journal of the Grassland Society of Southern Africa* **7**: 16–23.

Novellie, P. & Gaillard, A. (2013) Long-term stability of grazing lawns in a small protected area, the Mountain Zebra National Park. *Koedoe* **55**: 1–7.

Olivier, R. C. D. & Laurie, W. A. (1974) Habitat utilization by hippopotamus in the Mara river. *African Journal of Ecology* **12**: 249–271.

Owen-Smith, N. (1973) The behavioural ecology of the white rhinoceros. PhD thesis, University of Wisconsin, Madison, WI.

Owen-Smith, N. (1988) *Megaherbivores: the influence of very large body size on ecology*. Cambridge University Press, Cambridge.

Owen-Smith, N. (2004) Functional heterogeneity in resources within landscapes and herbivore population dynamics. *Landscape Ecology* **19**: 761–771.

Person, B. T., Herzog, M. P., Ruess, R. W., *et al.* (2003) Feedback dynamics of grazing lawns: coupling vegetation change with animal growth. *Oecologia* **135**: 583–592.

Roberts, C. M. (2009) Marsupial grazing lawns in Tasmania: maintenance, biota and the effects of climate change. PhD thesis, University of Tasmania.

Scholes, R. J. (2003) Convex relationships in ecosystems containing mixtures of trees and grass. *Environmental and Resource Economics* **26**: 559–574.

Shrader, A. M. & Perrin, M. R. (2006) Influence of density on the seasonal utilization of broad grassland types by white rhinoceroses. *African Zoology* **41**: 312–315.

Stock, W. D., Bond, W. J., & Van de Vijver, C. A. D. M. (2010) Herbivore and nutrient control of lawn and bunch grass distributions in a southern African savanna. *Plant Ecology* **206**: 15–27.

Swemmer, T. (1998) The distribution and ecology of grazing lawns in a South African savannah ecosystem. Honours thesis, University of Cape Town, Cape Town.

Tracy, B. & McNaughton, S. J. (1995) Elemental analysis of mineral lick soils from the Serengeti National Park, the Konza Prairie and Yellowstone National Park. *Ecography* **18**: 91–94.

Valls-Fox, H., Bonnet, O., Cromsigt, J. P. G. M., Fritz, H., & Shrader, A. M. (2015) Legacy effects of different land-use histories interact with current grazing patterns to determine grazing lawn properties. *Ecosystems* **18**: 720–733.

Van de Vijver, C. A. D. M., Poot, P., & Prins, H. H. T. (1999) Causes of increased nutrient concentrations in post-fire regrowth in an East African savanna. *Plant and Soil* **214**: 173–185.

Van der Plas, F., Anderson, T. M., & Olff, H. (2012) Trait similarity patterns within grass and grasshopper communities: multitrophic community assembly at work. *Ecology* **93**: 836–846.

Van der Plas, F., Zeinstra, P., Veldhuis, M., *et al.* (2013) Responses of savanna lawn and bunch grasses to water limitation. *Plant Ecology* **214**: 1157–1168.

Van der Waal, C., Kool, A., Meijer, S. S., *et al.* (2011) Large herbivores may alter vegetation structure of semi-arid savannas through soil nutrient mediation. *Oecologia* **165**: 1095–1107.

Veldhuis, M. P., Howison, R. A., Fokkema, R. W., Tielens, E., & Olff, H. (2014) A novel mechanism for grazing lawn formation: large herbivore-induced modification of the plant–soil water balance. *Journal of Ecology* **102**: 1506–1517.

Veldhuis, M. P., Fakkert, H. F., Berg, M. P., & Olff, H. (2016). Grassland structural heterogeneity in a savanna is driven more by productivity differences than by consumption differences between lawn and bunch grasses. *Oecologia* **182**: 841–853.

Verweij, R. J. T., Verrelst, J., Loth, P. E., Heitkönig, I. M. A. & Brunsting, A. M. H. (2006) Grazing lawns contribute to the subsistence of mesoherbivores on dystrophic savannas. *Oikos* **114**: 108–116.

Vesey-Fitzgerald, D. F. (1960) Grazing succession among East African game animals. *Journal of Mammalogy* **41**: 161–172.

Vos, I. A. (2010) Spatial heterogeneity of resources – variation in space and time in the distribution and persistence of grazing lawns. MSc thesis, Utrecht University, Utrecht.

Waldram, M. S., Bond, W. J., & Stock, W. D. (2008) Ecological engineering by a megagrazer: white rhino impacts on a South African savanna. *Ecosystems* **11**: 101–112.

Walters, M., Midgley, J. J., & Somers, M. J. (2004) Effects of fire and fire intensity on the germination and establishment of *Acacia karroo*, *Acacia nilotica*, *Acacia luederitzii* and *Dichrostachys cinerea* in the field. *BMC Ecology* **4**: 1–13.

Wood, T. G. & Sands, W. (1978) The role of termites in ecosystems. In: *Production ecology of ants and termites* (ed. M. V. Brian), pp. 245–292. Cambridge University Press, Cambridge.

Young, T. P., Patridge, N., & Macrae, A. (1995) Long-term glades in acacia bushland and their edge effects in Laikipia, Kenya. *Ecological Applications* **5**: 97–108.

7 · *Demographic Bottlenecks and Savanna Tree Abundance*

WILLIAM J. BOND, A. CARLA STAVER, MICHAEL D. CRAMER, JULIA L. WAKELING, JEREMY J. MIDGLEY, AND DAVE A. BALFOUR

7.1 Introduction

In Hluhluwe-Mfolozi Park (HiP), tree abundance varies widely at a scale of just hundreds of metres both within savannas and between savannas and closed forests or thickets. Similar differences, but on much larger geographical scales, are characteristic of many savanna-dominated landscapes. Explanations for this variation in tree cover have long been divided into resource constraints on tree growth (bottom-up) and consumer (herbivores, fire) control of tree cover (top-down). A popular resource-based hypothesis for tree–grass coexistence in savannas, for example, assumes that grasses and trees partition soil resources, with grasses accessing surface layers and trees using deeper layers (Walter, 1971). Counter to such resource-based hypotheses are top-down arguments based, for example, on numerous fire experiments showing that frequent fires prevent saplings from growing into large trees. Heavy pruning of stems by mammalian browsers may act in a similar way, preventing trees from growing to mature tree heights for decades (reviewed by Bond, 2008).

The effects of resources and consumers can be integrated by considering the changing effects of disturbance as trees grow larger. Taller trees escape injury from surface fires if the trunk is insulated, and from herbivory if foliage is beyond the reach of browsers. Resources influence the rates at which plants grow to critical threshold sizes (Figure 7.1) and thus contribute to the probability of saplings escaping fire and browsers (Bond and van Wilgen, 1996; Higgins *et al.*, 2000; Silva *et al.*, 2013). Models have been important in helping to identify the relative role of resource

Conserving Africa's Mega-Diversity in the Anthropocene, ed. Joris P. G. M. Cromsigt, Sally Archibald and Norman Owen-Smith. Published by Cambridge University Press.

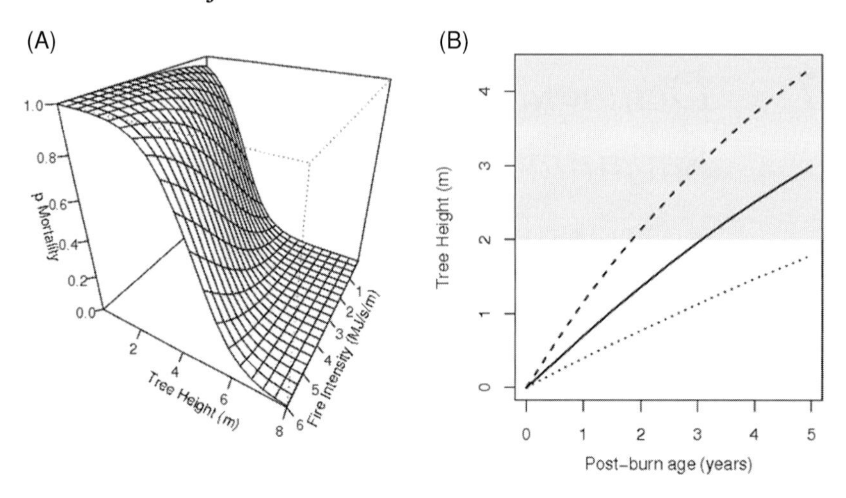

Figure 7.1 The probability of a sapling growing into a tree in frequently burned savannas depends on top-kill of stems and the rate of post-burn growth. (A) The relationship between top-kill (mortality), stem height, and fire intensity for an African savanna tree. The probability of top-kill decreases sharply once plants reach escape height, 2–4 m in this example. (B) How sapling growth rates determine the period needed to reach escape height between top-killing fires. This period varies from a minimum of 2, 3, and more than 5 years for the three growth rates illustrated. Growth rate varies among species and is influenced by resource supply, which is influenced by competition with grasses. The interaction between the fire regime and sapling growth rates determines juvenile recruitment to adult size classes. (Reprinted with permission from Annual Reviews; Bond, 2008.)

and consumer control and of key demographic stages (e.g. Higgins *et al.*, 2000). Higgins and colleagues (2000) developed a simulation model based on parameter values of South African savannas, and showed that tree–grass coexistence was possible over a wide rainfall gradient, given variable conditions for recruitment and 'storage' of successful recruitment events in long-lived adults. For example, fires in mesic savannas may be frequent enough and severe enough to top-kill saplings, preventing their growth into adult trees. For trees to exist in such savannas, there would need to be periodic escape opportunities, e.g. from longer intervals between fires or locally reduced fire severity: the 'variable conditions for recruitment'. However, for the population to persist, adult trees would also need to survive from one recruitment episode to the next, thereby 'storing' successful recruitment episodes until the next opportunity arises. In the Higgins *et al.* (2000) model, simulated adult tree densities were most limited by seedling establishment in arid savannas and by sapling escape in frequently burned savannas. Sensitivity analysis of this model showed that

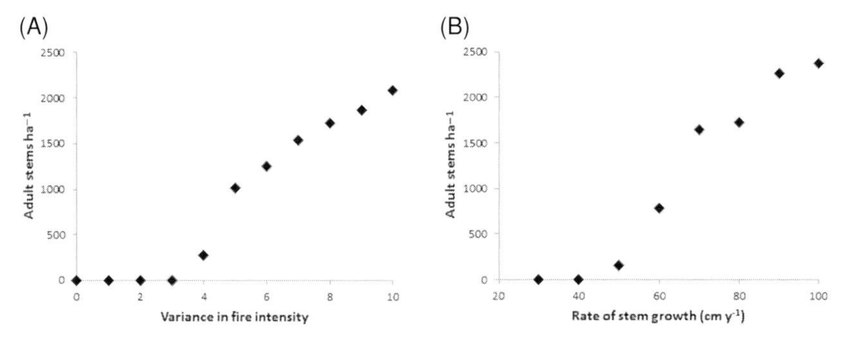

Figure 7.2 Sensitivity of number of adult trees to (A) variance in fire intensity, and (B) rate of sapling stem growth according to a demographic model of South African savanna trees (from Higgins *et al.*, 2000). The high sensitivity to these two model parameters indicates the necessity for combining (A) top-down controls with (B) bottom-up controls to understand tree populations in savannas.

tree cover is sensitive both to the frequency of top-killing fires and to growth rates to threshold sizes where saplings are too large to be top killed (Figure 7.2). The implication of the threshold models, or demographic bottleneck models (DBMs), is that tree densities in flammable grassy systems can only be understood if both the resource constraints on growth and the factors controlling the disturbance regime are integrated.

In this chapter we outline the interacting roles of resource constraints and consumers for the different life stages of a woody plant and show how studies from HiP have been central for developing the concept of demographic bottlenecks. Finally, we will show that this concept is increasingly influential globally for theory development and practical applications.

7.2 Life-History Stages and Bottom-Up and Top-Down Controls on Demographic Transitions

There are several potential demographic bottlenecks (labelled 'hurdles' by Midgley and Bond, 2001) in savanna tree populations starting with pollination and the probability of seed set. No studies have analysed all of these hurdles for a single population in any single locality. While such a study would no doubt reveal interesting information, the complexity of tree demography in savannas, and generalities among species and environments, may be better understood by focusing on just the key life-history stages in a particular population. Here we consider the role of resources and consumers in influencing seedling establishment, sapling transitions to adult size classes, and adult tree mortality. Our focus is on trees, woody plants that produce most of their flowers and fruits above the flame zone and antelope browse height (~3 m). Shrubs, which mature within flame and browse height, are relatively poorly studied, but are discussed where information is available.

7.2.1 Seedlings

Seedling establishment is the period from post-germination to a 1- or 2-year old young plant. During this time, the plant will have survived its first dry season and, in mesic savannas, probably its first burn. It is also likely to have encountered insect and vertebrate herbivores.

Resources. Resource constraints on seedling establishment have been explored in a series of field experiments conducted at Hluhluwe (Cramer *et al.*, 2007, 2010, 2012; Cramer and Bond, 2013). These consisted primarily of nutrient addition treatments, with and without grass, applied to *Acacia* and broadleaved thicket species. The results consistently showed very strong competitive effects of grass on seedling growth: seedling biomass with grass removed was typically > 10 times that of treatments with grass present (Figure 7.3). Above-ground grass-clipping experiments showed that competition was overwhelmingly below-ground. There were significant (between 67% and 140%) increases in growth with seedlings of three tree species supplied with NPK (Cramer *et al.*, 2012), but not when supplied with N, P, or K separately, indicating that competition is for general nutrition. The implication is that grass competition is a formidable obstacle to early seedling growth and that any resource addition, such as atmospheric nitrogen deposition, will promote grasses at the expense of tree seedlings. The results of these studies, at least for the establishment phase, are opposite to those expected from root niche differentiation. Instead of trees and grasses avoiding competition by exploiting different soil layers, tree establishment and sapling growth are strongly limited by root competition with grasses (see also February *et al.*, 2013; Tedder *et al.*, 2014).

Consumers. The demographic effects of fire vary with plant life-history stages. Mortality as a result of burning is most likely at the vulnerable seedling and subsequent early establishment phases of growth. In a common garden experiment comparing fire survival rates of 1-year-old acacias and broadleaved species from different habitat types, there were striking differences among species. Mesic savanna acacias survived an experimental fire with greater than 70% survival rates, in striking contrast to arid savanna species, all of which were killed by burning (Table 7.1). Two acacia species occurring in thicket vegetation, which seldom burns,

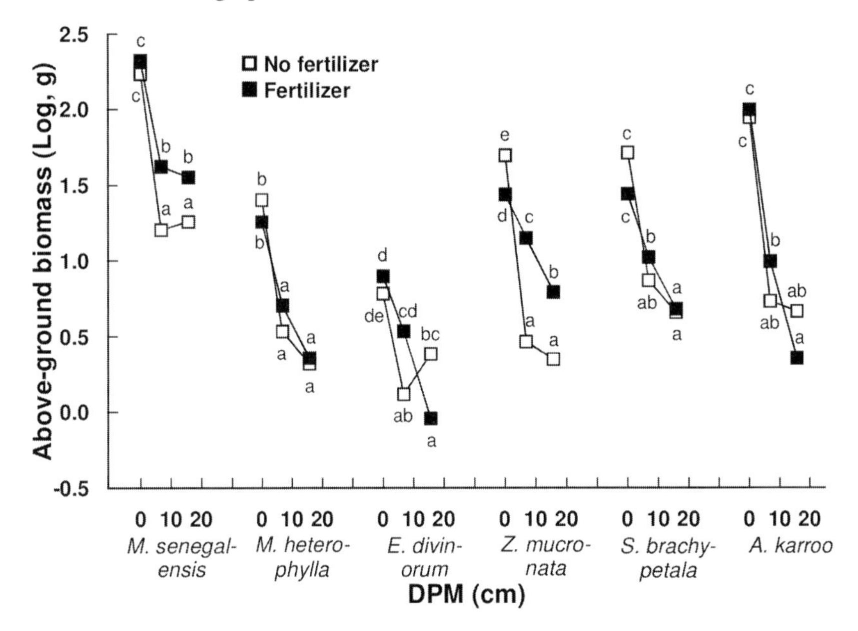

Figure 7.3 Competitive suppression of broadleaved thicket species and *Acacia karroo* by clipped and unclipped grass either with or without N-fertilizer addition. Grass sward height was measured by a disc pasture meter (DPM, cm) as a proxy for grass biomass. Different letters indicate statistically significant differences between grass treatments (three-way GLMM). (Reprinted with permission from Springer International Publishing; Cramer and Bond (2013).)

were both killed by fire. Broadleaved thicket and forest species were mostly killed by fire with one or two exceptions (Table 7.1). Thus fires during the early establishment phase would act as major selective filters on which species occur where. Differences in fire tolerance at establishment might explain the remarkable acacia species turnover from Hluhluwe, with frequent predictable fires, to Mfolozi with less-frequent and less-predictable fires especially in drought years (Balfour and Howison, 2002).

The same set of species as used in the fire experiment was planted in a grazing lawn (Seme, Hluhluwe) where grazing and browsing pressure was assumed to be high. There was no difference in survival rate of seedlings exposed or protected by wire cages from browsers. However, seedling survival, especially of mesic savanna, forest, and thicket species, was far poorer in these short grasslands than in the tall grasslands used for the burning experiment. Thus, in addition to the direct effects of browsing, the microclimate on grazing lawns can inhibit tree seedling establishment (Table 7.1).

Table 7.1. *Tolerance to fire and browsing for* Acacia *species from different habitats at the seedling stage. Values are number of species in each group. For the fire response, seedlings were transplanted into a tall grassland, left to establish for a year, and then burnt. All species survived in an adjacent control area over the same period. For the 'browse' response, seedlings were transplanted into a grazing lawn. In the grazing lawn, there was no difference in response in seedlings exposed, or protected by wire cages, from browsers. Thus 'browse' response is primarily to the exposed lawn microclimate. There were 10 replicate seedlings of each species for each treatment. Species are defined as tolerant (t) = > 70% survival, intolerant (n) = < 30% survival, intermediate (i) = 30–60% survival. ZLTP unpublished.*

	Fire Tolerant	Fire Intolerant	Fire Inter-mediate	Browse Tolerant	Browse Intolerant	Browse Inter-mediate
Arid savanna acacias	0	4		4	0	
Mesic savanna acacias	2	0		2	0	
Thicket acacias		2		2		
Broad-leaved thicket, forest species	2	4	1	2	3	1

Arid savanna acacias: *borleae, grandicornuta, nigrescens, tortilis.*
Mesic savanna acacias: *gerrardii, karroo.*
Thicket acacias: *ataxacantha, robusta.*

Broadleaved thicket and forest species: *Apodytes dimidiata* (n), *Bridelia cathartica* (i), *Ekebergia capensis* (t), *Englerophytum natalense* (n), *Erythrina lysistemon* (t), *Schotia capitata* (n), *Ziziphus mucronata* (n).

7.2.2 Saplings (Gullivers)

A striking feature of many higher rainfall savannas is the large number of stunted 'saplings' especially relative to the few adult trees. Bond and van Wilgen (1996) coined the word 'gulliver' for this common life stage after the famous sailor held captive by minute Lilliputians (= grass) but towering over them when released (the adult tree). Plants can be stuck in this gulliver phase for decades, top-killed repeatedly by fire but very seldom killed by burning. The probability of emerging from this stage to grow into a large tree is strongly influenced by consumers but also by growth rates to safe size thresholds.

7.2.2.1 Resources and Sapling Growth Rates

Whereas the Walter hypothesis emphasized niche differentiation of trees and grasses through resource partitioning of soil rooting space, the emphasis of the DBM is on how resources determine growth rates to threshold sizes. The sensitivity analyses of Higgins *et al.* (2000) showed very large effects of sapling growth rates on tree population size by facilitating sapling escape from the fire trap. But little or no data on growth rates, or on the effects of environmental factors on growth rates, existed in the literature at the time. Since then, considerable research effort has been expended on exploring factors influencing juvenile growth rates of trees. The magnitude of a growth response to resources is measured by its effect on the probability of advancing from one demographic stage to the next within a given disturbance regime (e.g. Bond *et al.*, 2003; Silva *et al.*, 2013). Experimental studies on various growth factors were generally single factor manipulations because of logistic constraints on multifactorial experiments. Many of the studies were conducted on *Acacia* species drawn from HiP with additional experiments on acacias and broadleaved species from Kruger National Park. The results are summarized in Table 7.2. Thus far, the most important factors influencing sapling growth rates were grass competition (February *et al.*, 2013), changes in atmospheric CO_2 (Kgope *et al.*, 2010; Bond and Midgley, 2012) and climate constraints on the length of the growing season as experienced across an altitudinal gradient (Wakeling *et al.*, 2012). In the presence of grass, manipulation of rainfall (50 to 150% of the control) had negligible effects on sapling growth (February *et al.*, 2013) while soil nutrients had a statistically significant, but biologically minor effect on growth and escape probabilities (Wakeling *et al.*, 2010).

7.2.2.2 Consumers and Sapling Growth

Considerable progress has been made in understanding how consumers of plant biomass, fire and herbivory, can influence vegetation structure at a variety of scales in savannas and other biomes.

Fire The effects of a single fire on woody plants depend on fire properties, especially intensity but also fire season, and on the properties

Table 7.2. *The effects of various environmental variables on savanna tree seedling and sapling growth. Growth difference is expressed relative to the treatment reported as '1' in the table. For example, a value of 4.5 for* Acacia karroo *grown at 370 ppm CO_2 indicates that plant biomass was 4.5 times greater than plants grown at 180 ppm. Cases refer to species except where indicated. G+ with grass, G− without grass. N+ nitrogen added, N− no nitrogen added. All acacia seeds sourced from HiP except for* A. sieberana *and* Acacia karroo *(Bloem) from a high frost area near Bloemfontein.*

Variable	Cases	Treatments			
(a) Rainfall		G+	G+	G+	G−
(% of natural)		50	100	150	100
	semi-arid	1	1.5	1.1	2.1
	Mesic	1	0.9	0.9	1.7
(b) CO_2 (ppm)[2]		G−	G−	G−	G−
		180	280	370	450
	Acacia karroo	1	1.4	4.5	4.5
	Acacia nilotica	1	1.3	2.0	4.3
	Terminalia sericea	1	1.2	2.7	3.5
(c) Elevation[3]		G−	G−		
		High grassland	Low savanna		
	Acacia karroo	1	1.8		
	Acacia karroo (Bloem)	1	3.2		
	Acacia gerardii	1	1.5		
	Acacia tortilis	1	3.1		
	Acacia sieberana	1	5.0		
	Mean	1	2.9		
(d) Soil nutrients[4]		G−	G−		
		grassland	savanna		
	Acacia karroo	1	1.4		
	Acacia sieberana	1	1.4		
(e) N fertilization[5]		G+	G+	G−	G−
		N−	N+	N−	N+
	Maytenus senegalensis	1	1.8	10.0	12.6
	Maytenus heterophylla	1	1.0	14.1	10.0
	Euclea divinorum	1	0.4	2.5	3.5
	Ziziphus mucronata	1	2.5	20.0	12.6
	Schotia capitata	1	1.0	10.0	6.3
	Acacia karroo	1	0.6	20.0	20.0
	Mean	1	1.2	12.8	10.8

Sources: 1, February *et al.* (2013); 2, Bond and Midgley (2012); Kgope *et al.* (2010); 3, Wakeling *et al.* (2012); 4, Wakeling *et al.* (2010); 5, Cramer and Bond (2013).

of plants being burnt. The response of plants changes with plant size and varies among species and their traits (e.g. Higgins *et al.*, 2012). The maximum potential intensity of fire is limited by the grass fuel load, but the realized intensity also varies with grass species composition, degree

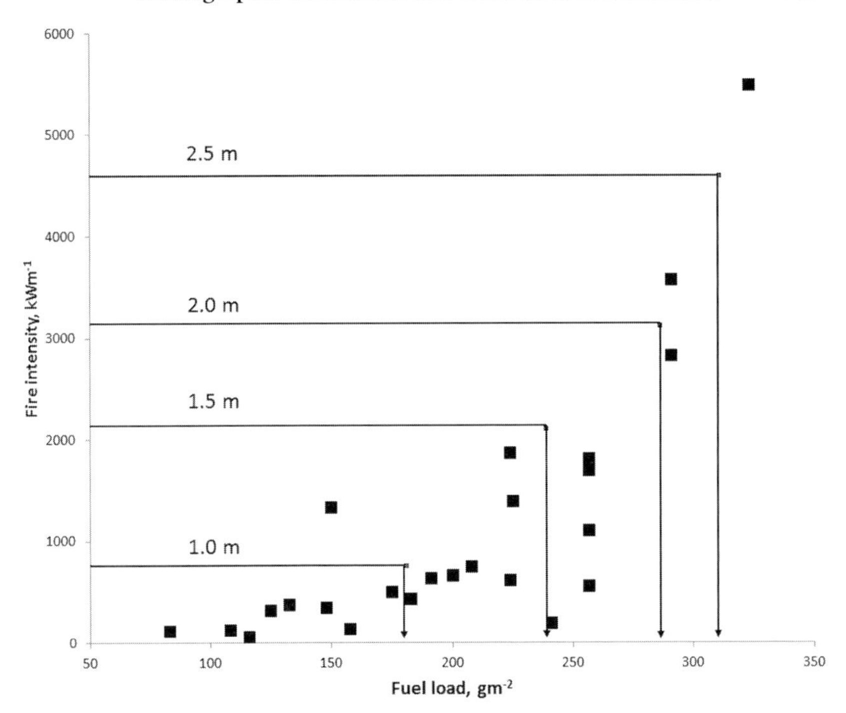

Figure 7.4 The relationship between fire intensity and fuel load in HiP and probability of top-kill. The horizontal lines indicate fire intensities needed to top-kill saplings of increasing height and the vertical arrows indicate the minimum grass fuel loads necessary to achieve these fire intensities. Note that fire intensities for a given fuel load vary with fire weather conditions and this will influence top-kill. Estimates of fire intensities for achieving top-kill (75–85% of stems killed) from Trollope *et al.* (1995) for Kruger National Park. Saplings > 2.5 m were rarely top-killed in this data set. Fire intensity and fuel load data ex ZLGP unpublished.

of curing, and weather conditions at the time of fire (Trollope, 1984). The data on fire intensity from HiP shown in Figure 7.4 indicate a triangular relationship between fuel load and fire intensity, with considerable variability for any given fuel load. Maximum fire intensity for a given fuel load increased exponentially with grass biomass over this sample of fires. The minimum grass biomass for fires to spread in HiP is in the order of 200 g/m^2 (see also Trollope, 1993). Almost any fire will top-kill woody plants < 1.0 m tall (Figure 7.4). To generate fire intensities that will top-kill 2-m tall saplings, a minimum fuel load of ~ 300 g/m^2 would be required. In addition, the fires should be burnt under suitable weather conditions and with cured grass fuels to realize the potential for top-kill. Because fire intensity depends on grass fuel load, grazers can strongly influence fire patterns depending on how much grass they consume. Waldram *et al.* (2008) showed how the removal of white rhinoceros, a grazer, had a considerable influence on fire spread. Fire and herbivory have to be considered together when evaluating options for managing vegetation (trees and grasses) in this savanna (see Chapter 10 for further details).

Once tree species have reached the pole phase of juvenile growth (Bond *et al.*, 2001; Archibald and Bond, 2003), they are very resilient to burning. Top-killed saplings can survive repeated fires for decades without being killed in both arid and mesic savannas (Higgins *et al.*, 2000; Balfour and Midgley, 2008). However, there is some evidence that if saplings have escaped and grown into trees, they lose the capacity to sprout and will die if top-killed in very severe fires (Maze, 2001). In the case of *Acacia karroo*, Maze (2001) hypothesized that the mortality of larger size classes was linked to the redistribution of carbon stores from the roots into the above-ground stems after the transition from sapling to adult tree, but this has not been tested.

Herbivores Herbivores, like fire, can impose a demographic bottleneck on tree establishment. Woody species responses to herbivory can vary widely, and browsing pressure itself is variable as well, depending on overall herbivore populations (see Chapter 4), herbivore size (Wilson and Kerley, 2003), and local herbivore use intensity. In general, herbivore pressure is inversely related to fire frequency (Archibald *et al.*, 2005), and browsing on trees appears to be most intense in grazing lawns and analogous bare patches (e.g. in Mfolozi; O'Kane *et al.*, 2012) and least in the high-rainfall, tall-grass savannas of Hluhluwe. However, the effects of herbivory can be significant even on species and in areas not usually considered susceptible to herbivory (Staver *et al.*, 2009). Exclosures erected throughout herbivore- and fire-prone areas of HiP have demonstrated that herbivores can strongly limit the recruitment of saplings into adult trees. Under a biannual fire regime, saplings do not recruit into trees at all but, when fire is absent, browser exclusion increased recruitment rates from zero to as much as 3.5% per year (Staver and Bond, 2014).

Susceptibility to browsing also depends strongly on the size of the tree being browsed (Wilson and Kerley, 2003). Exclosure studies in HiP suggest that mixed feeders, including impala and nyala, which are small but numerous throughout the park, have large effects on sapling growth and can substantially limit large tree populations (Staver and Bond, 2014; see also O'Kane *et al.*, 2012 for correlative studies). These medium-bodied antelope can only browse within 1–2 m of the ground, and are limited in the diameter of shoot they can consume (Wilson and Kerley, 2003). Elsewhere, dominant herbivores are usually also what is most abundant – at Mpala Ranch in Kenya, for instance, where dik-dik abound, their impacts on seedling growth appear to limit tree emergence (Sankaran *et al.*, 2013), such that the height of the browse trap is shorter.

The browse trap differs from the (episodic) fire trap in that chronic, sustained browsing is required to prevent trees reaching escape size. Anecdotally, trees can only escape the effects of browsing during periods when browsing pressure is reduced, because of changing landscape use or even large-scale disease epidemics (Prins and van der Jeugd, 1993). In a rare long-term study at HiP, acacias were protected from browsing for a decade and then exposed to browsers for 3 years after exclosure fences were removed (Staver and Bond, 2014). After the re-introduction of browsers, trees retained the height they had reached in the exclosures and, where they had grown above escape size, also survived without being top-killed by subsequent fires. Evidence from Kruger National Park (Moncrieff *et al.*, 2011) and Ithala Game Reserve (Bond and Loffell, 2001) suggests that the effects of herbivory are similarly height-structured; above the browse line, trees are no longer as susceptible to browsing as they are as saplings. However, there are very few formal studies of the browse trap in African savannas and its analogies to the fire trap. More studies are needed to understand the interaction between herbivore densities, fire, and the controls that both these consumers exert on savanna structure.

Because they are limited to impacting only relatively small trees, small browsers and mixed feeders very rarely drive increases in tall tree mortality (Sankaran *et al.*, 2013; Staver and Bond, 2014). However, elephants are not so limited, and can in fact be a major factor influencing the mortality of established adult trees (e.g. Boundja and Midgley, 2010; White and Goodman, 2010). They can alter the structure of the savanna by snapping off branches and even the main trunk of large trees such as *Sclerocarya birrea*. They kill trees directly by pushing them over or, less directly, by ring barking and bark stripping, which then exposes stems to fire damage (Moncrieff *et al.*, 2008). At least in theory, these elephant impacts on tree mortality can drive major reductions in tree cover at the landscape level (Dublin *et al.*, 1990), although these system-level impacts have to date only been observed where elephant densities are very high (Skarpe *et al.*, 2004).

Giraffe are often overlooked as animal agents that kill trees. However, the introduction of giraffe into Ithala Game Reserve in KwaZulu-Natal led to the extirpation of adult trees of *Acacia davyi* and *A. caffra* and the heavy mortality of *A. karroo*, all of which are tree species characteristic of frequently burnt savanna. Acacias from more arid savannas, such as *A. tortilis*, were better defended and experienced no mortality (Bond and Loffell, 2001).

7.2.2.3 Simply the Best

Demographic models typically estimate transition probabilities from one size class to the next from mean growth rates within a size class. Mean growth rates in HiP proved to be so slow that, under typical fire return intervals, saplings would never emerge out of the fire trap. Based on mean growth rates, intervals between top-killing fires for some species would be in the order of decades, long enough for savannas to be replaced by forests. Field observations revealed highly left-skewed frequency distributions of sapling growth rates. Thus, mean values are heavily biased towards the slowest-growing individuals (Figure 7.5). Wakeling *et al.* (2011) suggested using only the top percentiles of growth rates to estimate escape probabilities. If growth rates were based only on the top 5% of plants, time to escape height would be 3–4 years for *A. karroo* versus 13 years for mean rates. For *A. gerrardii* and *A. tortilis*, time to escape height would be 5–6 years compared to > 20 years for mean rates. The maximum rates are a feasible time interval for occasional sapling recruitment to the tree layer under typical fire frequencies. Bond *et al.* (2012) recognized the same problem in their study assessing which species would be most likely to escape the fire trap and dominate the tree layer in Australian savannas. Instead of using growth rates of individuals to estimate transition probabilities, they estimated sapling transitions to adult sizes directly by counting the density of tagged saplings emerging from below to above escape height over a 13-year period. Of the hundreds of fire-trapped juveniles per hectare, only 75 eucalypts/ha grew to adult sizes and a paltry 10.8 non-eucalypts/ha. The difference in escape rates matched the 6:1 dominance of large tree eucalypts versus non-eucalypts in this savanna woodland. The radical ecological inference is that sapling escape from the fire trap, and not interspecific competition among adult trees, accounts for eucalypt dominance in this savanna (Bond *et al.*, 2012). In support of this, Bond *et al.* (2012) noted that the density and biomass of non-eucalypt savanna trees increased several fold after 20 years of fire suppression, whereas eucalypt populations were stable or declined. Eucalypts

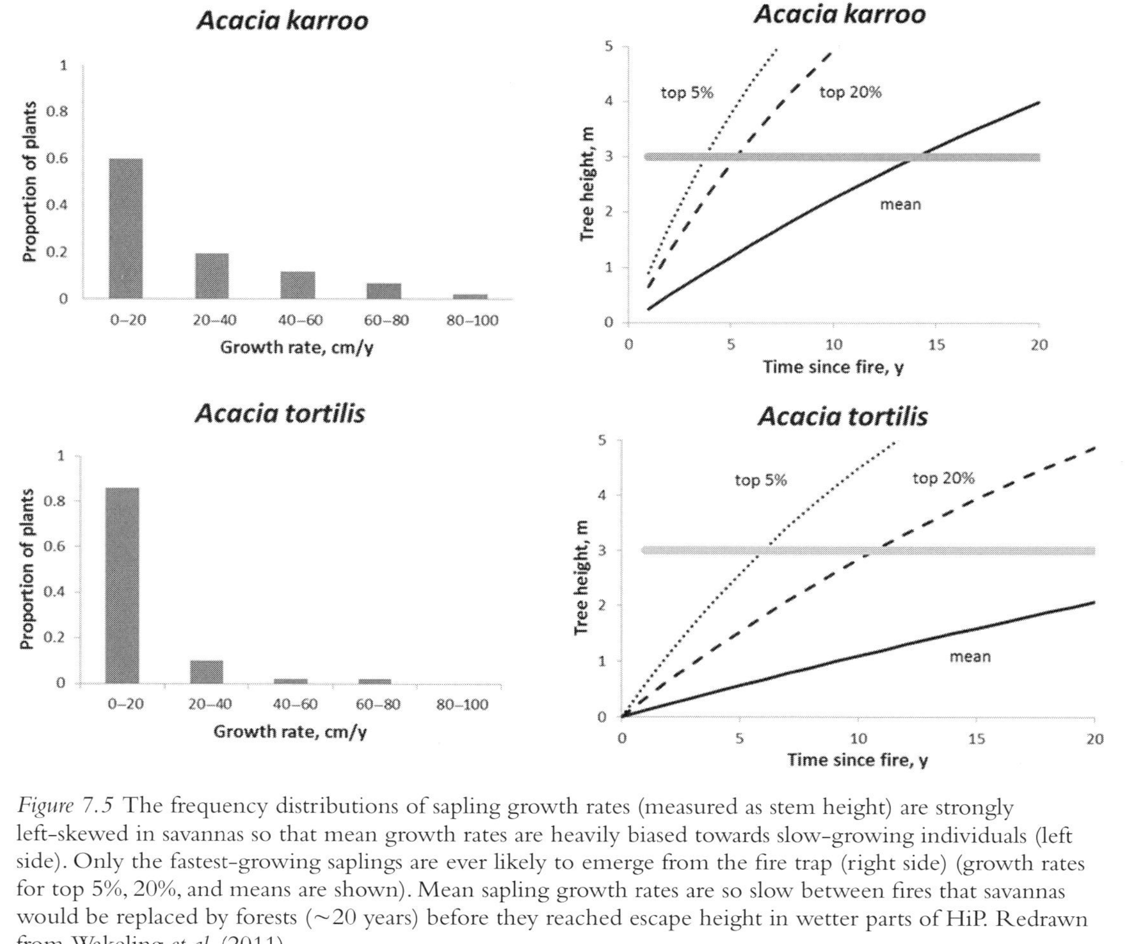

Figure 7.5 The frequency distributions of sapling growth rates (measured as stem height) are strongly left-skewed in savannas so that mean growth rates are heavily biased towards slow-growing individuals (left side). Only the fastest-growing saplings are ever likely to emerge from the fire trap (right side) (growth rates for top 5%, 20%, and means are shown). Mean sapling growth rates are so slow between fires that savannas would be replaced by forests (~20 years) before they reached escape height in wetter parts of HiP. Redrawn from Wakeling *et al.* (2011).

are fire-weeds in this system and would be lost without very frequent fires.

7.2.3 Which Demographic Stage: Seedling Establishment or Sapling Release?

As lamented by Midgley *et al.* (2010), there are no complete demographic studies of all life-history stages of any savanna tree in any locality anywhere. Nevertheless, to help determine if seedling recruitment or sapling release is the primary constraint on large tree densities, a demographic bottleneck index, n/K, has been proposed, where n is the density of established juvenile plants in a sample and K is the maximum number of trees that space filling allows (Wakeling *et al.*, 2015). This maximum is estimated for a site from the density of non-overlapping canopies using the mean canopy area of large trees. Thus for a tree with a mean canopy diameter of 5 m, K is ~ 500 trees/ha, and for a broad-canopied tree of diameter 10 m, K is ~ 127 trees/ha. Where the demographic bottleneck index (n/K) is greater than one, there are sufficient juveniles to fully occupy a site with adults and the population is assumed to be restricted by processes influencing sapling release (release-limited). Where the demographic bottleneck index is less than one, the key bottlenecks are assumed to be related to the establishment of new individuals (establishment-limited).

7.2.3.1 A Case Study: The Savanna–Grassland 'Treeline'

Tree-less grasslands in HiP crown the highest hills, while savannas occupy lower elevations. This savanna 'treeline' is widespread in the higher-rainfall eastern parts of South Africa and also more generally in upland tropical and subtropical grassy landscapes elsewhere in the world. Among possible explanations for the absence of trees is that juveniles are killed by frost or that sapling growth rates are too slow in the cooler upland climates for saplings ever to escape the fire trap where grasslands burn frequently. We tested these hypotheses by growing *Acacia* species, mostly from HiP, across an elevation gradient from 40 to 1700 m (Wakeling *et al.*, 2012). Despite a severe frost over the region during the study, frost damage was negligible except at the highest grassland site. Savanna saplings showed positive growth across nearly the entire elevation gradient, but growth rates were markedly slower in cooler grassland versus warmer savanna sites. Thus in cooler, upland climates saplings would very seldom escape the fire trap in frequently burnt grasslands (Table 7.2).

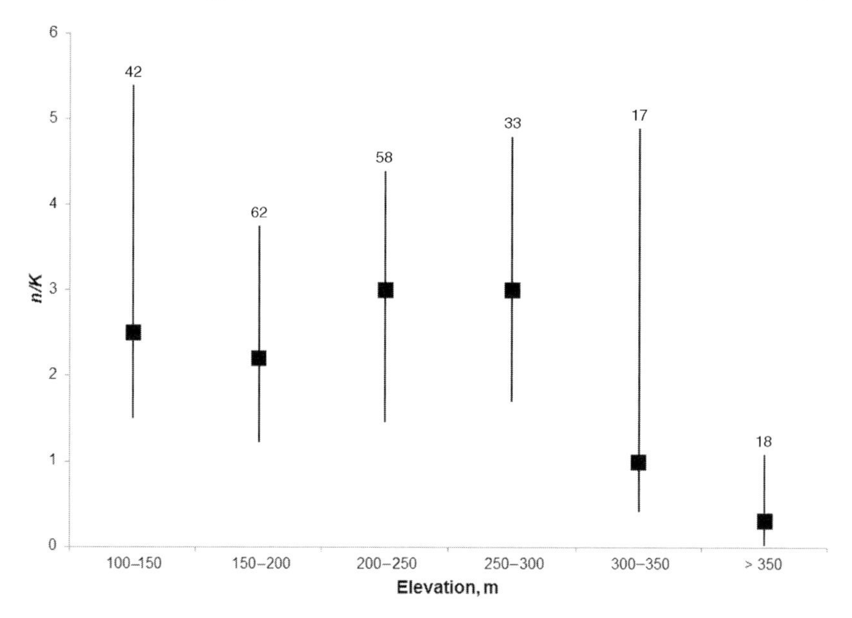

Figure 7.6 The demographic bottleneck index for *Acacia* spp. along an elevation gradient. The boxes indicate median values, whiskers upper and lower quartiles with sample size above each class. Sites either have a demographic establishment bottleneck ($n/K < 1$) or a release bottleneck ($n/K > 1$). n refers to juvenile density per hectare and K is the maximum potential number of adult trees per hectare given the mean canopy area of adult trees. Where the demographic bottleneck index (n/K) is greater than one there are sufficient juveniles to fully occupy a site and the population is assumed to be restricted by processes influencing sapling release (release bottleneck). Where the demographic bottleneck index is less than one the tree population is assumed to be restricted by the establishment of new individuals (establishment bottleneck). (Reprinted with permission from Elsevier; Wakeling *et al.*, 2015.)

However, the absence of trees in grasslands could also reflect earlier demographic bottlenecks at the seedling establishment phase. Figure 7.6 shows the demographic bottleneck index (DBI = n/K) estimated from surveys of juvenile plants at sites from HiP to high montane grasslands. The DBI results point to a release bottleneck at lower, warmer, drier elevations changing to an establishment bottleneck in upland grasslands (Figure 7.6). Experiments at HiP show that grass competition is the likely cause of the upland bottleneck, adding a twist to the assumption of Higgins *et al.* (2000) that the establishment bottleneck would dominate in arid savannas but be replaced by the release bottleneck at higher-rainfall

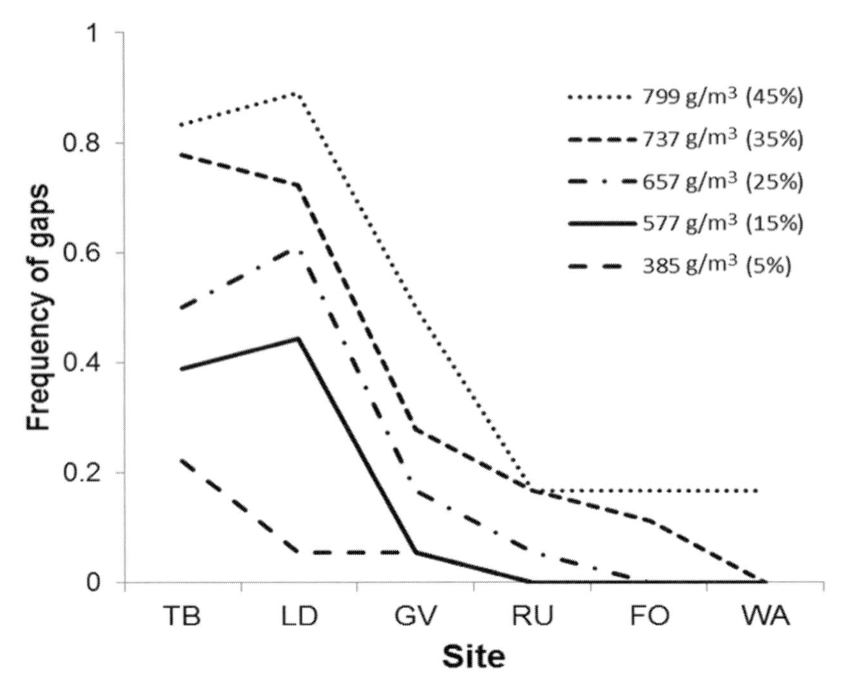

Figure 7.7 The frequency of 'root gaps' in savannas versus grasslands. A root gap is defined by grass root biomass in holes dug to 40 cm at a grid of 18 holes at each site. Values of root biomass per 40 cm hole (g/m^3) are equivalent to percentiles of 5%, 15%, 25%, 35%, and 45% of biomass for the total sample of 108 holes. (Reprinted with permission from Elsevier; Wakeling *et al.*, 2015.)

sites. Jurena and Archer (2003), working in the Texas rangelands, found that the establishment of *Prosopis* seedlings depended on a 'root gap', a micro-site where seedlings can rapidly extend roots to deeper moist soil layers with minimal interference from grass roots. Could the lack of root gaps in upland grasslands explain the puzzling absence of savanna trees in this biome?

We tested the idea by sampling grass roots in a grid of auger holes dug to 40 cm depth at each of six sites spread across an elevation gradient from semi-arid savannas to upland grasslands (Wakeling *et al.*, 2015). The frequency of root gaps indeed declined steeply with increasing elevation, dwindling to none at all in upland grassland sites (Figure 7.7). Thus the savanna–grassland biome boundary may be explained by the failure of tree seedlings to establish in grasslands due to intense competition from grass roots. Savannas would then occur where root gaps are more frequent, presumably due to grass mortality from factors such as drought, animal activity, or self shading in the absence of fire.

7.2.4 Adult Mortality

Savanna tree lifespans vary considerably among species. *Acacia karroo* and *A. nilotica*, the most common acacia species of Hluhluwe, are thought to have lifespans as adult trees of little more than 50 years. There has been some debate as to whether lifespans are intrinsically short (programmed senescence) or whether the trees are killed by a complex of accumulated fire damage and bark removal by mammals (Midgley and Bond, 2001). *Acacia nigrescens* trees in Mfolozi mostly date from the early twentieth century with no trees exceeding 100 years old (Staver *et al.*, 2011). Bark stripping of this species by elephants is widespread and extensive tree mortality is occurring in many Mfolozi populations. It seems reasonable to expect that as trees die they will be replaced by new recruitment from saplings. However, pole-sized saplings of *A. nigrescens* were extremely rare in Mfolozi during our studies from 2000 to 2012. Furthermore, after protection of *A. nigrescens* 'seedlings' in exclosures for 10 years, including 5 years with no fires, almost no seedlings had grown to pole-sized saplings capable of escaping the fire trap. This degree of protection from browsers and fire is extremely unlikely under prevailing conditions in the park. Thus the loss of large trees of *A. nigrescens* can be considered permanent for the 'foreseeable future' in Mfolozi.

HiP experienced severe droughts in the early 1980s and again in the early 1990s. However, drought deaths of mature trees were not documented. In Kruger National Park, the same 1990s drought caused negligible adult tree mortality with the lowest rates, perhaps surprisingly, in the arid savannas (Viljoen, 1995). High-intensity fires periodically kill adult savanna trees. Species that sprout vigorously as juveniles are often weak sprouters, or fail to sprout altogether, after they have grown into adult trees (Schutz *et al.*, 2009). Ageing *A. nilotica* trees which dominated Hluhluwe savannas until the 1990s are particularly vulnerable to fire death when flames from tall-grass fires consume their low canopies (Midgley *et al.*, 2001).

Because of generally high rates of adult tree mortality, and the short lifespan of adult acacias in HiP, the tree layer is particularly dynamic. Tree populations have been shown to have shifted in space, time and composition (Bond *et al.*, 2001; Staver *et al.*, 2011). Hence, an understanding of how new trees are recruited has been critical in this ecosystem.

7.3 Bush Encroachment

Bush encroachment is the term commonly used in South Africa for an increase in tree densities in savannas and is synonymous with 'woody thickening' used elsewhere (Scholes and Archer, 1997; O'Connor *et al.*, 2014). It has long been of concern in HiP (Watson and Macdonald, 1983; Whateley and Porter, 1983; Wigley *et al.*, 2010; Chapter 3). Two distinct forms of bush encroachment have been observed in HiP and elsewhere. The first is an increase in savanna trees and shrubs but without loss of grasses. The second is colonization of savannas by forest and thicket trees and shrubs which shade out the grass layer (Wigley *et al.*, 2010; Gordijn *et al.*, 2012). The consequence, in this case, is a biome shift from savannas to thicket: an alternative closed non-savanna of densely wooded vegetation (Parr *et al.*, 2012; Chapter 8).

7.3.1 Woody Thickening in Savannas

Given the nature of tree population dynamics in savannas, a local increase in trees is an entirely natural part of system dynamics. Larger-scale (park-wide) increases in gulliver release to adult trees might also be expected following regional climate events such as severe drought, especially when droughts persist for several years. Severe droughts promote adult tree recruitment by suppressing grass production, reducing its competitive influence on saplings and also reducing the fuel for fires, and reducing herbivore densities (February *et al.*, 2013). The current cohort of *Acacia karroo* in Hluhluwe dates to the early 1990s when there was a protracted drought in the park (Staver *et al.*, 2011). More worrying from a management perspective is sustained woody thickening over large areas with no obvious climate signal (Section 6.3).

In tall-grass areas of the park, woody thickening can be effectively managed by the application of frequent (1–3-year) high-intensity burns ('trollopeing' trees after Winston Trollope, 1984, who pioneered this approach). These intense burns maximize top-kill and, while seldom killing trees, will alter savanna structure, producing an open, grassy ecosystem. The effectiveness of this approach has been shown experimentally for *Acacia karroo*-dominated savannas in the northern part of the park (Zululand Tree Project, unpublished data) and in management burns in the Corridor (D. Robertson, personal communication, 2010; see also Balfour and Midgley, 2008).

Woody thickening by savanna shrubs, especially by *Dichrostachys cinerea* var *nyassana*, is a much more difficult problem to deal with than thickening of savanna trees. *Dichrostachys* can spread vegetatively from root suckers (Wakeling and Bond, 2007), bypassing the vulnerable seedling stage in high-rainfall savannas. The shrub is typically top-killed by fire, but resprouts vigorously, especially after dry season fires. This species, and similar shrubs such as *Acacia caffra*, cannot be suppressed by high-intensity dry season burns because their root reserves of carbohydrates are at their maximum during the dry season. Thus, fires in the dry season that are effective for controlling the structure of savanna trees are ineffective for controlling shrubs. Attempts to burn *Dichrostachys* in early summer (late November/early December) when root reserves are exhausted from supporting new growth and reproduction showed some success in reducing *Dichrostachys* (ZLTP unpublished). However, fires burnt at this time are of low intensity and spread slowly because of the high moisture content of the grasses. Early summer burns would therefore only be practical for reducing *Dichrostachys* populations in small areas.

7.3.2 Forest Colonization of Savannas

The second form of bush encroachment, replacement of savannas by closed forests or thickets (thus no longer savannas; Chapter 8), has much more severe biodiversity consequences. The loss of grasses, coupled with a compositional shift in trees, shrubs, and herbs, causes a biome shift to an alternative biome state (Parr *et al.*, 2012). Large areas of Hluhluwe were colonized by thicket from the 1930s to the 2000s and forests also expanded into grasslands over this period (Skowno *et al.*, 1999; Wigley *et al.*, 2010; Chapter 3). Although not formally studied as yet, Mfolozi does not seem to have undergone the same degree of thicket transformation. The causes of forest expansion are not well understood, but, as it is occurring across vastly different land management systems (communal farms, commercial cattle ranches and conservation areas), Wigley *et al.* (2010) inferred a global driver. Climate change, whether of rainfall or temperature, has been negligible in the Hluhluwe area over the period and Wigley *et al.* suggested that increasing atmospheric CO_2 was the most likely candidate. Gordijn *et al.* (2012) suggested that changes in rainfall distribution (more rainfall events > 10–20 mm and fewer < 10 mm) were important in promoting thicket invasion at Ithala, in northern KwaZulu-Natal. Thicket invasion is promoted by fire suppression or by low-intensity early season burns. *Chromolaena* invasion seems to have been a by-product of the invasion process because this invasive weed is excluded by frequent fires but establishes readily under the relatively open canopy of thicket vegetation (see Chapter 15).

7.3.3 CO_2 and the Causes of Bush Encroachment

Bush encroachment is not a phenomenon restricted to HiP and surroundings. A general increase in savanna trees and shrubs is occurring more widely in South Africa and elsewhere. *Acacia karroo* is a prominent encroaching species with radical increases reported for this and other species in the Eastern Cape (O'Connor and Crow, 1999; Buitenwerf *et al.*, 2012), KwaZulu-Natal (Russell and Ward, 2014), and the northern and north-eastern parts of South Africa (Stevens, 2014). Local land use such as overgrazing or fire suppression is undoubtedly a factor in the rate and spread of savanna trees (O'Connor *et al.*, 2014), but simulations, glasshouse experiments, and long-term field studies, in addition to the widely observed trends of increasing trees, indicate that increasing CO_2 has been a significant factor in bush encroachment over the last few decades. Experiments across a gradient of CO_2 have shown drastically increased growth rates of species such as *Acacia karroo* and *Terminalia sericea* from pre-industrial to late twentieth century CO_2 levels (Kgope *et al.*, 2010; Bond and Midgley, 2012). CO_2 fertilization is most likely to stimulate plant growth when there is a large carbon sink to utilize the extra carbon gain from photosynthesis. Carbon sinks include starch-storing lignotubers in savanna trees such as *A. karroo* (Wigley *et al.*, 2009; Schutz *et al.*, 2009), root suckers of clonal shrubs such as *Dichrostachys*, or simply woody plants top-killed by fire with large underground root systems maintained by initially small resprouting shoots.

The important management implication of the effects of CO_2 on the tree–grass balance is that methods that were effective for controlling woody plants in the past will be much less effective now and in the future. In HiP, this might imply, for example, that cooler early-season burns that were effective in the past should be replaced by more intense late-season burns to manage tree populations and to prevent thicket and forest expansion in the future. The problem of thickening of shrubs is unresolved.

It is interesting to note that the least problematic areas for bush encroachment are heavily grazed patches: grazing lawns and related very heavily denuded grazing patches in Mfolozi. According to classic rangeland science, heavy grazing should promote bush encroachment, but it seems this is not necessarily the case where the herbivores are indigenous African mammal fauna. The 'browse trap' is far less well understood than the fire trap, but clearly exists in parts of HiP. The recent (2010–2014) decline in ungulate populations (see Chapter 4), and especially of mixed feeders such as impala, can be expected to result in the release of shrubs and trees in many heavily grazed patches in HiP. Mfolozi, where the grass layer is sparse and seasonal in grazing patches, may be particularly vulnerable to widespread bush encroachment as a result of decreased grazing and browsing.

7.4 Evaluation of the Demographic Bottleneck Concept and its Applications

The demographic bottleneck (DB) approach has been successfully applied in Africa (e.g. Holdo *et al.*, 2009), Australia (Werner and Prior, 2013), and South America (Hoffmann *et al.*, 2009). It provides a means of integrating the effects of consumers (fire, herbivores) and resources. While conceptually simple, the ecology is inevitably more complex. An important contribution is that the effects of resources can now be explicitly linked to demographic transitions, thereby allowing direct evaluation of the ecological importance of different resources. One such resource is CO_2 and the DB model led directly to recognition of the contribution of changing atmospheric CO_2 to increasing tree cover in savannas. The DB concept predicts that species composition is determined by sapling escape rates and not by competition among established trees, a radical suggestion but consistent with eucalypt dominance in Australian savannas (Bond *et al.*, 2012). It also explains the otherwise puzzling landscape-scale replacement of one dominant tree by another in HiP. Where soil type is usually invoked as determining species composition, the *Acacia* species turnover on the same soil in HiP is caused by a change in the dominant consumer from frequent fires to heavy browsing or vice versa (Bond *et al.*, 2001). If the interaction between resources and consumers determines species distribution, then species distribution models designed to predict the effects of climate change will be poor tools for predicting future range changes in savanna trees. Instead, it will be essential to develop predictive models that include changes in the fire/herbivore regime (Stevens, 2014).

The DB concept has been influential in explaining the distribution of savannas and their boundaries with other biomes. New analytical models, for example, explore the demographic effects of fire as a key positive feedback maintaining alternative biome states (Staver and Levin, 2012). Moreover, differential growth rates to minimum thresholds for escaping top-kill have been incorporated in general models of forest/ savanna boundaries (Hoffmann *et al.*, 2012). These are being extended to include ways in which edaphic differences in soil nutrients might determine the location of biome boundaries (Silva *et al.*, 2013). The DB hypothesis has also been used to explore grassland/savanna boundaries by suggesting explicit predictions, which led to the first experimental test of the mechanisms accounting for this biome boundary in South Africa (Wakeling *et al.*, 2012). These formal conceptual models open the way to explicit prediction of future changes in biome boundaries as a result of global change. They are a major advance over the correlative methods widely used in the past.

The importance of sapling escape, and adult tree resistance, especially in frequently burnt savannas, has led to new studies of functional trait differences both within and between savannas and forests (see Chapter 8 on plant traits). The influence of consumer type on tree species distribution has emerged in studies of acacias. Different species with different architectures ('cage' vs 'pole') occupy heavily browsed versus frequently burnt savannas (Staver *et al.*, 2012). In areas of the world where the megafauna is extinct, these trait syndromes can be used to reconstruct the evolutionary legacy of the big vertebrates.

At large regional scales, the DB concept has provided the key mechanistic processes for a simulation model, the aDGVM, which can analyse interacting effects of climate, CO_2, and fire on tree–grass mixtures to explore global change futures for Africa (Scheiter and Higgins, 2009). The importance of this development can be gauged by comparing simulated futures of a deterministic climate model (Bergengren *et al.*, 2011) with those of the aDGVM. Whereas the climate model predicts that Africa will be the most ecologically stable region in the world over the next century or two, the aDGVM predicts massive increases in trees and a retreat of C4 savannas in the coming century (Higgins and Scheiter, 2012). The simulated trend towards a woody plant increase is largely because of CO_2 effects on trees recovering from fire. The climate model, in contrast, incorporates neither fire nor CO_2 in its projections (Bergengren *et al.*, 2011). The aDGVM has also been used to explore the past in an attempt to resolve causes of the extraordinary late Miocene emergence and explosive spread of C4 savannas (Scheiter *et al.*, 2012). By using a series of elegant simulation experiments, this study showed that fire and low CO_2 promoting C4 grasses were key ingredients of the savanna revolution. The aDGVM has also been used to explore the legacy of the last glacial on contemporary vegetation in Africa. Simulations showed that the frequent fires in the grassy biomes that dominated Africa during the last glacial, when atmospheric CO_2 dropped to less than half its present level, inhibited the development of forests in the Holocene as climates warmed and CO_2 increased. The legacy of the last glacial, in promoting savannas, persists in contemporary Africa, where large areas of savannas occur in climates warm enough and wet enough for forests (Moncrieff *et al.*, 2014).

7.5 Conclusion

Perhaps the largest unanswered question in African ecology is the extent to which tree cover within savannas, and the distribution of savannas versus forests, are influenced by the African megafauna. Answers on the role of large mammals in constructing large-scale vegetation patterns are of interest worldwide in those regions where the megafauna went extinct (e.g. Owen-Smith, 1987; Svenning, 2002; Weigl and Knowles, 2014). That herbivores do influence tree populations is obvious at least for local patches such as grazing lawns. The critical question concerns the scale of that impact. Do large mammals alter savanna structure and composition on a scale visible on a map of, say, Kruger National Park or South Africa or Africa? Would the vegetation maps change if the megafauna went extinct, or is their effect merely a local phenomenon? Some of the key information needed includes browse effects on demographic transitions from saplings to trees and how these relate to animal type and densities. Demographic effects of browsers on the tree layer, both with and without fire, are beginning to emerge and studies in HiP have helped lay a foundation for analysing the browse trap (Staver *et al.*, 2009; Staver and Bond, 2014). However, answers to the critical question of the scale of large mammal impacts on savanna structure remain elusive.

The DB hypothesis implies that edaphic effects on savanna composition are primarily through influencing demographic transitions from seedling to adult. A major challenge for the hypothesis is explaining the two major savanna types of southern Africa, the acacia savannas such as those of HiP versus the miombo woodlands on the old land surfaces in higher-rainfall regions of south-central Africa (Scholes and Walker, 1993). Comparative studies of the different savannas of Africa would be highly informative on the relative role of bottom-up soil resources and their interaction with top-down consumers, especially fire, on the major vegetation types of the continent. Without such studies the generality of concepts developed in the acacia-dominated savannas of HiP must be questioned.

7.6 References

Archibald, S. & Bond, W. J. (2003) Growing tall vs growing wide: tree architecture and allometry of *Acacia karroo* in forest, savanna, and arid environments. *Oikos* **102**: 3–14.

Archibald, S., Bond, W. J., Stock, W. D., & Fairbanks, D. H. K. (2005) Shaping the landscape: fire–grazer interactions in an African savanna. *Ecological Applications* **15**: 96–109.

Balfour, D. A. & Howison, O. E. (2002) Spatial and temporal variation in a mesic savanna fire regime: responses to variation in annual rainfall. *African Journal of Range and Forage Science* **19**: 45–53.

Balfour, D. A. & Midgley, J. J. (2008) A demographic perspective on bush encroachment by *Acacia karroo* in Hluhluwe-Imfolozi Park, South Africa. *African Journal of Range and Forage Science* **25**: 147–151.

Bergengren, J. C., Waliser, D. E., & Yung, Y. L. (2011) Ecological sensitivity: a biospheric view of climate change. *Climatic Change* **107**: 433–457.

Bond, W. J. (2008) What limits trees in C4 grasslands and savannas? *Annual Review of Ecology, Evolution, and Systematics* **39**: 641–659.

Bond, W. J. & Loffell, D. (2001) Introduction of giraffe changes acacia distribution in a South African savanna. *African Journal of Ecology* **39**: 286–294.

Bond, W. J. & Midgley, G. F. (2012) CO_2 and the uneasy interactions of trees and savanna grasses. *Philosophical Transactions Royal Society B* **367**: 601–612.

Bond, W. J. & Van Wilgen, B. W. (1996) *Fire and plants*. Chapman and Hall, London.

Bond, W. J., Smythe, K. A. & Balfour, D. A. (2001) Acacia species turnover in space and time in an African savanna. *Journal of Biogeography* **28**: 117–128.

Bond, W. J., Midgley, G. F., & Woodward, F. I. (2003) The importance of low atmospheric CO_2 and fire in promoting the spread of grasslands and savannas. *Global Change Biology* **9**: 973–982.

Bond, W. J., Cook, G., & Williams, R. J. (2012) Which trees dominate in savannas? The escape hypothesis and eucalypts in northern Australia. *Austral Ecology* **37**: 678–685.

Boundja, R. P. & Midgley, J. J. (2010) Patterns of elephant impact on woody plants in the Hluhluwe-Imfolozi Park, KwaZulu-Natal, South Africa. *African Journal of Ecology* **48**: 206–214.

Buitenwerf, R., Bond, W. J., Stevens, N., & Trollope, W. S. W. (2012) Increased tree densities in South African savannas: > 50 years of data suggests CO_2 as a driver. *Global Change Biology* **18**: 675–684.

Cramer, M. D. & Bond, W. J. (2013) N-fertilization does not alleviate grass competition induced reduction of growth of African savanna species. *Plant and Soil* **366**: 563–574.

Cramer, M. D., Chimphango, S. B. M., Van Cauter, A., Waldram, M. S., & Bond, W. J. (2007) Grass competition induces N_2 fixation in some species of African *Acacia*. *Journal of Ecology* **95**: 1123–1133.

Cramer, M. D., van Cauter, A., & Bond, W. J. (2010) Growth of N_2-fixing African savanna *Acacia* spp. is constrained by below-ground competition with grass. *Journal of Ecology* **98**: 156–167.

Cramer, M. D., Wakeling, J. L., & Bond, W. J. (2012) Belowground competitive suppression of seedling growth by grass in an African savanna. *Plant Ecology* **213**: 1655–1666.

Dublin, H. T., Sinclair, A. R., & McGlade, J. (1990) Elephants and fire as causes of multiple stable states in the Serengeti–Mara woodlands. *Journal of Animal Ecology* **59**: 1147–1164.

February, E. C., Higgins, S. I., Bond, W. J., & Swemmer, L. (2013) Influence of competition and rainfall manipulation on the growth responses of savanna trees and grasses. *Ecology* **94**: 1155–1164.

Gordijn, P. J., Rice, E., & Ward, D. (2012) The effects of fire on woody plant encroachment are exacerbated by succession of trees of decreased palatability. *Perspectives in Plant Ecology, Evolution and Systematics* **14**: 411–422.

Higgins, S. I. & Scheiter, S. (2012) Atmospheric CO_2 forces abrupt vegetation shifts locally, but not globally. *Nature* **488**: 209–212.

Higgins, S. I., Bond, W. J., & Trollope, W. S. (2000) Fire, resprouting and variability: a recipe for grass–tree coexistence in savanna. *Journal of Ecology* **88**: 213–229.

Higgins, S. I., Bond, W. J., Combrink, H., *et al.* (2012) Which traits determine shifts in the abundance of tree species in a fire-prone savanna? *Journal of Ecology* **100**: 1400–1410.

Hoffmann, W. A., Adasme, R., Haridasan, M., *et al.* (2009) Tree topkill, not mortality, governs the dynamics of savanna–forest boundaries under frequent fire in central Brazil. *Ecology* **90**: 1326–1337.

Hoffmann, W. A., Geiger, E. L., Gotsch, S. G., *et al.* (2012) Ecological thresholds at the savanna–forest boundary: how plant traits, resources and fire govern the distribution of tropical biomes. *Ecology Letters* **15**: 759–768.

Holdo, R. M., Holt, R. D., & Fryxell, J. M. (2009) Grazers, browsers, and fire influence the extent and spatial pattern of tree cover in the Serengeti. *Ecological Applications* **19**: 95–109.

Jurena, P. N. & Archer, S. (2003) Woody plant establishment and spatial heterogeneity in grasslands. *Ecology* **84**: 907–919.

Kgope, B. S., Bond, W. J., & Midgley, G. F. (2010) Growth responses of African savanna trees implicate atmospheric [CO_2] as a driver of past and current changes in savanna tree cover. *Austral Ecology* **35**: 451–463.

Maze, K. E. (2001) Fire survival and life histories of Acacia and Dichrostachys species in a South African savanna. MSc thesis, University of Cape Town, Cape Town.

Midgley, J. J. & Bond, W. J. (2001) A synthesis of the demography of African acacias. *Journal of Tropical Ecology* **17**: 871–886.

Midgley, J. J., McLean, P., Botha, M., & Balfour, D. (2001) Why do some African thorn trees (*Acacia* spp.) have a flat-top: a grazer–plant mutualism hypothesis? *African Journal of Ecology* **39**: 226–228.

Midgley, J. J., Lawes, M. J., & Chamaillé-Jammes, S. (2010) Savanna woody plant dynamics: the role of fire and herbivory, separately and synergistically. *Australian Journal of Botany* **58**: 1–11.

Moncrieff, G. R., Kruger, L. M., & Midgley, J. J. (2008) Stem mortality of *Acacia nigrescens* induced by the synergistic effects of elephants and fire in Kruger National Park, South Africa. *Journal of Tropical Ecology* **24**: 655–662.

Moncrieff, G. R., Chamaillé-Jammes, S., Higgins, S. I., O'Hara, R. B., & Bond, W. J. (2011) Tree allometries reflect a lifetime of herbivory in an African savanna. *Ecology* **92**: 2310–2315.

Moncrieff, G. R., Scheiter, S., Bond, W. J., & Higgins, S. I. (2014) Increasing atmospheric CO_2 overrides the historical legacy of multiple stable biome states in Africa. *New Phytologist* **201**: 908–915.

O'Connor, T. G. & Crow, V. R. T. (1999) Rate and pattern of bush encroachment in Eastern Cape savanna and grassland. *African Journal of Range and Forage Science* **16**: 26–31.

O'Connor, T. G., Puttick, J. R., & Hoffman, M. T. (2014) Bush encroachment in southern Africa: changes and causes. *African Journal of Range & Forage Science* **31**: 67–88.

O'Kane, C. A., Duffy, K. J., Page, B. R., & Macdonald, D. W. (2012) Heavy impact on seedlings by the impala suggests a central role in woodland dynamics. *Journal of Tropical Ecology* **28**: 291–297.

Owen-Smith, N. (1987) Pleistocene extinctions: the pivotal role of megaherbivores. *Paleobiology* **13**: 351–362.

Parr, C. L., Gray, E. F., & Bond, W. J. (2012) Cascading biodiversity and functional consequences of a global change-induced biome switch. *Diversity and Distributions* **18**: 493–503.

Prins, H. H. T. & van der Jeugd, H. P. (1993) Herbivore population crashes and woodland structure in East Africa. *Journal of Ecology* **81**: 305–314.

Russell, J. M. & Ward, D. (2014) Remote sensing provides a progressive record of vegetation change in northern KwaZulu-Natal, South Africa, from 1944 to 2005. *International Journal of Remote Sensing* **35**: 904–926.

Sankaran, M., Augustine, D. J., & Ratnam, J. (2013) Native ungulates of diverse body sizes collectively regulate long-term woody plant demography and structure of a semi-arid savanna. *Journal of Ecology* **101**: 1389–1399.

Scheiter, S. & Higgins, S. I. (2009) Impacts of climate change on the vegetation of Africa: an adaptive dynamic vegetation modelling approach. *Global Change Biology* **15**: 2224–2246.

Scheiter, S., Higgins, S. I., Osborne, C. P., *et al.* (2012) Fire and fire-adapted vegetation promoted C4 expansion in the late Miocene. *New Phytologist* **195**: 653–666.

Scholes, R. J. & Archer, S. R. (1997) Tree–grass interactions in savannas. *Annual Review of Ecology and Systematics* **28**: 517–544.

Scholes, R. J. & Walker, B. H. (1993) *An African savanna: synthesis of the Nylsvley study*. Cambridge University Press, Cambridge.

Schutz, A. E. N., Bond, W. J., & Cramer, M. D. (2009) Juggling carbon: allocation patterns of a dominant tree in a fire-prone savanna. *Oecologia* **160**: 235–246.

Silva, L. C., Hoffmann, W. A., Rossatto, D. R., *et al.* (2013) Can savannas become forests? A coupled analysis of nutrient stocks and fire thresholds in central Brazil. *Plant and Soil* **373**: 829–842.

Skarpe, C., Aarrestad, P. A., Andreassen, H. P., *et al.* (2004) The return of the giants: ecological effects of an increasing elephant population. *Ambio* **33**: 276–282.

Skowno, A. L., Midgley, J. J., Bond, W. J., & Balfour, D. (1999) Secondary succession in *Acacia nilotica* (L.) savanna in the Hluhluwe Game Reserve, South Africa. *Plant Ecology* **145**: 1–9.

Staver, A. C. & Bond, W. J. (2014) Is there a 'browse trap'? Dynamics of herbivore impacts on trees and grasses in an African savanna. *Journal of Ecology* **102**: 595–602.

Staver, A. C. & Levin, S. A. (2012) Integrating theoretical climate and fire effects on savanna and forest systems. *American Naturalist* **180**: 211–224.

Staver, A. C., Bond, W. J., Stock, W. D., van Rensburg, S. J., & Waldram, M. S. (2009) Browsing and fire interact to suppress tree density in an African savanna. *Ecological Applications* **19**: 1909–1919.

Staver, A. C., Bond, W. J., & February, E. C. (2011) History matters: tree establishment variability and species turnover in an African savanna. *Ecosphere* **2**: art49.

Staver, A. C., Bond, W. J., Cramer, M. D., & Wakeling, J. L. (2012) Top-down determinants of niche structure and adaptation among African Acacias. *Ecology Letters* **15**: 673–679.

Stevens, N. (2014) Exploring the potential impacts of global change on the woody component of South African savannas. PhD thesis, University of Cape Town, Cape Town.

Svenning, J. C. (2002) A review of natural vegetation openness in north-western Europe. *Biological Conservation* **104**: 133–148.

Tedder, M., Kirkman, K., Morris, C. & Fynn, R. (2014) Tree–grass competition along a catenal gradient in a mesic grassland, South Africa. *Grassland Science* **60**: 1–8.

Trollope, W. S. W. (1984) Fire in savanna. In: *Ecological effects of fire in South African ecosystems* (eds P. de V. Booysen & N. M. Tainton), pp. 149–176. Springer-Verlag, Berlin.

Trollope, W. S. W. (1993) Fire regime of the Kruger National Park for the period 1980–1992. *Koedoe* **36**: 45–52.

Trollope, W. S. W., Potgieter, A. L. F., & Zambatis, N. (1995) Effect of fire intensity on the mortality and topkill of bush in the Kruger National Park in South Africa. *Bulletin of the Grassland Society of Southern Africa* **6**: 66.

Viljoen, A. J. (1995) The influence of the 1991/92 drought on the woody vegetation of the Kruger National Park. *Koedoe* **38**: 85–97.

Wakeling, J. L. & Bond, W. J. (2007) Disturbance and the frequency of root suckering in an invasive savanna shrub, *Dichrostachys cinerea*. *African Journal of Range & Forage Science* **24**: 73–76.

Wakeling, J. L., Cramer, M. D., & Bond, W. J. (2010) Is the lack of leguminous savanna trees in grasslands of South Africa related to nutritional constraints? *Plant and Soil* **336**: 173–182.

Wakeling, J. L., Staver, A. C., & Bond, W. J. (2011) Simply the best: the transition of savanna saplings to trees. *Oikos* **120**: 1448–1451.

Wakeling, J. L., Cramer, M. D., & Bond, W. J. (2012) The savanna–grassland 'treeline': why don't savanna trees occur in upland grasslands? *Journal of Ecology* **100**: 381–391.

Wakeling, J. L., Bond, W. J., Ghaui, M., & February, E. C. (2015) Grass competition and the savanna–grassland 'treeline': a question of root gaps? *South African Journal of Botany* **101**: 91–97.

Waldram, M. S., Bond, W. J., &, Stock, W. D. (2008) Ecological engineering by a mega-grazer: white rhino impacts on a South African savanna. *Ecosystems* **11**: 101–112.

Walter, H. (1971) *Ecology of tropical and subtropical vegetation.* Oliver and Boyd, Edinburgh.

Watson, H. K. & Macdonald, I. A. W. (1983) Vegetation changes in the Hluhluwe-Umfolozi Game Reserve Complex from 1937 to 1975. *Bothalia* **14**: 265–269.

Weigl, P. D. & Knowles, T. W. (2014) Temperate mountain grasslands: a climate–herbivore hypothesis for origins and persistence. *Biological Reviews* **89**: 466–476.

Werner, P. A. & Prior, L. D. (2013) Demography and growth of subadult savanna trees: interactions of life history, size, fire season, and grassy understory. *Ecological Monographs* **83**: 67–93.

Whateley, A. & Porter, R. N. (1983) The woody vegetation communities of the Hluhluwe–Corridor–Umfolozi Game Reserve Complex. *Bothalia* **14**: 745–758.

White, A. M. & Goodman, P. S. (2010) Differences in woody vegetation are unrelated to use by African elephants (*Loxodonta africana*) in Mkhuze Game Reserve, South Africa. *African Journal of Ecology* **48**: 215–223.

Wigley, B. J., Cramer, M. D., & Bond, W. J. (2009) Sapling survival in a frequently burnt savanna: mobilisation of carbon reserves in *Acacia karroo*. *Plant Ecology* **2003**: 1–11.

Wigley, B. J., Bond, W. J., & Hoffman, M. T. (2010) Thicket expansion in a South African savanna under divergent land use: local vs. global drivers? *Global Change Biology* **16**: 964–976.

Wilson, S. L. & Kerley, G. I. (2003) Bite diameter selection by thicket browsers: the effect of body size and plant morphology on forage intake and quality. *Forest Ecology and Management* **181**: 51–65.

8 · *Woody Plant Traits and Life-History Strategies across Disturbance Gradients and Biome Boundaries in the Hluhluwe-iMfolozi Park*

LAURENCE M. KRUGER, TRISTAN CHARLES-DOMINIQUE, WILLIAM J. BOND, JEREMY J. MIDGLEY, DAVE A. BALFOUR, AND ABEDNIG MKHWANAZI

8.1 Introduction

A key aim of community ecology is to classify plants according to life-history traits and establish general principles (McGill *et al.*, 2006) in terms of life forms (Raunkiær, 1937), plant vital attributes (Noble and Slatyer, 1980), strategies (e.g. Grime, 1997; Tilman, 1990), and functional types (McIntyre *et al.*, 1999), to name but a few. The challenge is to find convergence in multiple traits, thereby define ecological strategy schemes (Westoby *et al.*, 2002; McGill *et al.*, 2006), and ultimately develop a periodic table for plants. This would allow us to understand important opportunities and selective forces that shape the ecologies of plants, and also to describe ecosystems in terms of a limited number of ecological component types (Westoby, 1998) or 'vital attributes' (Noble and Slatyer, 1980). With this foundation, we could predict how species will respond to perturbations and/or changing environments (e.g. Pausas, 1999) and draw comparisons between regions at global scales (Rusch *et al.*, 2003).

Substantial research has been done on life-history strategy schemes and trait variation in temperate systems and forests, which are strongly

Conserving Africa's Mega-Diversity in the Anthropocene, ed. Joris P. G. M. Cromsigt, Sally Archibald and Norman Owen-Smith. Published by Cambridge University Press.

controlled by competitive forces (Grime, 1997; Loehle, 2000; Westoby *et al.*, 2002). The selection pressures faced by plants in systems, such as savannas, that are driven by consumers (fire, herbivory) are likely to result in very different suites of plant traits, focused on resistance of disturbance or recovery after disturbance (Bond and Midgley, 2001). For example, for fire, thick bark is a resistance trait which protects sensitive cambium from the heat of the flame, and the ability to resprout after the above-ground biomass has been damaged is a key fire-recovery trait (see Chapter 7). However, the traits required to resist or recover from fire do not necessarily confer tolerance to herbivory. The rapid elongation of sprouts of *Acacia karroo* following a fire, for example, results in an architecture that is vulnerable to herbivory (Archibald and Bond, 2003). Therefore, we expect to see changes in the composition of plant communities and corresponding changes in suites of plant functional traits across gradients in fire and herbivory due to differences among species in their ability to withstand these disturbances. In contrast, within closed-canopy forest or thicket formations where the roles of fire and mammalian herbivory are reduced, competition becomes the main influence on plant traits and species composition. Bond (2005) used the following terminology for systems that differ in their disturbance regime: 'Black World' (fire-controlled system), 'Brown World' (herbivore-controlled) and 'Green World' (climate- or competition-controlled).

In this chapter, we explore how woody plant trait sets differ among Black World, Brown World and Green World systems in Hluhluwe-iMfolozi Park (HiP). Hluhluwe-iMfolozi provides great opportunities to study woody plant traits in relation to variation in fire and herbivory. Rainfall varies from around 1000 mm on the highest hills within Hluhluwe to under 550 mm in low-lying regions of Mfolozi (Figure 2.10C), which has corresponding effects on vegetation distribution (Chapter 2), fire regime (Chapter 10), and herbivore distributions (Chapter 4). The savanna vegetation in HiP is exposed to steep disturbance gradients representative of almost the full range of disturbances found in African savannas as a whole.

In this chapter we compare woody plant trait variation at a broad scale among forest, thicket, and savanna communities, representing the three main woody biomes found in HiP (Figure 8.1). Then, at a finer scale, we compare and contrast the variation in traits between different types of acacia savannas across the park in relation to patterns of fire and herbivory.

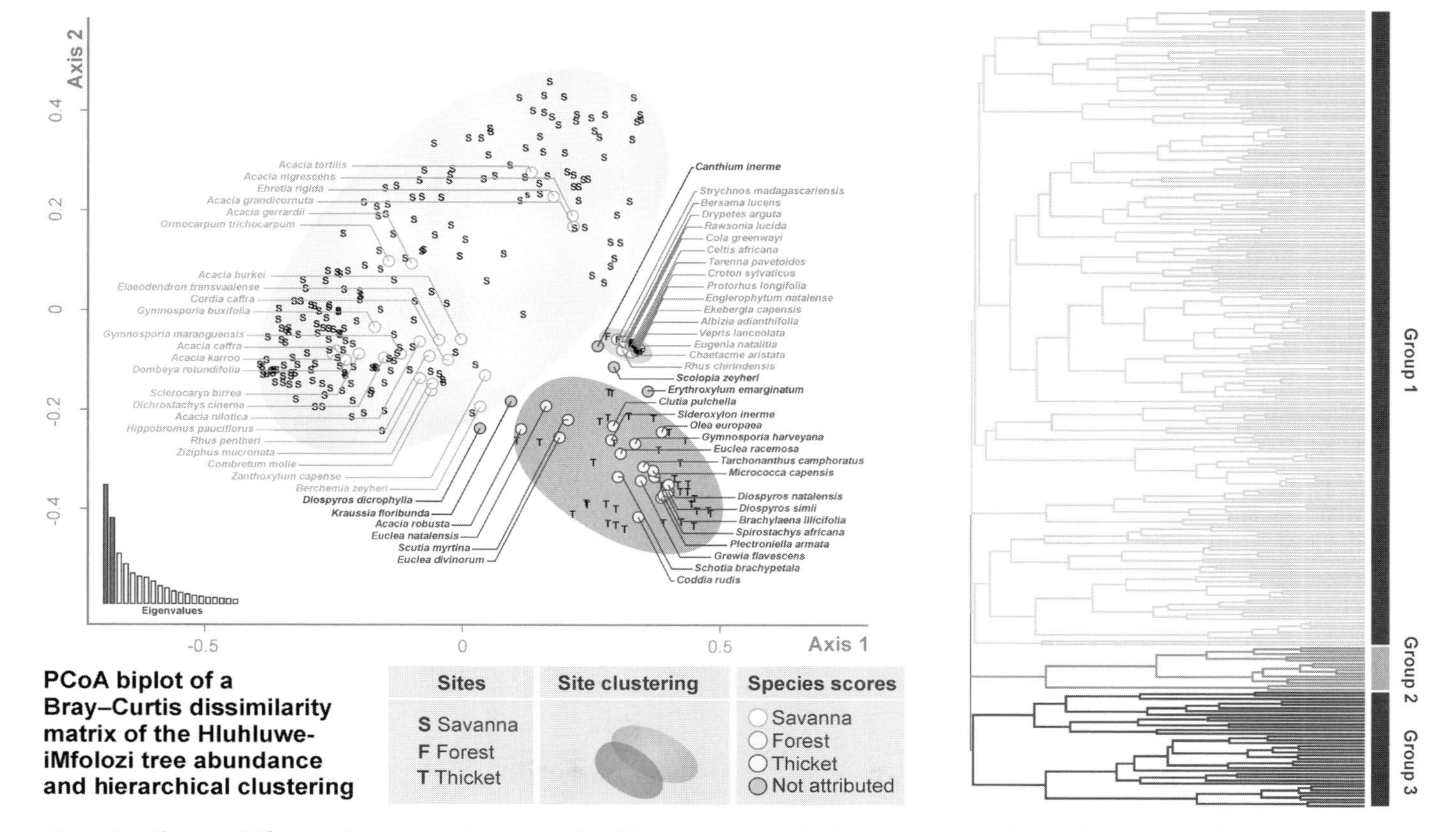

Figure 8.1 Floristic differentiation among the biomes in HiP. The graph on the left shows the position of sites (shown by letters) and species (circles) on the two first axes of the Principal Coordinate Analysis. The three envelopes correspond to the grouping made by hierarchical clustering. Each species has been attributed to one of the three biomes (see colour code) depending on its score. (Reprinted with permission from Elsevier; Charles-Dominique *et al.*, 2015b.) (For the colour version, please refer to the plate section. In some formats this figure will only appear in black and white.)

8.2 Biome Shifts at a Local Scale: Trait Variation across Forests, Thickets, and Savannas

The biome concept was introduced to characterize ecologically similar but floristically distinct vegetation types at large geographic scales. Initially there were few data on local climates. Hence, biomes were used to characterize climate and, in turn, climate classifications were designed to align with major vegetation patterns (reviewed by Keddy, 2007). Woodward *et al.* (2004) developed a new definition of biomes, which allows identification of major vegetation types without first characterizing the climate or geographic region in which the biome occurs. This definition emphasizes biomes as assemblages dominated by a particular growth form, or mix of growth forms (such as trees and grasses in savannas) typically mapped at continental scales. The emphasis on major growth forms underlies the most recent biome classification for South Africa (Mucina and Rutherford, 2006).

This growth-form definition of biomes frees us from the assumption that biomes and climates are linked so that associations between biomes and climate can be independently assessed. As a consequence, the presence of 'anomalous' vegetation in the 'wrong' climate is now seen as information, not noise (Bond, 2005). An important consequence is the recognition of alternative biome states: distinct vegetation forms occurring under the same climate and, often, on the same soils (Wilson and Agnew, 1992; Scheffer and Carpenter, 2003; Warman and Moles, 2009). Alternative states imply that plant species have the capacity to construct their own environmental conditions, in contrast to notions that the physical environment explains the presence of particular plant forms. Forest and savanna are classic examples found in the same landscape and, in HiP, on the same soil types but showing distinct water infiltration and water retention rates as well as differences in nutrient cycling (see Eldridge *et al.*, 2011). The presence of a C4 grassy layer is a key functional feature of savanna, where the grasses fuel the fires and feed the grazers that maintain savannas (Chapters 3 and 7). The shade cast by forest trees and the forest microclimate they generate are positive feedbacks for forest that inhibit the entry of fire by restricting C4 grass biomass (reviewed in Baudena *et al.*, 2015) and reducing wind speed (Hoffman *et al.*, 2012). Feedbacks on soil properties have also been reported (Pellegrini *et al.*, 2014).

HiP supports vegetation types representative of four distinct biomes: forest, thicket, savanna, and grassland (Figure 8.1). Mesic savanna, thicket, and forest communities dominate the wetter north-east of Hluhluwe, interspersed with scattered patches of C4 grassland on the highest hilltops. Semi-arid savanna predominates in the drier Mfolozi region in the southwest. Broadleaf thicket (found both in Mfolozi and Hluhluwe) is floristically and functionally distinct from both savanna and evergreen forest (Parr *et al*., 2012). It has affinities with the Albany Thicket biome of the South African Eastern Cape province, sharing dominance by mostly evergreen woody plants in genera such as *Euclea*, *Diospyros*, *Berchemia*, *Schotia*, and *Grewia*, but lacks succulents. It diverges from forests by having shorter canopy trees (4–6 m vs 10 m⊗), no intermediate layer of shade-tolerant trees, a more open canopy, and a dense understorey of subshrubs such as *Isoglossa*, *Justicia* and species within the Malvaceae. A continuous grass layer is lacking, although grassy patches may occur, typically of shade-tolerant grazing lawn species (e.g. *Dactyloctenium australe*).

Given the presence of these distinct biomes in HiP, we assessed plant functional traits in each by quantifying woody species composition in 253 transects (40 × 10 m) across HiP. Sites were selected based on the relative proportions of the three biomes in the park: 202 sites in savanna, 37 in thicket, and 14 in forest. Six functional traits were described for five saplings (2–5 m tall) of 58 tree species representative of the three biomes: 16 species in forest, 23 in savanna, and 19 in thicket. We ignored the grassland biome, due to the lack of woody plants. Table 8.1 describes the traits that we selected and, briefly, how they were assessed. For more detail on this study and the methods used, see Charles-Dominique *et al*. (2015a,b). Traits were analysed across vegetation types to ascertain whether they differed among woody species specific to savanna, forest, and thicket. In order to quantify the prevalent disturbance in each biome, we tracked fire history (using the park's annual fire maps) and herbivore pressure (using dung counts as a proxy) for each of 253 transects.

We found both floristic and functional differences among biomes. Sites separated a priori based on structural differences of their vegetation (presence/absence of C4 grass layer, mono-/multi-layered understorey, stem density, and height of the canopy) and had different tree species assemblages (Figure 8.2). The thicket biome in HiP is rich in species, with 19 of the 65 most abundant tree species occurring preferentially in this biome. The relationship between thicket and savanna in HiP is dynamic: thicket may replace savanna following fire suppression and/or heavy grazing (Watson, 1995; Staver *et al*., 2009; Wigley *et al*., 2010). During field observations, savanna patches with many thicket species were sometimes found in Hluhluwe and represented former thicket patches destroyed by recent firestorms (see Chapter 10). All of the thicket tree species found

Table 8.1. *Plant life-history traits assessed in Hluhluwe-iMfolozi Park*

Trait Across biomes:	Black World (fire-driven) Savannas	Brown World (herbivore-driven) Thickets	Green World (competition-driven) Forests
Height of first fork: an indication of early vs late set up of canopy. A low fifth-fork height indicates that a species forms a canopy quickly and invests less in vertical growth	Low	Low	Tall
Specific leaf area: SLA is considered to be a good indicator of growth rate (Reich *et al*., 1998).	Low	High	High
Wood growth rate: wood growth rate was described at the base of the trunk (between 10 and 20 cm height). Cuttings of transverse sections were photographed under a binocular microscope, and aged by counting wood rings	Low	Low	Med
Spinescence: we calculated the proportion of constituent species that bear spines. Greater proportion of spinescence indicates greater allocation to physical defences against herbivory	High	Low	Low
Architecture (cage vs pole): the index of cage architecture is calculated as follows: $ICA = \prod_{i=1}^{n} (1 + 2s_i + c_{i+1})$ with n the rank of the most peripheral axis category (*sensu* Barthélémy and Caraglio, 2007), s is the presence (1) or absence (0) of spine on a given axis category and c describes the conicity of the axis category (1 = conic; 0 = cylindrical)	Cage	Pole	Pole
Bite size: the bite size index, quantifying the phytomass a herbivore can remove in one bite, estimated following Wigley *et al*. (2014)	Low	Low	High
Bud protection and bark growth rate: cuttings of transverse sections of shoots were photographed under a binocular microscope. The thickness of the bark layer was measured using ImageJ© and divided by the age of the plant estimated from morphological markers (Barthélémy and Caraglio, 2007). We checked that dead epidermis was still visible at the periphery of the stem in order to avoid underestimation of bark growth rate due to bark shedding	High Medium	Medium Medium	Low Low

Accessory buds: indicative of sprouting ability and assessed by the presence/absence of accessory buds by longitudinal cuttings passing by the centre of the leaf scar	High	Low	Low
Acacias	Mesic savannas	Semi-arid savannas	Riparian forests
Height at the top of the canopy indicates whether plants escape both the flame and/or herbivory zone	Short to tall	Medium to tall	Tall
Seed size: a crude measure for both reproduction allocation and potential establishment success (Westoby *et al*., 1996), providing an indication of the reliance on seeds for regeneration. Dry seed mass was calculated for 15 mature, alive seeds collected from each individual)	Sprouters: Small Seeders: Large	Large	Large
Multi-stemmedness: a surrogate measure of sprouting ability as adults (Midgley, 1996; Kruger *et al*., 1997), calculated as the numbers of stems per individual originating from below the ground	Sprouters: Multi-stemmed, Seeders: Single	Single	Single
Bark thickness: key in determining resistance to fire. Stem thickness vs bark thickness regression analyses of each individual were used to determine the predicted bark at 100 mm of stem. Bark thickness values represented as a % of stem diameter to standardize the measurement across a range of stem diameters	Thick	Thin	Thin
Resprouting following experimental felling: stems were cut at ground level and resprout length and abundance were enumerated after 6 months (both adults and juveniles)	*Juveniles*: Few, long = pole *Adults*: Elongation	*Juveniles* Many, short = cage *Adults* None	*Juveniles* Few, long = pole *Adults* Few, long = pole
Acacias (juveniles): Staver *et al*. (2012)			
Sapling architecture: number of secondary branches emerging from the primary axis (density of branches per unit of height increment)	Pole	Cage	
Root to shoot ratios: dry weight of total shoot mass and below-ground root mass of saplings to a depth of 0.5 m	Variable	Variable	
Starch content: determined calorimetrically by phenol–sulphuric method (Dubois *et al*., 1956) and expressed as glucose concentration. Assessed in dry season when starch reserves are at their highest	High	Low	
Bark thickness: measured for height (controlled between 25 and 50 cm)	Thin	Thick	

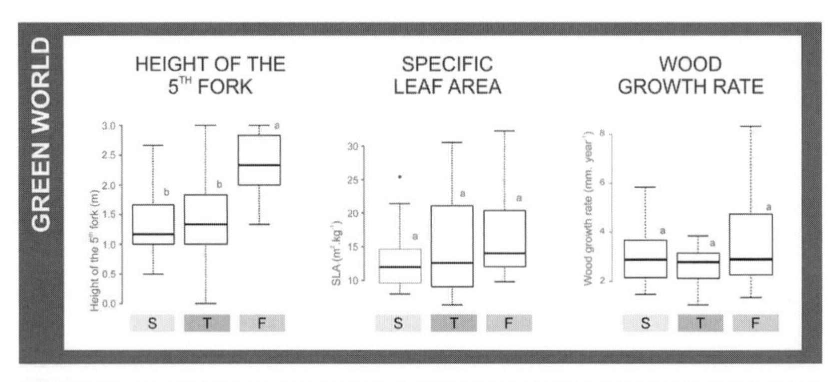

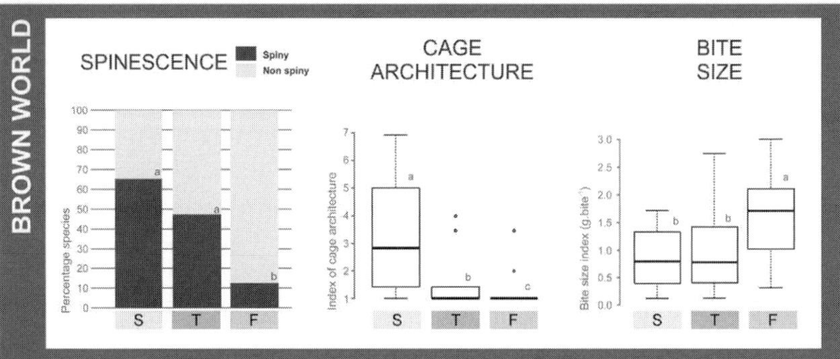

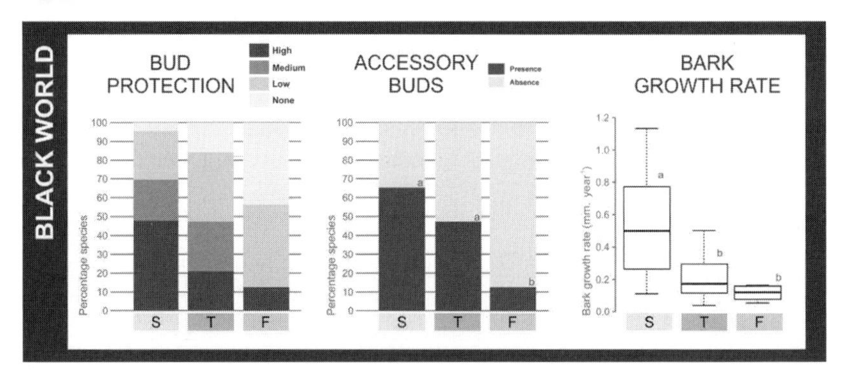

Figure 8.2 Functional distinctions species from the three biomes. The values shown in boxplots and barplots correspond to the average values/count on species attributed to each of the three biomes ('S' for savanna, 'T' for thicket, and 'F' for forest) with three replicates per species for anatomical traits (wood growth rate, bud protection, accessory buds, and bark growth rate) and five replicates for the other traits. Letters show the grouping after pairwise Wilcoxon–Mann–Whitney rank sum tests and pairwise comparisons of proportions with a Holm's correction for multiple testing. (Reprinted with permission from Elsevier; Charles-Dominique *et al.*, 2015b.)

in these patches were from resprouts and no new individuals recruiting into the savanna were observed. As regards thicket/forest relationships, there is very little evidence that thicket is a successional stage to forest. In 37 transects covered in sites dominated by thicket, there were almost no stems of forest species (six forest saplings encountered in total).

The distinctions between savanna, thicket, and forest vegetation have major consequences for their respective ecosystem functioning. Forest transects had low light availability and very sparse C4 grass cover contributing to a very low incidence of burning (no fire for the selected forest sites during the last 20 years). There was also low browser use within forest transects (216 ± 210 dung piles mostly of red duiker and nyala per hectare, mean ± SD). Thicket sites had a lighter understorey with discontinuous patches of grasses, infrequent fires (0.14 ± 0.30 fires/decade) and high browser (mostly nyala, elephant, grey duiker) presence (655 ± 508 dung piles/ha). Savanna sites, characterized by a continuous layer of C4 grasses, had frequent fires (3.07 ± 1.97 fires/decade) and medium browser (predominantly impala and giraffe) presence (404 ± 600 dung piles/ha, probably underestimated because of lower detection of dung piles within the grass layer).

The three biomes in HiP are associated with major functional differences among dominant tree species, with functional traits differentiated in context of the Brown World (impact of herbivores) and Black World (impact of fire; Bond, 2005). Savanna trees occur in open environments maintained by regular fire and herbivory, and consequently set up their canopy early (sympodial growth form, low height of the fifth fork in Figure 8.2). As saplings, most of their branches are accessible to browsers and subject to high fire intensity. Savanna trees are well-defended structurally against browsers: they have the highest proportion of spiny trees, often develop 'cages' (i.e. dense structures made of many woody and spiny axis categories), and offer low bite sizes (Figure 8.2). The most striking differences between trees from savanna and the two other biomes lie in their fire-related traits. Savanna trees have a distinct combination of high bark production and thus a high degree of bud protection (similar to eucalypts and other Myrtaceae in Australia: Burrows *et al.*, 2010; Charles-Dominique *et al.*, 2015a) and accessory buds that provide increased capacity for resprouting after fire (Burrows *et al.*, 2008). Savanna trees thus clearly live in a Brown and Black World maintained by regular fire and herbivory. Thicket trees also set up their canopy early (height of the fifth fork as low as savanna species). They also incur significant herbivore pressure and have a higher proportion of spiny species, offering small bite sizes, and more regularly form cages than forest trees. The strongest difference in traits between savanna and thicket trees follows from differences in fire frequency between these two vegetation types. Thicket trees have relatively low bark production and little bud protection compared to savanna species (Figure 8.2). However, many thicket species have accessory buds and can resprout basally after rare intense fires such as those caused by 'firestorms', making them adapted to infrequent but intense burns (Charles-Dominique *et al.*, 2015a,b). Like savanna trees, thicket trees live in a Brown and Black World, although more shifted towards the Brown than the Black World. Forest species have their growth directed vertically and start forking at greater heights than savanna and thicket species (a more monopodial growth form). Few forest trees are spiny and almost all are unable to develop a cage restricting the access to leaves and buds of herbivores. This absence of structural defences is reflected in the large bite sizes available to herbivores. Forest species have very little bark production, almost no accessory buds, and low bud protection, similar to species found in mangroves (Burrows *et al.*, 2010). Forest species are thus poorly adapted to withstand fires and herbivory, and clearly live in a Green World.

8.3 Life-History Trait Variation among Acacias within Savannas in HiP

The strong rainfall gradient in HiP creates the rather unique situation that mesic and semi-arid savannas occur within a 50-km transect. Both savanna types differ substantially in the relative role of fire and herbivory. Mesic savanna communities are characterized by high grass productivity and hence frequent, intense burns (Chapter 10), whereas relatively low grass productivity in semi-arid savannas leads to lower fire frequency. In HiP, this pattern may be enhanced by white rhino-derived grazing lawns, which are more widespread in the semi-arid savannas of Mfolozi and may act as firebreaks (Waldram *et al.*, 2008; Chapter 6). Given the inverse relationship between fire frequency and herbivore pressure (Archibald *et al.*, 2005; Staver *et al.*, 2012), browsers should have a greater influence on woody plant demography and related traits in the semi-arid savanna (Sankaran *et al.*, 2005; Staver *et al.*, 2012). In this section, we explore how the difference in the relative role of fire versus herbivory between mesic and semi-arid savannas influences acacia life-history traits. We adopt the approach of Keeley and Zedler (1998), who grouped *Pinus* species into functional groups according to selected life-history traits, and then attempted to explain these in terms of different disturbance regimes. We ask whether traits such as sprouting ability, bark thickness, seed size, and plant height cluster species into different functional groups that represent adaptations to fire or herbivory. We consider both saplings and adults and concentrate on acacia species forming the most widespread group of trees in HiP, with substantial variation in life forms between (Midgley and Bond, 2001) and within (Archibald and Bond, 2003) species. In identifying functional guilds of acacias, we assessed variation in key life-history traits (Table 8.1), then sought correlations among these traits, and finally used multi-dimensional scaling (MDS) to ordinate key life-history traits (based on five adult trees). To test sprouting ability experimentally, we cut down 10 individuals (five juveniles and five adults) and measured the length of regrowth after a year.

8.3.1 Variation in Selected Traits

The thickness of bark increased with stem diameter for all of the species and all species displayed early allocation to defences, indicating that defence against fire is critical for saplings (see also Chapter 7). *Acacia karroo*, *A. nilotica*, and *A. burkei* have inherently thick bark. Others, such as *A. caffra*, *A. gerrardii*, and *A. davyi*, develop thick bark by having rapid bark accumulation. Age and/or size strongly influence sprouting ability. We found that juveniles of all species can resprout, consistent with most intense fires occurring 2–3 m above ground. Although sprouting by juveniles is well known (Gignoux *et al.*, 1997; Hodgkinson, 1998; Higgins *et al.*, 2000; Bond and Midgley, 2003; Vesk, 2006; Clarke *et al.*, 2013), little knowledge is available on the ability of adult trees to sprout from the base. We found that, among adults, only the two mesic savanna species, *A. caffra* and *A. davyi*, resprouted strongly basally. Thus, strong sprouting as adults is important in mesic savannas where the probability of losing all aboveground biomass to fires of high intensity is greatest. We found relationships of plant height with seed size and number of stems only (Figure 8.3). Much like forest trees (Kruger *et al.*, 1997), there was a trade-off between plant height and number of stems, suggesting that resprouting trees are height-limited, probably as a consequence of the sharing of resources between stems (Midgley, 1996). Taller acacias had substantially larger seeds than shorter species, aligned with a life history shaped by competition rather than disturbance (Westoby *et al.*, 1996). In shaded habitats, stored reserves in larger seeds facilitate establishment (Leishman *et al.*, 2000).

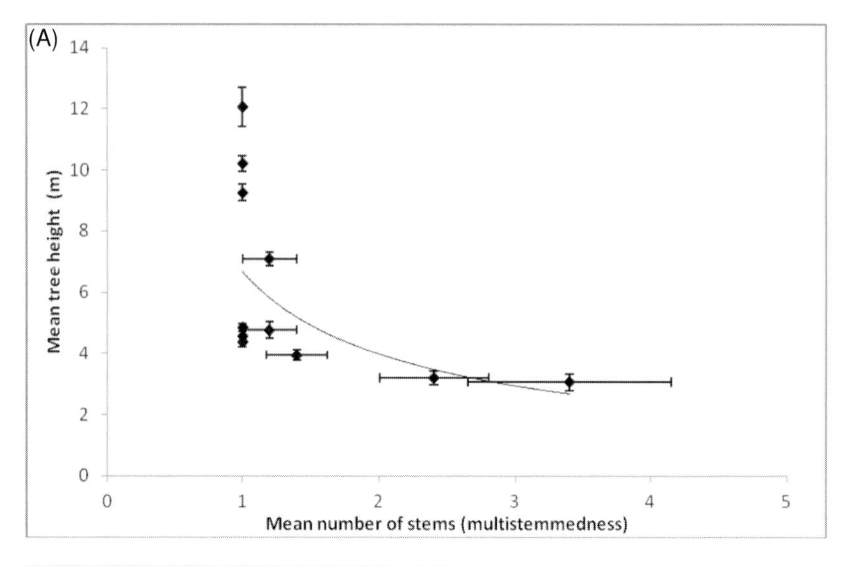

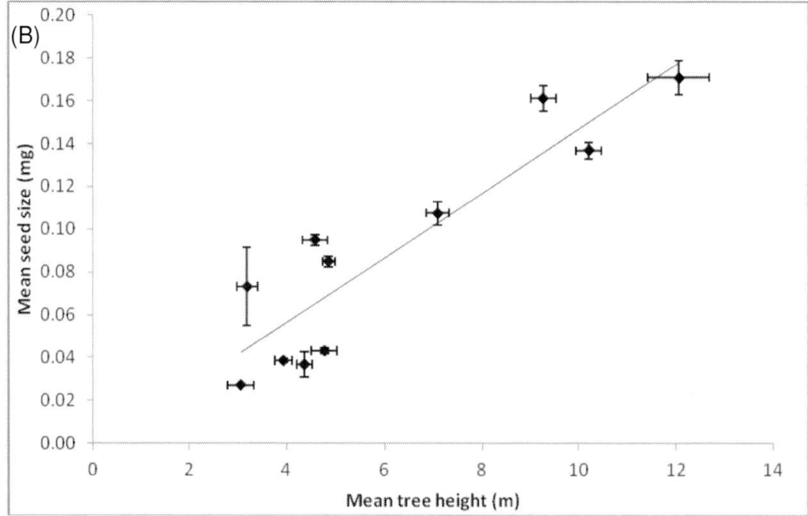

Figure 8.3 Relationships between (A) number of stems per individual ('multi-stemmedness') and plant height ($r^2 = 0.44$, $P < 0.01$, $y = 6.66x^{-0.78}$), and (B) plant height and seed weight ($r^2 = 0.82$, $P < 0.005$, $y = 0.02x + 0.003$) for 11 species of acacia trees in HiP.

8.3.2 Multi-Trait Analyses: Grouping Species

The multi-trait MDS analysis of the traits yielded three distinct groups (Figure 8.4, see Table 8.2 for species trait values). Species were ordinated

Table 8.2. *Life-history traits measured for each of 11* Acacia *species in HiP: including seed mass, wood density, height, predicted bark thickness at 100 mm of stem, and sprouting after experimental cutting (percentage of individuals sprouting, mean count of sprouts and vigour, i.e. mean length of sprouts, SE in parentheses)*

Species	Seed mass (mg)	Wood density (g/cm^3)	Height (m)	No. stems	Bark (mm)	Sprouting Juvenile % spr	Adult mean count	mean length (cm)	% spr	mean count	mean length (cm)
Mesic savannas											
Seeders											
A. burkeii	0.13 (0.01)	0.84 (0.04)	10.0 (0.37)	1.0 (0.01)	10.3	100	9.2 (1.11)	24.6 (3.86)	60	6.3 (2.23)	47.0 (8.89)
A. karroo	0.04 (0.002)	0.84 (0.038)	5.3 (0.44)	1.2 (0.02)	9.8	89	2.2 (0.35)	88.7 (8.05)	80	4 (0.73)	46.6 (1.39)
A. nilotica	0.10 (0.002)	0.93 (0.034)	4.7 (0.38)	1.0 (0.01)	10.4	100	1.4 (0.55)	92.0 (9.33)	0	0	0
A. gerrardii	0.09 (0.002)	0.87 (0.023)	5.4 (0.06)	1.1 (0.01)	10.9	100	3.5 (1.18)	64.1 (2.11)	0	0	0
Sprouters											
A. caffra	0.09 (0.001)	1.01 (0.034)	3.6 (0.15)	2.2 (0.01)	12.4	100	5.5 (0.81)	110.5 (5.72)	100	4.3 (0.56)	108.8 (8.59)
A. davyi	0.03 (0.001)	0.86 (0.023)	2.5 (0.20)	4.5 (0.45)	15.9	100	6.4 (1.83)	119.6 (11.81)	100	5.3 (1.48)	105.6 (4.71)
Riverine habitats											
A. robusta	0.17 (0.006)	0.78 (0.034)	12.4 (0.14)	1.0 (0.01)	7.6	100	9.2 (1.56)	99.9 (7.86)	75	13.3 (1.81)	109.3 (4.89)
Xeric savannas											
Seeders											
A. grandicornuta	0.11 (0.004)	0.97 (0.07)	7.4 (0.08)	1.2 (0.02)	7.2	100	10.2 (2.16)	38.8 (4.71)	40	17 (3.02)	30.9 (6.46)
A. nigrescens	0.16 (0.005)	1.07 (0.103)	10.4 (0.10)	1.1 (0.01)	8.9	66	13.3 (2.26)	26.9 (1.99)	50	22 (4.42)	25.6 (0.55)
A. tortilis	0.05 (0.009)	0.93 (0.91)	4.7 (0.91)	1.1 (0.02)	8.2	100	11.0 (2.19)	56.8 (11.35)	40	10.5 (2.75)	19.3 (6.14)
A. luederitzii	0.04 (0.002)	1.03 (0.023)	3.9 (0.02)	1.5 (0.13)	7.6	100	8.4 (1.98)	38.7 (6.50)	40	9 (2.35)	38.9 (5.72)

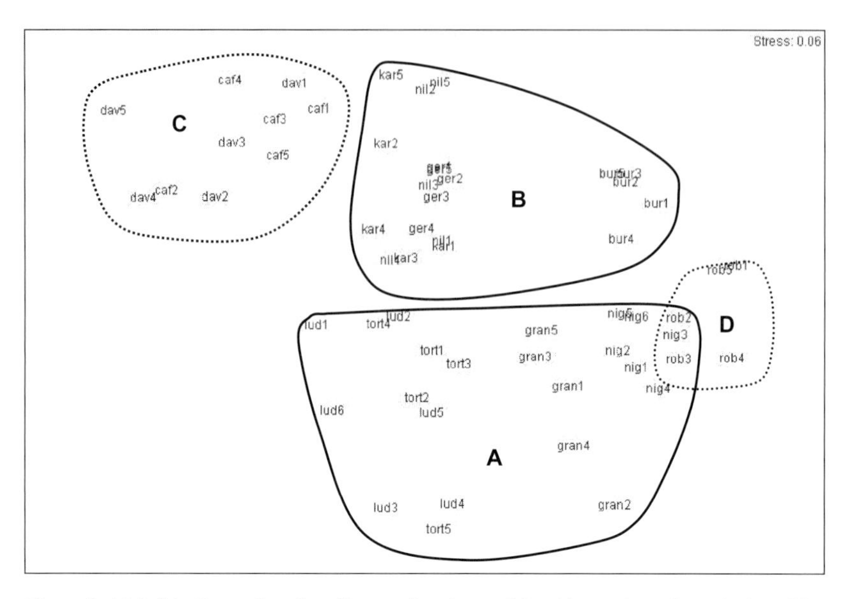

Figure 8.4 Multi-dimensional scaling ordination of the 11 species of acacia in HiP based on square root-transformed data and Euclidean distances. (A) Brown World Acacias – semi-arid reseeding species (lud: *Acacia luederitzii*; tort: *A. tortilis*; gran: *A. grandicornuta*; nig: *A. nigrescens*); (B) Black World Acacias – mesic reseeding species (kar: *A. karroo*; nil: *A. nilotica*; ger: *A. gerrardii*; bur: *A. burkeii*); (C) Black World Acacias – basal resprouting species (dav: *A. davyi*; caf: *A. caffra*); (D) Green World Acacias – reseeding riverine species (rob: *A. robusta*). All data were equally weighted but square root-transformed to ensure that larger trait values (e.g. plant height vs seed mass) are not overemphasized by the analysis.

into semi-arid and mesic savanna species, and within the mesic group separated between reseeders (low probability of resprouting as adults) and basal resprouters (Figure 8.5). Only one species straddled the trait space of two groups, *A. robusta*, which separated as a riverine species and was found across the range of savanna types. The functional groups, based on the MDS ordination, are graphically depicted in the context of both disturbance regimes (fire- versus herbivore-driven systems) and species characteristics in Figure 8.5.

8.3.2.1 Group A: 'Brown World Acacias' – Semi-Arid Reseeders

The species found in the semi-arid savannas of Mfolozi (*Acacia luederitzii*, *A. tortilis*, *A. grandicornuta*, and *A. nigrescens*) are sensitive to fire, as indicated by their relatively thin bark as juveniles, and do not resprout as strongly (basally) as adults (Table 8.1). They are generally well-defended, as indicated by their high spinescence indices (e.g. *A. tortilis* and *A. grandicornuta*;

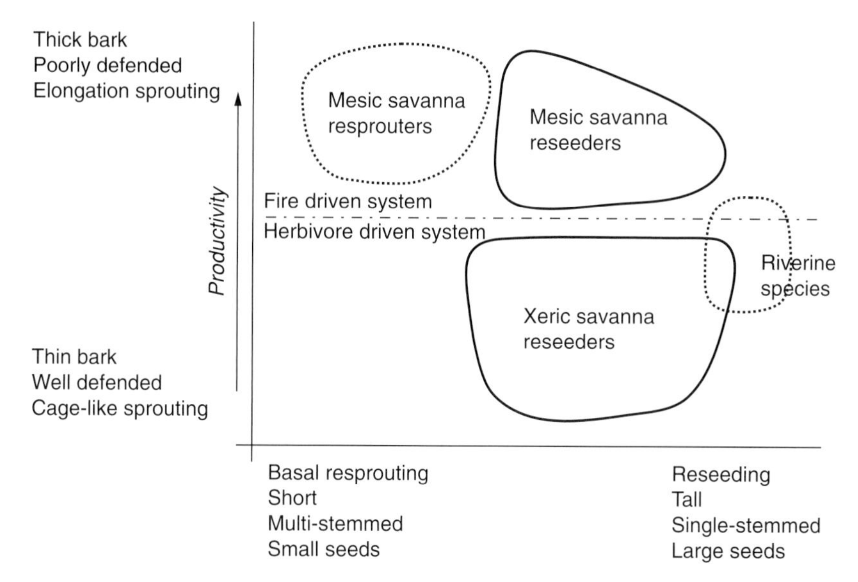

Figure 8.5 Stylized representation of functional groupings of various acacia species in HIP based on multi-dimensional scaling of character traits (Figure 8.3 for MDS plot and Table 8.2 for trait values).

Midgley *et al.*, 2001) and some display a cage-like sprouting structure. This is an effective architectural defence against herbivory in that, in conjunction with spines, the inner parts of the plant are well-defended (Brown, 1960). *A. luederitzii*, which is the shortest of the Mfolozi species (mean height: 3.89 m) and therefore remains in the zone of highest herbivory, has a highly branched architecture even as an adult (in many species the juvenile forms are more strongly defended against browsing). *A tortilis*, with both hooked and straight spines, has been recorded as highly resistant to herbivory (du Toit *et al.*, 1990; Bond and Loffell, 2001). *Acacia nigrescens* makes up for its small hooks by growing tall and thereby escaping much browsing impact (Midgley *et al.*, 2001). Moreover, *A. nigrescens* can adequately compensate for losses through browsing by strong shoot regrowth (i.e. by sprouting epicormically; du Toit *et al.*, 1990).

8.3.2.2 Group B; 'Black World Acacias' – Mesic Reseeders

As a group, these reseeders (*A. karroo*, *A. burkeii*, and *A. gerrardii*, and perhaps not as strongly *A. nilotica*) appear adapted to intense and/or frequent fires rather than herbivory. These species are fire survivors (thick bark as juveniles) and at least two of them (*A. karroo* and *A. gerrardii*) appear somewhat poorly defended against herbivory. When sprouting after the loss of above-ground stems, all juveniles and most adults (except for *A. nilotica*) of these species produce a few but vigorous resprouts. Rapid height gain increases the chances of pushing the new shoots above the firetrap and recruiting into the canopy (Archibald and Bond, 2003; Dantas and Pausas, 2013). Exactly how fast plants need to grow to escape fire depends on the frequency distribution of fire-free intervals, and there is much intraspecific variability (Wakeling *et al.*, 2011; Chapter 7). However, Higgins *et al.* (2000) used annual sprouting growth rates of 0.6 m, and resprouting shoots (after mechanical decapitation) sampled in HiP were on average this length or longer, suggesting that these species are adapted to a fire-prone existence. However, the same elongated nature of the resprouts, in conjunction with poor mechanical defences, may render them vulnerable to browsing, especially by giraffe and black rhino (Archibald and Bond, 2003). These species have larger seeds than the mesic sprouting species, and this feature may confer an advantage in the seedling establishment phase (Westoby *et al.*, 1996), a critical stage of acacia demography (Midgley and Bond, 2001). Note that *A. nilotica*, which appears somewhat exceptional in its impressive spines and low tree height, is also widespread in the semi-arid Mfolozi region.

8.3.2.3 Group C: 'Black World Acacias' – Basal Resprouters

Given the short adult height of *A. caffra* and *A. davyi* (mean of about 3.5 and 2.5 m, respectively) in Hluhluwe, these species remain within the flame zone for the duration of their lifespan, and thus require a strong ability to persist. These species could be considered as supreme fire-survivors as they are well protected by their relatively thick bark and ability to resprout vigorously, both as juveniles and adults. Sprouting after damage requires a substantial allocation to stored reserves (Hodgkinson, 1998; Schutz *et al.*, 2009). This prior allocation of resources to ensure fire survival represents a major cost, which may result in a trade-off between storage versus growth and reproduction (Bond and van Wilgen, 1996). Due to their weak spinescence (Midgley *et al.*, 2001) and the elongated nature of their resprouts, these species are very poorly defended against herbivory and have been found to be particularly vulnerable to browsing by giraffes (Bond and Loffell, 2001).

8.3.2.4 Group D: 'Green World Acacias' – Riverine Species

Acacia robusta is an outlier in the MDS analysis: it is tall and single-stemmed and bears large seeds but weak spines. As a juvenile it is both fire-sensitive (thin bark) and vulnerable to herbivory. This species seems to have traits associated with forest environments and is commonly found in riverine forest margins (Whateley and Porter, 1983). This would explain its rapid height gain (to gain access to light) as well as its large seed size (Westoby *et al*., 1996). Juvenile *A. robusta* nevertheless resprout strongly when damaged, indicating that they are able to survive damage from fires and herbivory.

8.3.3 Summary Overview

Bark thickness, plant height, seed size, and sprouting ability are all important in discriminating among savanna guilds. While some traits are well understood (e.g. architecture, resprouting, and bark thickness in relation to fire), other traits (seed size and defence against herbivory) remain poorly understood. Similarly to trait convergence along light gradients among biomes, we observed convergence in traits across disturbance gradients in savanna vegetation in HiP. The patterns for adult acacias echo those of Staver *et al.* (2012), although we found contradictory evidence for bark thickness, possibly because of different measurements used (bark thickness versus plant height as opposed to bark thickness and stem diameter; see Hempson *et al*., 2014 for review).

Despite these contradictions, our findings strongly support the contention of Staver *et al.* (2012) that top-down consumer controls, i.e. fire and herbivory, are critical in determining the realized niches of plants. Our results suggest that changes in disturbance regimes may have a substantial influence on the survivorship and distribution of trees. Indeed, in HiP we see that as the disturbance regime changes from fire at higher altitudes to herbivory at lower elevations, there is a concurrent shift in dominance from *A. karroo* to *A. nilotica*, a more herbivore-adapted species (Bond *et al*., 2001). Moreover, Bond and Loffel (2001) reported declines in *A. davii* and *A. caffra*, both fire-adapted species poorly defended against herbivory, in Ithala Game Reserve, due to herbivory by introduced giraffes. Balfour and Midgley (2008) suggested that changes in fire regime, in particular a lower frequency of intense fires, has facilitated *A. karroo* invasion in mesic areas. However, there are still large holes in our understanding concerning the ecology of shrub species, which remain exposed to both fire and browsing throughout their lifespans.

8.4 Conclusion

HiP presents an ideal natural laboratory for studying the effects of top-down controls on plant traits and species distributions and has been the source of novel methods to quantify disturbance-related traits (Table 8.1) as well as new ideas about how these traits are organized along gradients of disturbance. While climate is broadly important in determining biome distributions (Keddy, 2007), work in HiP on trait shifts across biomes contributes towards understanding biome boundary shifts at local scales. Our results demonstrate trait turnover at both biome (savanna, forests, and thickets) and within-biome (savannas) scales, distinguishing plants adapted to a Black World (fire-driven systems), Brown World (herbivore-driven systems), or Green World (climate-controlled systems; Bond, 2005). This greater appreciation for how changes in disturbance regimes could impact vegetation boundaries can guide managers in their efforts to counteract changing vegetation patterns. Moreover, while the shifts in vegetation in HiP are at a landscape scale, the insights from this work affect the way in which we assess the determinants and distributions of biomes globally.

Floristic and trait analyses across biome types highlight the difference between thicket vegetation and adjacent savanna and forest vegetation. Hoare *et al.* (in Mucina and Rutherford, 2006) provide a useful account of thicket vegetation types occurring at the drier end of the precipitation gradient in South Africa, and similar vegetation has been recognized elsewhere in Africa, Madagascar, Australia, India, and South and North America. Consequently, thicket should be considered independently from forest and savanna when assessing vegetation dynamics and ecosystem management. In particular, it is important to distinguish between bush encroachment within savanna (increases in density of savanna tree species) and the invasion of novel 'thicket' functional forms into formerly savanna regions (see Chapters 2, 3, and 7). The expansion of thicket at the cost of savanna that has occurred in Hluhluwe and elsewhere in KwaZulu-Natal and Africa over the last century is arguably the most insidious form of 'bush encroachment' because it leads to a non-grassy state excluding fire and results in cascading species turnovers (Parr *et al.*, 2012).

8.5 References

Archibald, S. & Bond, W. J. (2003) Growing tall vs. growing wide: tree architecture and allometry of *Acacia karroo* in forest, savanna and arid environments. *Oikos* **102**: 3–14.

Archibald, S., Bond, W. J., Stock, W. D., & Fairbanks, D. H. K. (2005) Shaping the landscape: fire–grazer interactions in an African savanna. *Ecological Applications* **15**: 96–109.

Balfour, D. A. & Midgley, J. J. (2008) A demographic perspective on bush encroachment by *Acacia karroo* in Hluhluwe-Infolozi Park, South Africa. *African Journal of Range and Forage Science* **25**: 147–151.

Barthélémy, D. & Caraglio, Y. (2007) Plant architecture: a dynamic, multilevel and comprehensive approach to plant form, structure and ontogeny. *Annals of Botany* **99**: 375–407.

Baudena, M., Dekker, S. C., van Bodegom, P. M., *et al.* (2015) Forests, savannas and grasslands: bridging the knowledge gap between ecology and Dynamic Global Vegetation Models. *Biogeosciences* **12**: 1833–1848.

Bond, W. J. (2005) Large parts of the world are brown or black: a different view of the Green World Hypothesis. *Journal of Vegetation Science* **16**: 261–266.

Bond, W. J. & Loffell, D. (2001) Introduction of giraffe changes acacia distribution in a South African savanna. *African Journal of Ecology* **39**: 286–294.

Bond, W. J. & Midgley, J. J. (2001) Ecology of sprouting in woody plants: the persistence niche. *Trends in Ecology and Evolution* **16**: 45–51.

Bond, W. J. & Midgley, J. J. (2003) The evolutionary ecology of sprouting in woody plants. *International Journal of Plant Sciences* **164**: 103–114.

Bond, W. J. & van Wilgen, B. A. (1996) *Fire and plants*. Chapman and Hall, London.

Bond, W. J., Smythe, K., & Balfour, D. A. (2001). Acacia species turnover in space and time in an African savanna. *Journal of Biogeography* **28**: 117–128.

Brown, W. (1960). Ants, acacias and browsing animals. *Ecology* **41**: 587–592.

Burrows, G. E., Hornby, S. K., Waters, D. A., *et al.* (2008) Leaf axil anatomy and bud reserves in 21 Myrtaceae species from northern Australia. *International Journal of Plant Sciences* **169**: 1174–1186.

Burrows, G. E., Hornby, S. K., Waters, D. A., *et al.* (2010) A wide diversity of epicormic structures is present in Myrtaceae species in the northern Australian savanna biome – implications for adaptation to fire. *Australian Journal of Botany* **58**: 493–507.

Charles-Dominique, T., Beckett, H., Midgley, G. F., & Bond, W. J. (2015a) Bud protection: a key trait for species sorting in a forest–savanna mosaic. *New Phytologist* **207**: 1052–1060.

Charles-Dominique, T., Staver, A. C., Midgley, G. F., & Bond, W. J. (2015b) Functional differentiation of biomes in an African savanna/forest mosaic. *South African Journal of Botany* **101**: 82–90.

Clarke, P. J., Lawes, M. J., Midgley, J. J., *et al.* (2013) Resprouting as a key functional trait: how buds, protection and resources drive persistence after fire. *New Phytologist* **197**: 19–35.

Dantas, V. D. L. & Pausas, J. G. (2013) The lanky and the corky: fire-escape strategies in savanna woody species. *Journal of Ecology* **101**: 1265–1272.

du Toit, J. T., Bryant, J. P., & Frisby, K. (1990) Regrowth and palatability of *Acacia* shoots following pruning by African savanna browsers. *Ecology* **71**: 149–154.

Dubois, M., Gilles, K. A., Hamilton, J. K., Rebers, P., & Smith, F. (1956) Colorimetric method for determination of sugars and related substances. *Analytical Chemistry* **28**: 350–356.

Eldridge, D. J., Bowker, M. A., Maestre, F. T., *et al.* (2011) Impacts of shrub encroachment on ecosystem structure and functioning: towards a global synthesis. *Ecological Letters* **14**: 709–722.

Gignoux, J., Clobert, J., & Menaut, J.-C. (1997) Alternative fire resistance strategies in savanna trees. *Oecologia* **110**: 576–583.

Grime, J. P. (1997) Biodiversity and ecosystem function: the debate deepens. *Science* **277**(5330) 1260–1261.

Hempson, G., Midgley, J. J., Lawes, M. J., Vickers, K. V., & Kruger, L. M. (2014) Comparing bark thickness: testing methods with bark-stem data from two South African fire-prone biomes. *Journal of Vegetation Science* **25**: 1247–1256.

Higgins, S. I., Bond, W. J., & Trollope, W. S. W. (2000) Fire, resprouting and variability: a recipe for grass–tree coexistence in savanna. *Journal of Ecology* **88**: 221–229.

Hodgkinson, K. C. (1998) Sprouting success of shrubs after fire: height-dependent relationships for different strategies. *Oecologia* **115**: 64–72.

Hoffmann, W. A., Jaconis, S. Y., McKinley, K. L., *et al.* (2012) Fuels or microclimate? Understanding the drivers of fire feedbacks at savanna–forest boundaries. *Austral Ecology* **37**: 634–643.

Keddy, P. A. (2007) *Plants and vegetation: origins, processes, consequences*. Cambridge University Press, Cambridge.

Keeley, J. E. & Zedler, P. H. (1998) Evolution of life histories in *Pinus*. In: *Ecology and biogeography of Pinus* (ed. D. M. Richardson), pp. 219–249. Cambridge University Press, Cambridge.

Kruger, L. M., Midgley, J. J., & Cowling, R. M. (1997) Resprouters versus reseeders in South African forest trees; a model based on canopy height. *Functional Ecology* **11**: 101–105.

Leishmann, M. R., Wright, I. A., Moles, A. T., & Westoby, M. (2000) The evolutionary ecology of seed size. In: *Seeds: the ecology of regeneration in plant communities* (ed. M. Fenner), pp. 31–57. CABI, Wallingford.

Loehle, C. (2000) Strategy space and the disturbance spectrum: a life-history model for tree species coexistence. *The American Naturalist* **156**: 14–33.

McGill, B. J., Enquist, B. J., Weiher, E., & Westoby, M. (2006) Rebuilding community ecology from functional traits. *Trends in Ecology and Evolution* **21**: 178–185.

McIntyre, S., Lavorel, S., Landsberg, J., & Forbes, T. D. A. (1999) Disturbance response in vegetation – towards a global perspective on functional traits. *Journal of Vegetation Science* **10**: 621–630.

Midgley, J. J. (1996) Why the world's vegetation is not totally dominated by resprouting plants. *Ecography* **19**: 92–95.

Midgley, J. J. & Bond, W. J. (2001) A synthesis of the demography of African acacias. *Journal of Tropical Ecology* **17**: 871–886.

Midgley, J. J., Botha, M. A., & Balfour, D. A. (2001) Patterns of thorn length, density, type and colour in African acacias. *African Journal of Range & Forage Science* **18**: 59–61.

Mucina, L. & Rutherford, M. C. (2006) *The vegetation of South Africa, Lesotho and Swaziland*. South African National Biodiversity Institute, Pretoria.

Noble, I. R. & Slatyer, R. O. (1980). The use of vital attributes to predict successional changes in plant communities subject to recurrent disturbances. In: *Succession*, pp. 5–21. Springer, Dordrecht.

Parr, C. L., Gray, E. F., & Bond, W. J. (2012) Cascading biodiversity and functional consequences of a global change-induced biome switch. *Diversity and Distributions* **18**: 493–503.

Pausas, J. G. (1999) Response of plant functional types to changes in the fire regime in Mediterranean ecosystems: a simulation approach. *Journal of Vegetation Science* **10**: 717–722.

Pellegrini, A. F. A., Hoffmann, W. A., & Franco, A. C. (2014) Carbon accumulation and nitrogen pool recovery during transitions from savanna to forest in central Brazil. *Ecology* **95**: 342–352.

Raunkiær, C. (1937) *Plant life forms*. The Clarendon Press, Oxford.

Reich, P. B., Walters, M. B., Ellsworth, D. S., *et al.* (1998) Relationship of leaf dark respiration to leaf nitrogen, specific leaf area, and leaf life-span: a test across biomes and functional groups. *Oecologia* **114**: 471–482.

Rusch, G. M., Pausas, J. G., & Lepš, J. (2003) Plant functional types in relation to disturbance and land use: introduction. *Journal of Vegetation Science* **14**: 307–310.

Sankaran, M., Hanan, N. P., Scholes, R. J., *et al.* (2005) Determinants of woody cover in African savannas. *Nature* **438**(7069): 846–849.

Scheffer, M. & Carpenter, S. R. (2003) Catastrophic regime shifts in ecosystems: linking theory to observation. *Trends in Ecology & Evolution* **18**: 648–656.

Schutz, A. E. N., Bond, W. J. & Cramer, M. D. (2009) Juggling carbon: allocation patterns of a dominant tree in a fire-prone savanna. *Oecologia* **160**: 235–246.

Staver, A. C., Bond, W. J., Stock, W. D., van Rensburg, S. J., & Waldram, M. S. (2009) Browsing and fire interact to suppress tree density in an African savanna. *Ecological Applications* **19**: 1909–1919.

Staver, A. C., Bond, W. J., Cramer, M. D., & Wakeling, J. L. (2012) Top-down determinants of niche structure and adaptation among African acacias. *Ecology Letters* **15**: 673–679.

Tilman, D. (1990) Constraints and tradeoffs: toward a predictive theory of competition and succession. *Oikos* **58**: 3–15.

Vesk, P. A. (2006) Plant size and resprouting ability: trading tolerance and avoidance of damage? *Journal of Ecology* **94**: 1027–1034.

Waldram, M. S., Bond, W. J., & Stock, W. D. (2008) Ecological engineering by a mega-grazer: white rhino impacts on a South African savanna. *Ecosystems* **11**: 101–112.

Warman, L. & Moles, A. T. (2009) Alternative stable states in Australia's wet tropics: a theoretical framework for the field data and a field-case for the theory. *Landscape Ecology* **24**: 1–13.

Wakeling, J. L., Staver, A. C., & Bond, W. J. (2011) Simply the best: the transition of savanna saplings to trees. *Oikos* **120**: 1448–1451.

Watson, H. K. (1995) Management implications of vegetation changes in Hluhluwe-Umfolozi Park. *South African Geographical Journal* **77**: 77–83.

Westoby, M. (1998) A leaf–height–seed (LHS) plant ecology strategy scheme. *Plant and Soil* **199**: 213–227.

Westoby, M., Leishman, M. R., & Lord, J. M. (1996) Comparative ecology of seed size and seed dispersal. *Philosophical Transactions of the Royal Society* **351**: 1309–1318.

Westoby, M., Falster, D. S., Moles, A. T., Vesk, P. A., & Wright, I. J. (2002) Plant ecological strategies: some leading dimensions of variation between species. *Annual Review of Ecology, Evolution and Systematics* **33**: 125–159.

Whateley, A. & Porter, R. N. (1983) The woody vegetation communities of the Hluhluwe–Corridor–Umfolozi Game Reserve Complex. *Bothalia* **14**: 745–758.

Wigley, B. J., Bond, W. J., & Hoffman, M. (2010) Thicket expansion in a South African savanna under divergent land use: local vs. global drivers? *Global Change Biology* **16**: 964–976.

Wigley, B. J., Fritz, H., Coetsee, C., & Bond, W. J. (2014) Herbivores shape woody plant communities in the Kruger National Park: lessons from three long-term exclosures. *Koedoe* 56: art. #1165.

Wilson, J. B. & Agnew, A. D. (1992) Positive-feedback switches in plant communities. *Advances in Ecological Research* **23**: 263–336.

Woodward, F. I., Lomas, M. R., & Kelly, C. K. (2004) Global climate and the distribution of plant biomes. *Philosophical Transactions of the Royal Society of London B*, **359**: 1465–1476.

9 · *Contributions of Smaller Fauna to Ecological Processes and Biodiversity*

NORMAN OWEN-SMITH, CLEO M. GOSLING, NICOLE HAGENAH, MARCUS J. BYRNE, AND CATHERINE L. PARR

9.1 Introduction

The Hluhluwe-iMfolozi Park (HiP) supports a huge diversity of smaller organisms besides the large mammals emphasized in other chapters. Besides 23 ungulates and seven large carnivores, the mammal species listed include four primates, 10 small carnivores, 16 rodents, 13 bats, seven shrews or moles, and two hares (Bourquin *et al.*, 1971). Over 400 bird species have been recorded, encompassing two-thirds of the species represented in KwaZulu-Natal province, among them 20 southern African endemics (Macdonald and Birkenstock, 1980). Within the thorn savanna alone, about 90 bird species were recorded weekly from either sightings or calls (Owen-Smith, 1980). The herpetofauna comprises minimally 59 reptiles and 26 amphibians (Bourquin *et al.*, 1971). The diversity of invertebrates present within various taxa has yet to be catalogued. The occurrence of some of the smaller mammals listed is based on only one or two records, and there have been no recent surveys of them. Furthermore, certain bird species may have shifted their distributions since the earlier records. For example, the grey go-away-bird (*Corythaixoides concolor*) is listed by Bourquin *et al.* (1971), but was not recorded during a three-year survey (1969–1971) in the savanna vegetation where they typically occur (Owen-Smith, 1980), although there have been occasional sightings in the region surrounding HiP (Macdonald and Birkenstock, 1980). Hence a comprehensive survey of faunal diversity cannot be provided in this chapter.

Conserving Africa's Mega-Diversity in the Anthropocene, ed. Joris P. G. M. Cromsigt, Sally Archibald and Norman Owen-Smith. Published by Cambridge University Press.

Nevertheless, there have been some notable studies on a few functional groups. These concerned (1) the effects of the vegetation changes induced by large herbivores on small rodents; (2) the dependence of particular bird species on the short grasslands promoted by large grazers, or associated with white rhinos; (3) the contribution of HiP towards conserving vultures and large raptors threatened by human impacts outside the protected area; (4) the role of termites as contributors to savanna heterogeneity; (5) the activities of ants as supreme gatherers of animal remains and seeds; (6) how dung beetles from HiP became candidates to cope with cow dung accumulations in Australia; and (7) the role of tsetse flies as vectors of parasites transmitted from wild ungulates to domestic livestock. This chapter outlines some of the main findings from these studies.

9.2 Rodents

Most rodents are primarily granivores consuming various plant seeds, although some species do feed on green plant tissues, and insects typically also form an important dietary component. Rodents serve as a food resource for many small carnivores, both mammalian and avian. Certain rodents are responsible for the spread of diseases such as typhus and plague via the fleas that they carry. Populations of small rodents may fluctuate widely in abundance, especially in temperate, boreal, and arctic regions where their densities can reach several hundred animals per hectare at times (e.g. Smit *et al.*, 2001; Bakker, 2003; Table 9.1). Repeated outbreaks of rodents are a feature of mid-latitudes in Europe, with peaks in rodent abundance associated with pulses of seed production by oaks and other forest trees (Jedrzejewski and Jedrzejewska, 1996). Within African savannas, the multimammate mouse (*Mastomys natalensis*; Leirs *et al.*, 1996a) and grass rat (*Arvicanthis niloticus*; Sinclair *et al.*, 2013; Byrom *et al.*, 2014) can attain very high densities during years of high rainfall. Wide fluctuations in rodent numbers have been recorded in savannas in parts of South America (Emmons, 2009). Spatial variation in rodent numbers can also be quite large (Korn, 1987).

Population densities of small rodents recorded in HiP are towards the low end of those reported for other African savannas (Bowland and Perrin, 1989; Hagenah, 2006; Hagenah *et al.*, 2009), but not much different from those found in several grassland habitats in North America (Table 9.1). An important question is the extent to which the abundance of rodents is limited by competition for food with large herbivores or restricted by the habitat modification brought about by large

Table 9.1. *Densities (number of animals per ha) of small mammals in various regions and range in these estimates over time or space*

			Large mammals				
			Present		Absent		
Group	Place	Habitat	Mean	Range	Mean	Range	Reference
Voles and mice	De Hoge Veluwe, Netherlands	Woodlands and heathland	261	169–353	278	305–332	Smit *et al.*, 2001
Voles	Junner Koeland, Netherlands	Floodplain grassland	115	54–201	302	196–406	Bakker, 2003
Voles and mice	Bialowieza, Poland	Deciduous forest	78	11–315			Jedrzejewski and Jedrzejewska, 1998
Rodents	Portal, Arizona	Chihuahuan desert	23				Brown and Zeng, 1989
Murid rodent	Yukon, Canada	Boreal forest	20	4–43			Galindo and Krebs, 1987
Small mammals	Lawrence, Kansas	Old field	19	5–48			Schweiger *et al.*, 2000
Small mammals	Illinois, USA	Restored tallgrass prairie	14	2–40			Yunger, 2004
Small mammals	Oklahoma	Tall grassland	7		45	30–60	Yunger, 2004
Small mammals	Colorado, Montana, and Washington	Various grasslands	5	3–9	5.5	2–8	Grant and Birney, 1979; Grant *et al.*, 1982
Rodents	Rwenzori, Uganda	Grassland	40	17–63			Cheeseman and Delaney, 1979
Small mammals	Mpala, Kenya	Savanna	37	5–54	57	70–107	Keesing, 2000
Small rodents	Nylsvley, South Africa	Savanna	27	11–41			Korn, 1987
Small mammals	S. Turkana, Kenya	Arid savanna	22				
Rodents	Swaziland	Grassland			22	14–28	Monadjem and Perrin, 2003
Rodents	Kenya	Moist grassland	16.5				Oguge, 1995
Rodents	Kalahari	Arid savanna	15	10–20			Blaum *et al.*, 2007
Rodents	West Africa	Savanna	15	8–21			Bellier, 1967
Small mammals	Mfolozi, South Africa	Savanna	8	5–10			Bowland and Perrin, 1989
Murid rodents	Hluhluwe-iMfolozi, South Africa	Savanna	4	0–12	25	5–47	Hagenah *et al.*, 2009
Small mammals	Mlawula, Swaziland	*Acacia* woodland	2.8	0.02–11			Mahlaba and Perrin, 2003

No fence	Rhino fence	Zebra fence	Impala fence	Hare fence
White rhino	-	-	-	-
Buffalo	Buffalo	-	-	-
Zebra	Zebra	-	-	-
Warthog	Warthog	Warthog	-	-
Impala	Impala	Impala	-	-
Nyala	Nyala	Nyala	-	-
Duiker	Duiker	Duiker	Duiker	-
Hares	Hares	Hares	Hares	-
Rodents	Rodents	Rodents	Rodents	Rodents

Figure 9.1 The 'Russian Doll' experiment, successively excluding herbivores differing in size by height and mesh diameter of fencing (from Hagenah, 2006).

ungulate grazers. In central Kenya, small mammals exhibited substantially greater abundance when cattle and other large herbivores were excluded by fencing (Keesing, 1998, 2000). Although this was interpreted as release from competition with large herbivores for green food of high nutritional value, this response was manifested largely by increased numbers of the pouched mouse *Saccostomys campenstris*, a small murid that is typically seed-eating. Possibly, grasses not eaten by large herbivores produced more seeds. A similar study conducted in Uganda also found an increase in small mammal abundance following the exclusion of large herbivores. However, it was not established whether this was due to release from competition for high-quality food, as suggested by Monadjem and Perrin (1996, 1998), or a response to lowered predation due to increased vegetation cover (Leirs *et al.*, 1996b), or some combination of both (Okullo *et al.*, 2013).

To investigate relationships between the grazing impacts of the large herbivores and rodent abundance, an elaborate 'Russian doll' experiment was set up in HiP (Hagenah *et al.*, 2009). Herbivores of diminishing size were successively excluded by reducing the height and mesh size of fencing. The final interior fence allowed only small mammals to penetrate (Figure 9.1). Findings showed that in the centre where all larger herbivores were excluded, rodents attained much higher densities, up to 50 animals per hectare, than they exhibited outside the experiment where the grass was kept short by grazing. This increase was mainly by the single-striped mouse *Lemniscomys rosalia*, which is largely diurnal and feeds primarily on grass leaves rather than seeds. Other rodent species captured more frequently where larger herbivores were excluded were the Natal multimammate mouse (drier sites in Mfolozi only), pouched mouse (wetter sites in Hluhluwe only), and bush rat (*Aethomys* spp.). The varied diets of these rodents indicates that their increased abundance was due to the greater height of the grass, and hence better cover provided against avian predators, rather than to changes in food abundance or quality. Radio-tracking showed that grass height was indeed the main factor influencing the habitat selection of single-striped mice (Hagenah, 2006). An earlier study undertaken in the Mfolozi section likewise found that plant cover was the main influence on spatial variation in rodent abundance (Bowland and Perrin, 1989). Trapping success was very low where heavy grazing had severely reduced the grass cover in a drought year, but higher where some grass remained.

Periodic rodent outbreaks comparable to those of grass rats in Serengeti (Sinclair *et al.*, 2013) have not been recorded in HiP. Their occurrence in Serengeti may be a result of the seasonal exodus of large grazers from the short-grass plains, leaving grass high in nutrients but low in biomass behind. In Serengeti, rodent outbreaks are associated with greatly increased numbers of black-shouldered kites (*Elanus caeruleus*), a small raptor that is a rodent specialist. The low rodent densities in HiP may be a consequence of the high grazing pressure by white rhinos and other ungulates, exposing rodents to high predation by the numerous small raptors and owls that occur there. One of the chapter authors (NO-S) noted the low abundance of snakes in the Mfolozi region – about one sighting every two months in an area walked almost daily during the late 1960s. This might also be a reflection of severe predation by raptors on snakes. The low abundance of rodents could possibly limit populations of small carnivores, including black-backed jackals (*Canis mesomelas*, formerly abundant but now seemingly extinct in HiP; see Chapter 12), honey badgers (*Mellivora capensis*, rarely sighted and possibly locally extinct), and large-spotted genets (*Genetta tigrina*).

9.3 Birds

HiP supports a wide diversity of bird species through the diversity of habitats present (Macdonald and Birkenstock, 1980; Owen-Smith, 1980; Table 9.2). Particularly notable is the conspicuousness of migrants from Europe during the wet season. Among the species most commonly recorded

Table 9.2. *The 20 most commonly recorded bird species in Mfolozi thorn savanna in order of their sighting frequency index during 1969–1971 (from Owen-Smith, 1980)*

Common name	Scientific name
Year-round:	
Cape turtle dove	*Streptopelia capicola*
Rattling cisticola	*Cisticola chiniana*
Red-billed oxpecker	*Buphagus erythrorhynchus*
Golden-breasted bunting	*Emberiza flaviventris*
Common bulbul	*Pycnonotus barbatus*
Blue waxbill	*Uraeginthus angolensis*
Cape starling	*Lamprotornis nitens*
White-backed vulture	*Gyps africanus*
Hadedah ibis	*Bostrychia hagedash*
Emerald-spotted dove	*Turtur chalcospilos*
Pied crow	*Corvus albus*
Bushveld pipit	*Anthus caffer*
Chinspot batis	*Batis molitor*
Helmeted guineafowl	*Numida meleagris*
Southern yellow-billed hornbill	*Tockus leucomelas*
Summer only:	
Red-backed shrike	*Lanius collurio*
Spotted flycatcher	*Muscicapa striata*
Yellow-billed kite	*Milvus aegyptius*
Barn swallow	*Hirundo rustica*
European roller	*Coracias garrulus*

at that time of the year were the spotted flycatcher (*Muscicapa striata*), red-backed shrike (*Lanius collurio*) and European roller (*Coracias garrulus*), along with barn swallows (*Hirundo rustica*), all of which greatly outnumbered their resident congeners (Owen-Smith, 1980). However, the only detailed study of habitat relationships among the avifauna was directed at the effects of the grassland changes brought about by white rhinos and other grazers on bird species associated with grasslands (Krook *et al.*, 2007; see also Chapter 6). Findings showed that three bird species are narrowly restricted to short grasslands, although able to use bunch grasslands that have been recently burned as an alternative habitat (Table 9.3). These are African pipit *Anthus cinnamomeus*, crowned lapwing *Vanellus coronatus*, and sabota lark *Calendulauda sabota*. In order to support a viable local breeding group (i.e. several pairs) of all three species, a minimum patch area of ~8 ha was needed. Two other species were associated with mosaics of short

Table 9.3. *Association of bird species that are grassland specialists with particular grassland habitats (from Krook* et al., *2007)*

Grassland type	Common name	Scientific name	Local density (birds/ha)
Short grass	African pipit	*Anthus cinnamomeus*	1.18
	Crowned plover	*Vanellus coronatus*	0.31
	Sabota lark	*Calendulauda sabota*	0.23
Mixed tall and short grass	Yellow-throated longclaw	*Macronyx croceus*	0.63
	Rufous-naped lark	*Mirafra africana*	0.25
Tall grass	Zitting cisticola	*Cisticola juncidis*	1.46
	Croaking cisticola	*C. natalensis*	1.26
	Fan-tailed widowbird	*Euplectes axillaris*	0.35
	Red-collared widowbird	*E. ardens*	0.26
	White-winged widowbird	*E. albonotatus*	0.16
	African stonechat	*Saxicola torquatus*	0.12
	Common buttonquail	*Turnix sylvaticus*	0.10

and tall grasses: yellow-throated longclaw *Macronyx croceus* and rufous-naped lark *Mirafra africana*. However, seven grassland specialists occurred only in tall bunch grasslands and could be threatened by the widespread conversion of these grasslands into grazing lawns (Table 9.3).

Short grasslands are not restricted to the protected area, because structurally similar short grass is generated by heavy grazing by domestic livestock in communal rangelands surrounding HiP. However, investigations extended to these regions found that local densities of these short-grass specialists as well as other bird species were greatly depressed because of the killing of birds for food by the rural people (Krook *et al.*, 2007). From questionnaires, it was estimated that each person consumed on average 50 birds per year. These findings emphasize how crucial the protected area is not only for the conservation of large mammals, but also for many other species adversely affected by the dense human settlements that have developed in the surrounding region.

A bird species characteristically associated particularly with white rhinos is the red-billed oxpecker (*Buphagus erythrorynchus*). It has become confined mostly to protected areas because of the poisons used to control ticks in cattle ranches. These birds settle on various ungulates to feed on ticks and nibble at wounds, but favour particularly white rhinos and buffalos because of the large area of hairless hide on these two hosts (Stutterheim, 1980). A mutualistic relationship seems to have evolved between white rhinos and oxpeckers. Birds that are settled on rhinos give strident churring calls when they detect the nearby presence of a human, before flying off. White rhinos react to this call by shuffling about apprehensively, seeking the source of the birds' alarm. Because of their poor eyesight, rhinos cannot identify an immobile human intruder beyond a range of about 20–30 m. If the person moves, the rhinos immediately run off. What is the benefit to these birds of warning rhinos of a nearby human? Their strident call is perhaps a flock rallying signal indicating that their host is about to run off. However, this signalling relationship is not obvious between oxpeckers and other ungulate hosts (Owen-Smith, 1973). The deep evolutionary fixation of the relationship is evident from the fact that rhinos continue to respond with alarm to the calls of oxpeckers, even though rhinos have not been hunted for many decades, beyond the memory of the individuals currently alive.

Pied crows (*Corvus albus*) also commonly settle on white rhinos, particularly while the latter are lying down sleeping. They walk around picking off food morsels, probably ticks. Rhinos show no response to the croaking calls of crows. Cape starlings (*Lamprocoleus nitens*) frequently walk near the feet of grazing rhinos, looking out for insects flushed by the rhinos. Cattle egrets (*Bubalis ibis*) typically associate with large ungulates in this same way, but are uncommon in HiP.

Other birds strongly dependent on the protected area for their regional persistence include vultures and large eagles. Rated as especially endangered are the white-backed vulture *Gyps africanus*, lappet-faced vulture *Torgos tracheliotis*, and white-headed vulture *Trigonoceps occipitalis*, while martial eagle *Polemaetus bellicosus*, tawny eagle *Aquila rapax*, and bateleur eagle *Terathopius ecaudatus* are regarded as vulnerable. All of these species breed within the confines of HiP (Table 9.4, Figure 9.2). Breeding populations of these species within HiP are quite small, and outside HiP they are found in only a few other protected areas in northern KwaZulu-Natal. The reason is not merely the paucity of ungulate carcases in land settled by humans, but because vultures are commonly killed by rural people to obtain their body parts for traditional remedies traded in distant cities. Vultures also get killed through feeding on poisoned baits placed to control predators on commercial livestock ranches. However, while vultures seemed to be maintaining their breeding populations within HiP, the declining trend in nests of various eagles recorded is cause for concern (Table 9.4).

Another large bird potentially dependent on the protected area for security is the southern ground hornbill *Bucorvus leadbeateri*. Only three

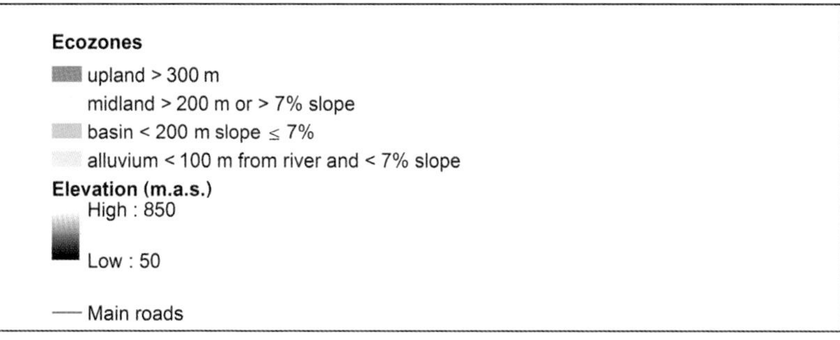

Figure 2.2 A three-dimensional view of model of Hluhluwe-iMfolozi Park showing the regional setting with respect to elevation in detail, and the main elevation-based landscape zones.

Figure 2.3 Rolling hills with mosaic of grassland, thicket, and forest in northern Hluhluwe (photo: Norman Owen-Smith).

Figure 2.4 Upland 'scarp forest' alternated with fire-dominated grassland patches in the northern part of the park (photo: Norman Owen-Smith).

Figure 2.5 Rolling fire-dominated upland grassy hills in the central (corridor) part of the park, with trees only occurring along the drainages (photo: Han Olff).

Figure 2.6 Basin of the White Mfolozi river in the southern part of the park looking towards the hills on the western boundary (photo: Norman Owen-Smith, taken in 1970).

Figure 2.7 Knob thorn savanna in Mfolozi (photo: Norman Owen-Smith).

Figure 2.8 Umbrella thorn savanna in western Mfolozi with grazing lawn grassland (photo: Norman Owen-Smith).

Figure 2.9 Black Mfolozi river. Much of the riverine woodland was removed by floods following cyclone Demoina in February 1984 (photo: Norman Owen-Smith).

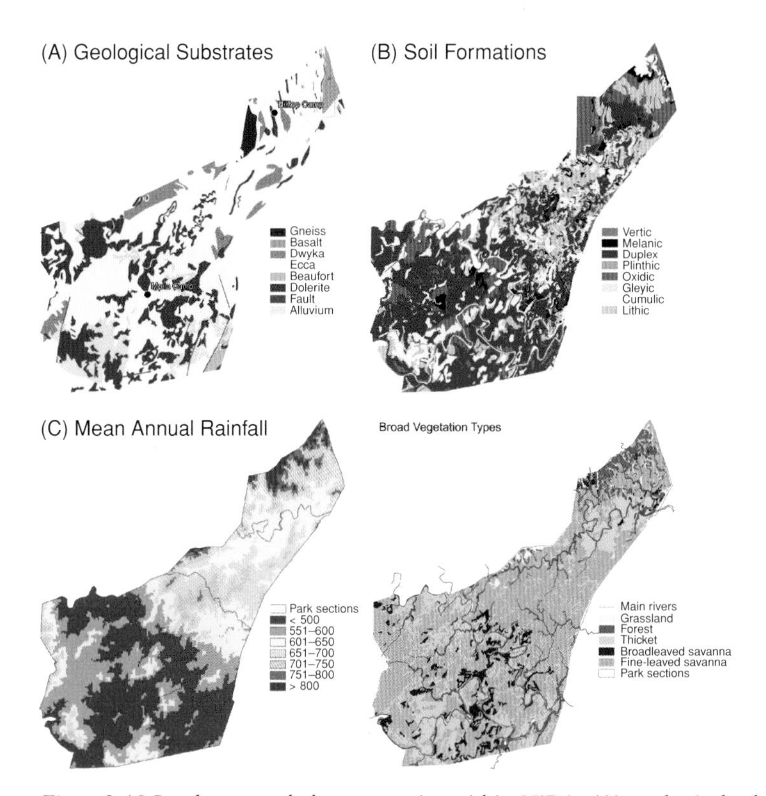

Figure 2.10 Landscape-scale heterogeneity within HiP in (A) geological substrates, (B) soil formations, (C) mean annual rainfall (from kriging of records from 17 recording stations during 2001–2007), and (D) broad vegetation types (adapted from Whateley and Porter, 1983 – see Chapter 8 for a more detailed classification of this map).

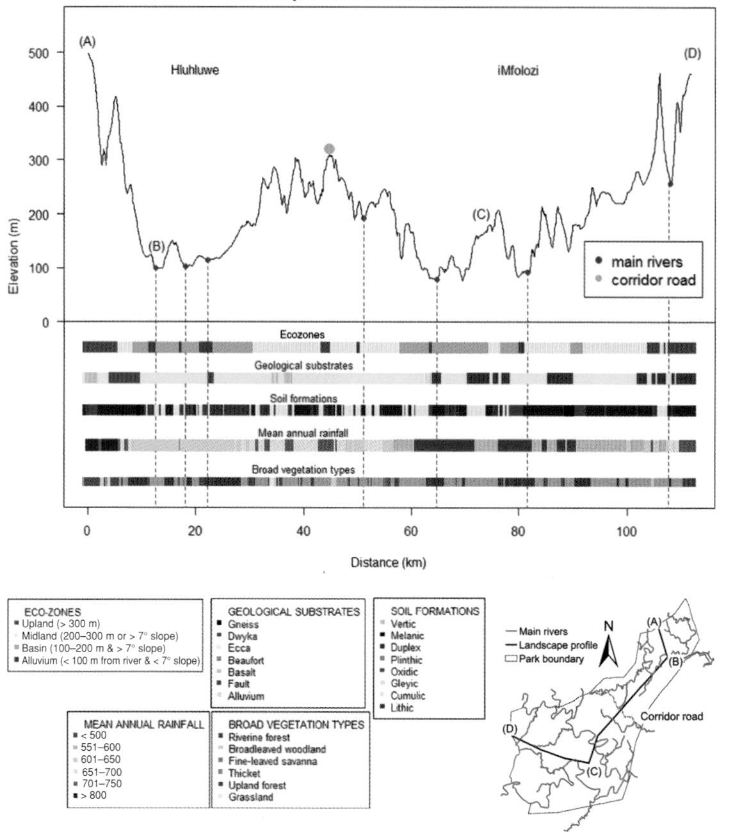

Figure 2.13 Landscape profile of Hluhluwe-iMfolozi Park, showing the general undulating profile from high elevation in the north through the basin of the Hluhluwe River, rising again through the central Corridor and through to the undulating basin of the Black and White Mfolozi Rivers, rising again in the south to an elevation almost equal to that of the upland forests in the north. The five coloured bars beneath the graph show the proportion of the landscape within the four ecozones and the different geological substrates, soil formations, mean annual rainfall regimes, and the broad vegetation types.

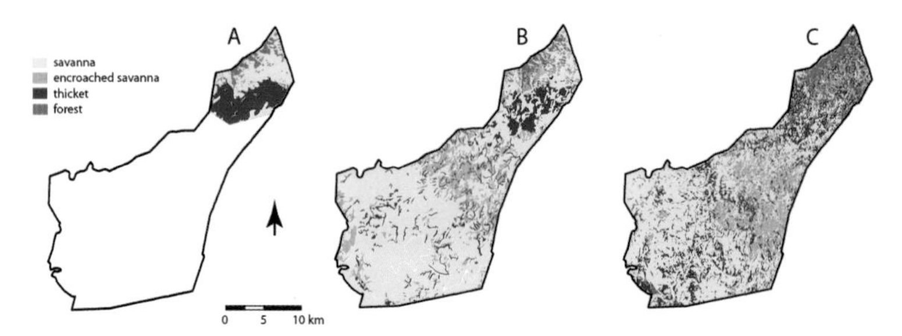

Figure 3.4 Vegetation maps for (A) 1937 (redrawn from Henkel, 1937), (B) 1979 (redrawn from data collected by Whateley and Porter, 1983), and (C) 2001 (redrawn from data in Dora, 2004).

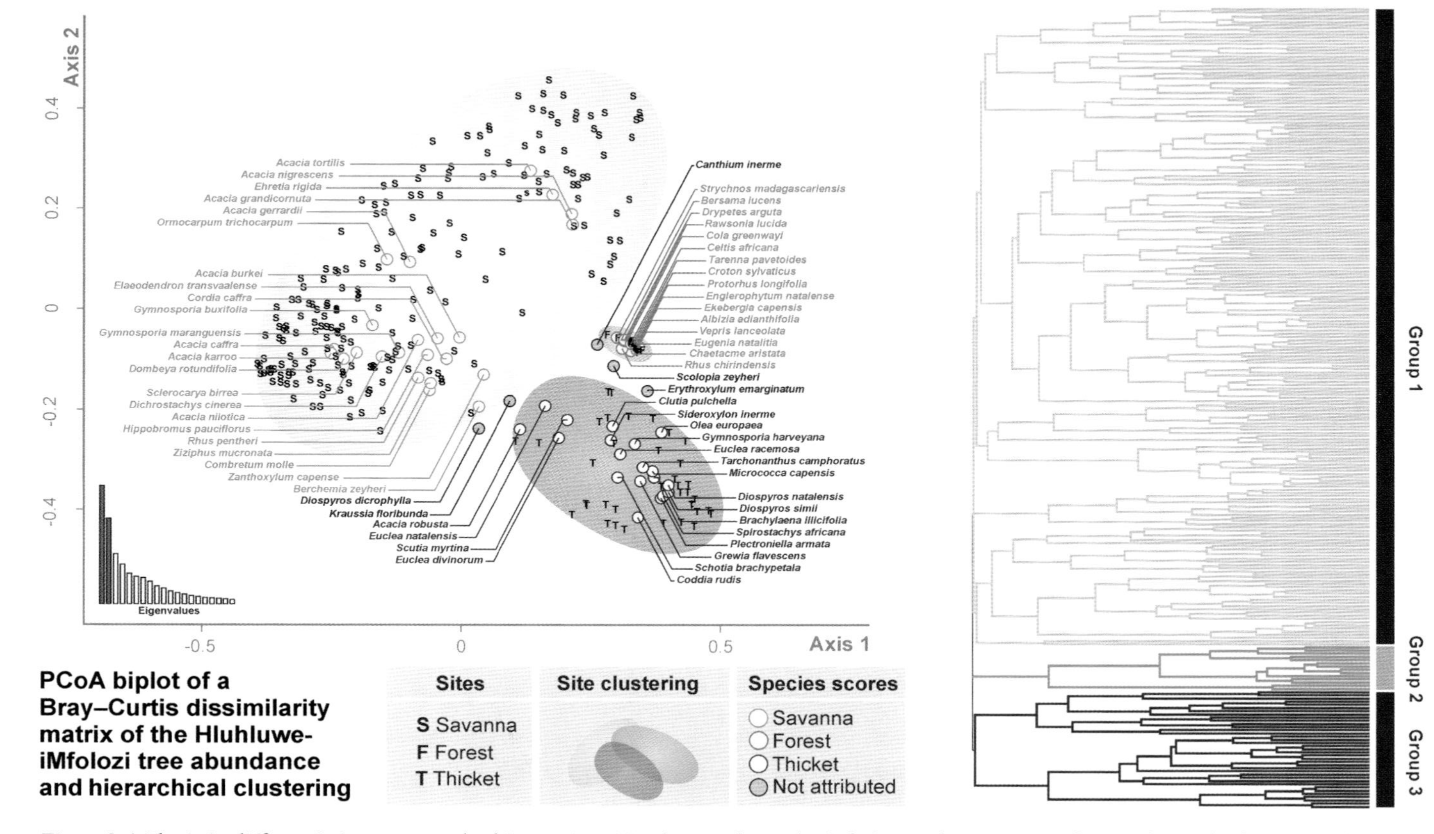

Figure 8.1 Floristic differentiation among the biomes in HiP. The graph on the left shows the position of sites (shown by letters) and species (circles) on the two first axes of the Principal Coordinate Analysis. The three envelopes correspond to the grouping made by hierarchical clustering. Each species has been attributed to one of the three biomes (see colour code) depending on its score. (Reprinted with permission from Elsevier; Charles-Dominique *et al.*, 2015b.)

Figure 15.1 (A) *Chromolaena odorata* infestation in a *Euclea racemosa* woodland in Maphumulo area of northern Hluhluwe showing the dense tangle of stems. The path through the Chromolaena monoculture was created by elephants (June 2003). (B) Chromolaena infestation in the same area during the wet season (March 2005). (C) Chromolaena infestation of the south bank of the Hluhluwe river opposite Maphumulo picnic site, preventing access for buffalo (August 2004). (D) Detail of Chromolaena plant (March 2005). (E) Cut-stump clearing treatment (August 2003). (F) Chromolaena stem residues remaining after clearing operation (August 2003). Photos: Mariska te Beest.

Table 9.4. *Number of breeding pairs or nests of vultures and large eagles recorded in HiP during annual aerial surveys*

Species	Period 1972–1980	2009	2010	2011	2012
White-backed vulture	106	398	392	381	366
Lappet-faced vulture	15	9	10	12	12
White-headed vulture	9	5	5	4	4
Wahlberg's eagle	20	1		2	1
Tawny eagle	10	1	2	3	3
Crowned eagle	9	1	2	1	
Bateleur	5				
Martial eagle	4				6
Fish eagle		2	1		2
Secretary bird	1			1	

family groups occur within HiP, but these also spend some time outside the park foraging on short grazed areas in communal rangelands. They are not generally killed by rural people, but their actual status in communal rangelands remains unknown. Their occurrence is threatened by the conversion of cattle farms into wildlife ranches, leading to a herbaceous

Figure 9.2 White-headed vulture on nest (photo: Dave Druce).

cover that is too tall and dense for the foraging method of these hornbills (Lucy Kemp, project@ground-hornbill.org.sa).

9.4 Insects

Numerous kinds of insects and other invertebrates occur within HiP. Many of these are clearly important ecologically, like the bees, beetles and flies that serve as pollinators, spiders that serve as major predators on insects, and grasshoppers of various forms acting as consumers of vegetation. In this section we consider merely the four groups of insects for which particular studies have contributed insights into their contributions to ecological functions in HiP.

9.4.1 Termites

Termites (Blattodea: Termitoidae) function as litter-feeders or detritivores in savanna vegetation. They are leading agents in the decomposition of plant litter in the form of leaves and wood, as well as dung deposited by large herbivores. In some places, the termite biomass outweighs the total biomass of all large herbivores combined (Ferrar, 1982; Deshmukh, 1989). Termites not only concentrate mineral nutrients in their nests (Jouquet *et al.*, 2005), but also alter soil properties by bringing clay fractions to the surface in the mounds that they construct (Bagine, 1984). Their activities contribute importantly to vegetation heterogeneity by locally excluding fire due to the lack of sufficient fuel on the mounds (Moe *et al.*, 2009; Okullo and Moe, 2012; Davies *et al.*, 2014; Joseph *et al.*, 2014). The short grasses associated with mounds provide nutritious forage that attracts grazing by white rhinos and other large herbivores (Loveridge and Moe, 2004; Mobaek *et al.*, 2005; Grant and Scholes, 2006; Chapter 6). Harvester termites may locally compete with large herbivores for remaining grass during the dry season, especially in drought years (Walker *et al.*, 1987; Moe *et al.*, 2009).

Some termites develop conspicuous mounds, while colonies of other species are obscurely hidden underground. Among the mound-builders, members of the subfamily Macrotermitinae, notable for their symbiotic relationship with fungi (*Termitomyces*), are most well studied in Africa because of their widespread occurrence and especially large mounds (Uys, 2002). Other mound-building termites also contribute to ecosystem pattern and function. Within HiP, the presence of 15 termite species representing two families, four subfamilies and 12 genera has been confirmed

Table 9.5. *Comparison of mounds formed by members of different subfamilies of Termitidae in terms of mean values for diameter, height, density, and proportion of the park area affected (from Gosling, 2014)*

Feature	Diameter (m)	Height (m)	Densities (per ha)	Area affected (% of HiP)
Macrotermes	4.18	1.21	1.25	1.18
Odontotermes	3.61	0.47	0.37	0.39
Trinervitermes	0.86	0.35	1.45	0.09

(Davies *et al.*, 2013). A study by Gosling (2014) addressed the contributions by different mound-building termites in the family Termitidae to savanna heterogeneity. Two subfamilies within the Termitidae cultivate fungus gardens in their nests and depend on the fungi to break down the lignocellulose into the products that they consume, while members of a third subfamily are not fungus-growing.

Members of the subfamily Macrotermitinae are the dominant litter-feeders in abundance, number of species, and extent of the area affected by their mounds in HiP (Gosling, 2014). Common mound-builders in the group include *Macrotermes natalensis* and *Odontotermes* spp. Mounds built by *M. natalensis* are the largest and most conspicuous (Table 9.5, Figure 9.3A). Growing on them are distinct communities of trees, especially species with broad, evergreen leaves that are less resistant to fire than the acacias prevalent in the surrounding savanna matrix (Van der Plas *et al.*, 2013; see also Chapter 8). Trees growing on mounds attract more browsing than those in the surrounding savanna woodland elsewhere in Africa (Loveridge and Moe, 2004; Mobaek *et al.*, 2005), but this pattern was not found in HiP (Gosling, 2014). The regional distinction is probably due to the nutritious foliage offered by the various acacia trees prevalent in HiP.

Species within the genus *Odontotermes* occurring in HiP could not be identified at species level because the genus is undergoing revision. Sampling indicated the presence of two or three distinct morphotypes (Davies *et al.*, 2013; Gosling, 2014). Mounds built by the termites in this genus are low in elevation, bare of vegetation, and are often large in diameter (Table 9.5, Figure 9.3B). A ring of lawn-forming grasses surrounds the bare central area. The density of these mounds in HiP is lower than those of *M. natalensis*. Mounds of species within these two genera tend to be

Figure 9.3 Comparison of mounds formed by termites in different subfamilies of the Termitidae. (A) *Macrotermes* (photo: Cleo Graf); (B) *Odontotermes* (photo: Jan Graf); (C) *Trinervitermes* (photo: Jan Graf).

negatively associated, perhaps because members of *Odontotermes* seem to utilize primarily grass litter (Jouquet *et al.*, 2004), while species within *Macrotermes* gather woody plant parts as well as other material. Nutrient levels in soils associated with *Odontotermes* mounds are less elevated than those around *M. natalensis* mounds, because less clay is needed to stabilize the lower structures (Gosling *et al.*, 2012).

The most numerous mounds in HiP are those built by *Trinervitermes trinervoides* in the subfamily Nasutitermitinae, but their diminutive size means that the area of HiP affected is quite small (Gosling, 2014; Table 9.5, Figure 9.3C). These termites are not fungus-growing, and their small mounds serve purely as a storage facility for grass. Although not enriched in micronutrients, mound soils have higher elevations of N and P compared with surrounding soils than those associated with mounds of termites in other subfamilies (Gosling *et al.*, 2012). Relationships between these various mounds and the formation of grazing lawns are discussed in Chapter 6.

Also abundant in HiP are harvester termites (*Hodotermes mossambicus*) in the family Hodotermitidae, which do not build mounds. Their distribution appears to be concentrated in the drier south-western region of the park. They can remove almost all grass over quite extensive areas, especially during years with low rainfall when they compete with mammalian herbivores for this food. However, no detailed studies on their activities and influence have been carried out in HiP.

Termites make a crucial contribution to the ecology of HiP by contributing along with fire and grazing by large herbivores to spatiotemporal heterogeneity in vegetation and grazing. The functional contributions by species besides the mound-builders have yet to be studied.

9.4.2 Ants

Ants (Hymenoptera: Formicidae) are diverse and abundant in the savannas and grasslands of Africa, including HiP. They perform a range of ecological functions from predators to scavengers to seed harvesters. Ant diversity in HiP is comparable with, if not greater than in, other savannas, with a total of 118 species and a mean of 27 species/0.25 ha plot recorded in open savanna and thicket (Parr *et al.*, 2012). This is similar to the 111 species and a mean of 28 species/0.25 ha plot collected in the Satara area of Kruger National Park with much greater sampling effort (Parr *et al.*, 2004). The most specious genera in these African savannas are *Tetramorium*, *Pheidole*, and *Monomorium*.

Ants are the dominant surface-active invertebrates in savannas; in HiP on average 70% of all ground active invertebrates are ants, with > 80% in some places. The dominance of ants in terms of numbers and biomass means they are likely to have significant effects on the structure and functioning of the ecosystem. Indeed, work from savannas elsewhere in South Africa has demonstrated they are important predators of termites and therefore indirectly regulate key processes such as decomposition (C. L. Parr and P. Eggleton, unpublished). Ants also respond to changes in habitat structure brought about by large mammal grazers. While a reduction in grass height and structural complexity does not affect species richness, it can significantly alter the composition of ant assemblages. Grazing lawns are dominated by the fast-moving, aggressive pugnacious ant (*Anoplolepis custodiens*) because it favours areas with abundant bare ground.

Ant assemblages differ strongly between open savanna and thicket habitats in Hluhluwe. In particular, the development of a well-established litter layer in thickets provides a critical habitat for cryptic leaf-litter ants, including *Amblyopone* sp. (undescribed), *Microdaceton exornatum*, *Prionopelta aethiopica*, and *Pristomyrmex cribrarius* (Parr *et al.*, 2012). There are also differences in functional representation in the ant assemblages. In thicket habitats, 80% of all individual ants and 44% of species were classified as predatory (e.g. *Pyramica*, *Leptogenys*, *Hypoponera*), presumably in response to the availability of prey such as isopods. This contrasts with only 6% and 8%, respectively, in savanna habitats (Parr *et al.*, 2012).

These findings provide only a narrow glimpse into the effects of changing savanna heterogeneity on the composition and functional diversity of ant assemblages, but do confirm their ecological importance.

9.4.3 Dung Beetles

Dung beetles (Coleoptera: Scarabaeoidae) are especially abundant and diverse in HiP, reflecting the numerous large herbivores present and variety of sizes and shapes of the dung piles that these animals produce. These beetles play an important role in burying the dung deposited by various vertebrates, thereby contributing to recycling the nutrients in this material. Dung contains not only undigested plant remains, but also gut microbes involved in degrading plant tissues plus cells lost from the gut wall and residues of the enzymes added. Large scarabeids are associated particularly with the dung boluses produced by white rhinos, and concentrations of these beetles and their buried larvae around rhino middens attract the attention of guinea fowl (*Numida meleagris*), spurfowl

(*Francolinus* spp.), yellow-billed hornbill (*Tockus leucomelas*), pied crow (*Corvus albus*), and banded mongoose (*Mungos mungo*).

Given its richness of dung beetles, it was appropriate that HiP was chosen as the field site for an investigation of potential candidates for introduction into Australia, where the dung pats produced by cattle were not being buried by beetles indigenous to Australia, which are adapted to handle merely kangaroo pellets. African beetles better able to deal with large dung pats were sought, the aim being to control the flies breeding in cattle dung. In 1970 the Dung Beetle Research Unit (DBRU) was established with HiP as its field station, and in 1983 a small building was constructed to replace the caravan that had previously housed dung beetle researchers. This building was later taken over as student accommodation for a wider range of studies, and remains affectionately known as 'Dung Beetle', as a legacy of this pioneering research programme.

Dung beetles in the subfamily Scarabaeinae, which sequester dung in discrete packages as larval food, formed the core of the research. These beetles remove dung either by rolling it away in balls, or by tunnelling directly beneath the pat. The buried dung can serve as food for adult beetles or to provision larval development. Investigations revealed different foraging strategies among the tunnellers: some rapidly buried dung within 24–48 h (*Copris* sp.), while others took many days to remove a complete dung pat (*Onitis* sp.; Giller and Doube, 1989).

The African buffalo fly *Haematobia thirouxi potans* was used as a surrogate for the Australian fly pest *H. irritans exigua*. In HiP, *H. thirouxi potans* larvae incurred 95% mortality in dung pats, dependent on soil type and associated beetles (Fay and Doube, 1987). This was significantly greater that the 67% mortality of *H. irritans exigua* measured in Australia (Doube *et al.*, 1988). These findings encouraged attempts to identify which members of the predatory dung beetle fauna were killing the flies (Davis *et al.*, 1988), shifting attention to three other families of dung beetles. These were the Histeridae, the Hydrophilidae (which are coprophagus as adults but predaceous as larvae), and the Staphylinidae (which are predators as adults and larvae, although some species in the Aleocharinae are pupal parasitoids of flies during the beetle's larval stage; Wright *et al.*, 1989). Five species of histerid were introduced into Australia, and three became established, although these were not particularly successful in controlling flies (Edwards, 2007).

Overall, 23 species of dung beetle were successfully established in Australia, among them nine species present in HiP, from where the parental stocks may have come (P. B. Edwards, personal communication). Some of the introduced species are now so abundant that they have common names in Australia, and contribute to removing a substantial portion of the dung dropped by cattle (Ridsdill-Smith and Edwards, 2011). This has improved pasture quality across much of Australia and all but eradicated the bush fly from high rainfall areas of southern Australia. Unfortunately, there has been little impact on the buffalo fly, which can tolerate moderate dung beetle activity (Doube, 2014).

9.4.4 Tsetse Flies

Tsetse flies (Diptera: *Glossina* spp.) played a major role in the ecological history of Zululand, first positively by restricting places of settlement of people with their livestock because of the parasites that these flies transmitted to cattle, and then negatively via mounting pressure from white settlers to eliminate the wildlife that served as the prime hosts of these flies (Chapter 1). Ultimately, *Glossina pallidipes*, the species serving as the main vectors of nagana, was completely eradicated within the area that become consolidated as HiP by aerial spraying with insecticides, although forest-dwelling species still survived in the surrounding region.

Ground-breaking research conducted near HiP established that the cattle disease locally called nagana was caused by a unicellular parasite (*Trypanosoma* spp.) transmitted between hosts by tsetse flies. Tsetse flies disappeared from the lowveld region of eastern South Africa following the decimation of wildlife by hunting coupled with the rinderpest epizootic towards the end of the nineteenth century. The main hosts of these flies had been buffalo, eland, kudu, and wildebeest. Thereafter, tsetse flies remained within the borders of South Africa only in parts of low-lying Zululand. This led to the pressure to deproclaim the game reserves and shoot the wildlife (see Chapter 1). Despite the shooting campaigns, large numbers of the more cryptic ungulates, particularly grey duiker and bushbuck, remained as potential hosts of the fly, leading to attempts to find an alternative solution to the nagana problem.

The person who took up this challenge was R. H. T. Harris, employed initially by the veterinary authority but later by the province of Natal. Harris established that tsetse flies detected ungulate hosts by sight rather than smell. This led him to develop the Harris fly trap in the form of a dark box with a gauze cover. Flies attracted to this shape entered via the opening below and became trapped in the gauze while attempting to fly towards the light above. By killing the flies captured in this way, the abundance of tsetse flies was greatly reduced by the traps. Harris also established the dependence of tsetse flies on shady places for resting, explaining their absence from regions with a low tree or bush cover. Another feature of biology that contributed to the later eradication of the flies was their slow reproduction, with just a single egg laid, indicating their dependence on the secure food supply provided by the blood of their ungulate hosts.

Once it became recognized that the fly traps could reduce but not eliminate tsetse flies, a more drastic solution was implemented: spraying of most of the HiP region with insecticides, initially DDT but later BHC, from aircraft. In order to ensure that the flies did not disperse beyond the area sprayed, a 3.2-km wide strip was cleared of all woody vegetation along the western boundary of the Umfolozi Game Reserve, and other regions of dense bush hosting pockets of tsetse were also cleared of trees and shrubs. This spraying operation succeeded in extirpating tsetse flies within the game reserves. The cost in the form of other organisms killed by the insecticides applied was not assessed.

Two forest-dwelling species (*G. brevipalpis* and *G. austeni*) survived in isolated pockets near HiP. *G. brevipalpis* is quite widely distributed through both dense indigenous forest and savanna habitats, while *G. austeni* is restricted to dense forests where its preferred hosts are bushpigs and duikers. These two fly species still remain in dense woody habitats in Hluhluwe (Gillingwater *et al.*, 2010). In 1990, a widespread outbreak of nagana occurred among cattle in areas surrounding the Hluhluwe section of HiP (van den Bossche *et al.*, 2006). The outbreak was brought under control by treating the infected cattle. Nevertheless, infections with the blood parasite *Trypanosoma congolense* remain a threat to livestock in communal rangelands adjoining the protected area, although not to wild ungulates within the park (see Chapter 13).

9.5 Summary Overview

Among the invertebrates present in HiP, the two groups of social insects, ants and termites, are most prominent in abundance and in the diverse range of ecological functions that they serve – from agents of decomposition to predators on other invertebrates, soil engineers, and nutrient cyclers. The enormous contribution of termites by moving soil and redistributing mineral nutrients, as well as by degrading plant detritus, is widely recognized, although difficult to study. Research in HiP identified the contributions of termites to savanna heterogeneity through interactions with grazing by white rhinos and other large herbivores, influenced by characteristics of the mounds formed by particular termite genera. Dung beetles are also important in the decomposer assemblage, burying plant material undigested by herbivores and thereby promoting the incorporation of the nutrients contained in the dung back into the soil. Dung beetles from HiP made an international contribution through being the source of some of the beetle species established in Australia and New Zealand to deal with pestilent flies breeding in cow dung.

Ants of various kinds serve a wider range of ecological roles, from scavenging on both animal and plant remains to acting as predators, notably on termites as well as other invertebrates. Their species composition depends on the vegetation structure including the changes in the herbaceous and litter layers.

The wide diversity of bird species within HiP is also an expression of the local habitat heterogeneity. Certain birds associated with short-grass lawns benefit from an abundance of large grazers, while other species frequenting tall grass might become threatened if the extent of this habitat became greatly reduced. The protected area also plays an important role in hosting breeding populations of vultures and large raptors, scarce or absent in surrounding communal rangelands. Oxpeckers have a close mutualistic relationship with rhinos, although not narrowly dependent on the presence of rhinos.

The abundance of small rodents within HiP seems to be unusually low, apparently due to exposure to predation as a result of the sparse grass cover generated by the grazing impacts of large herbivores. This restricts the contributions that some of these rodents make as seed-eaters or herbivores on green vegetation and as food for a number of small carnivores.

Lastly, the proclamation of the game reserves that became amalgamated into HiP was largely due to the presence of tsetse flies, which had limited human settlements because of the parasitic disease they transmitted to cattle. However, their continued threat to nearby cattle ranches almost brought about the deproclamation of the Umfolozi Game Reserve. The tsetse species primarily responsible for the transmission of nagana was locally exterminated by spraying insecticides, but two forest-dwelling species remain and still pose a threat to livestock in communal rangelands surrounding the protected area.

9.6 References

Bagine, R. K. N. (1984) Soil translocation by termites of the genus *Odontotermes* (Holmgren) (Isoptera: Macrotermitinae) in an arid area of Northern Kenya. *Oecologia* **64**: 263–266.

Bakker, E. S. (2003) Herbivores as mediators of their environment: the impact of large and small species on vegetation dynamics. PhD thesis, Wageningen University.

Bellier, L. (1967). Recherches écologiques dans la savane de Lamto (Côte d'Ivoire): densités et biomasses des petitsmammiféres. *Terre Vie* **2**: 319–329.

Blaum, N., Rossmanith, E., & Jeltsch, F. (2007) Land use affects rodent communities in Kalahari savanna rangelands. *African Journal of Ecology* **45**: 189–195.

Bourquin, O., Vincent, J., & Hitchins, P. M. (1971) The vertebrates of the Hluhluwe Game Reserve–Corridor (State land)–Umfolozi Game Reserve Complex. *Lammergeyer* **14**: 1–58.

Bowland, A. E. & Perrin, M. E. (1989) The effect of overgrazing on the small mammals in Umfolozi Game Reserve. *Zeitschrift fur Saugetierkunde* **54**: 251–260.

Brown, J. H. & Zeng, Z. (1989) Comparative population ecology of eleven species of rodents in the Chihuahuan Desert. *Ecology* **70**: 1507–1525.

Byrom, A. E., Ruscoe, W. A., Nkwabi, A. K., *et al.* (2014) Small mammal diversity and population dynamics in the Greater Serengeti Ecosystem. In: *Serengeti IV. Sustaining biodiversity in a coupled human–natural system* (eds. A. R. E. Sinclair, K. L. Metzger, S. A. R. Mduma, & J. M. Fryxell), pp. 323–358. University of Chicago Press, Chicago.

Cheeseman, C. L. & Delaney, M. J. (1979) The population dynamics of small rodents in a tropical African grassland. *Journal of Zoology* **187**: 451–476.

Davies, A. B., Eggleton, P., van Rensburg, B. J., & Parr, C. L. (2013) Assessing the relative efficiency of termite sampling methods along a rainfall gradient in African savannas. *Biotropica* **45**: 474–479.

Davies, A. B., Robertson, M. P., Levick, S. R., *et al.* (2014) Variable effects of termite mounds on African savanna grass communities across a rainfall gradient. *Journal of Vegetation Science* **25**: 1405–1416.

Davis, A. L. V., Doube, B. M., & McLennan, P. D. (1988) Habitat associations and seasonal abundance of coprophilous Coleoptera (Staphylinidae, Hydrophilidae and Histeridae) in the Hluhluwe region of South Africa. *Bulletin of Entomological Research* **78**: 425–434.

Deshmukh, I. (1989) How important are termites in the production ecology of African savannas? *Sociobiology* **15**: 155–168.

Doube, B. M. (2014) *Dung down under: dung beetles for Australia*. Dung Beetle Solutions, Australia.

Doube, B. M., Macqueen, A., & Fay, H. A. C. (1988) Effects of dung fauna on survival and size of buffalo flies (*Haematobia* spp.) breeding in the field in South Africa and Australia. *Journal of Applied Ecology* **25**: 523–536.

Edwards, P. B. (2007) *Introduced dung beetles in Australia 1967–2007: current status and future directions*. Sydney: Landcare Australia.

Emmons, L. H. (2009) Long term variation in small mammal abundance in forest and savanna of Bolivian cerrado. *Biotropica* **41**: 493–502.

Fay, H. A. C. & Doube, B. M. (1987) Aspects of the population dynamics of adults of *Haematobia thirouxi potans* (Bezzi) (Diptera: Muscidae) in southern Africa. *Bulletin of Entomological Research* **77**: 135–144.

Ferrar, P. (1982) Termites of a South African savanna. IV. Subterranean populations, mass determination and biomass estimations. *Oecologia* **52**: 147–151.

Galindo, C. & Krebs, C. J. (1987) Population regulation in deer mice: the role of females. *Journal of Animal Ecology* **56**: 11–23.

Giller, P. S. & Doube, B. M. (1989) Experimental analysis of inter- and intraspecific competition in dung beetle communities. *Journal of Animal Ecology* **58**: 129–142.

Gillingwater, K., Mamabolo, M. V., & Majiya, P. A. O. (2010) Prevalence of mixed *Trypanosoma congolense* infections in livestock and tsetse in KwaZulu-Natal, South Africa. *Journal of the South African Veterinary Association* **81**: 219–223.

Gosling, C. M. (2014) Biotic determinants of heterogeneity in a South African savannah. PhD thesis, Groningen University.

Gosling, C., Cromsigt, J., Mpanza, N., & Olff, H. (2012) Effects of erosion from mounds of different termite genera on distinct functional grassland types in an African savannah. *Ecosystems* **15**: 128–139.

Grant, C. C. & Scholes, M. C. (2006) The importance of nutrient hot-spots in the conservation and management of large wild mammalian herbivores in semi-arid savannas. *Biological Conservation* **130**: 426–437.

Grant, W. E. & Birney, E. C. (1979) Small mammal community structure in North American grasslands. *Journal of Mammalogy* **60**: 23–36.

Grant, W. E., Birney, E. C., French, N. R., & Swift, D. M. (1982) Structure and productivity of grassland small mammal communities related to grazing-induced changes in vegetation cover. *Journal of Mammalogy* **63**: 248–260.

Hagenah, N. (2006) Among rodents and rhinos. Interplay between small mammals and large herbivores in a South African savanna. PhD thesis, Wageningen University.

Hagenah, N., Prins, H. H. T., & Olff, H. (2009) Effects of large herbivores on murid rodents in a South African savannah. *Journal of Tropical Ecology* **25**: 483–492.

Jedrzejewska, B. & Jedrzejewski, W. (1998) *Predation in vertebrate communities*. Springer, Berlin.

Jedrzejewski, W. & Jedrzejewska, B. (1996) Rodent cycles in relation to biomass and productivity of ground vegetation and predation in the Palearctic. *Acta Theriologica* **41**: 1–34.

Joseph, G. S., Seymour, C. L., Cumming, G. S., Cumming, D. H. M., & Mahlangu, Z. (2014) Termite mounds increase functional diversity of woody plants in African savannas. *Ecosystems* **17**: 808–819.

Jouquet, P., Boulain, N., Gignoux, J. & Lepate, M. (2004) Association between subterranean termites and grasses in a West African savanna: spatial pattern analysis shows a significant role for *Odontotermes* n. *pauperans*. *Applied Soil Ecology* **27**: 135–164.

Jouquet, P., Barre, P., Lepage, M., & Velde, B. (2005) Impact of subterannean fungus-growing termites on chosen soil properties in a West African savanna. *Biology and Fertility of Soils* **42**: 365–370.

Keesing, F. (1998) Impacts of ungulates on the demography and diversity of small mammals in central Kenya. *Oecologia* **116**: 381–389.

Keesing, F. (2000) Cryptic consumers and the ecology of an African savanna. *BioScience* **50**: 205–214.

Korn, H. (1987) Densities and biomass of non-fossorial southern African savanna rodents during the dry season. *Oecologia* **72**: 410–413.

Krook, K., Bond, W. J., & Hockey, P. A. R. (2007) The effect of grassland shifts on the avifauna of a South African savanna. *Ostrich – Journal of African Ornithology* **78**: 271–279.

Leirs, H., Verhagen, R., Verheyen, W., *et al.* (1996a) Forecasting rodent outbreaks in Africa: an ecological basis for *Mastomys* control in Tanzania. *Journal of Applied Ecology* **33**: 937–943.

Leirs, H., Verhagen, R., Verheyen, W., *et al.* (1996b) Spatial patterns in *Mastomys natalensis* in Tanzania (Rodentia, Muridae). *Mammalia* **60**: 545–555.

Loveridge, J. P. & Moe, S. R. (2004) Termitaria as browsing hotspots for African megaherbivores in miombo woodland. *Journal of Tropical Ecology* **20**: 337–343.

Macdonald, I. A. W. & Birkenstock, P. J. (1980) Birds of the Hluhluwe-Umfolozi Game Reserve complex. *Lammergeyer* **29**: 1–56.

Mahlaba, T. A. M. & Perrin, M. R. (2003) Population dynamics of small mammals at Mlawula, Swaziland. *African Journal of Ecology* **41**: 317–323.

Mobaek, R., Narmo, A. K., & Moe, S. R. (2005) Termitaria are focal feeding sites for large ungulates in Lake Mburo National Park, Uganda. *Journal of Zoology* **267**: 97–102.

Moe, S., Mobaek, R., & Narmo, A. K. (2009) Mound building termites contribute to savanna vegetation heterogeneity. *Plant Ecology* **202**: 31–40.

Monadjem, A. & Perrin, M. (1996) The effects of additional food on the demography of rodents in a subtropical grassland in Swaziland. *Mammalia* **60**: 785–789.

Monadjem, A. & Perrin, M. (1998) The effect of supplementary food on the home range of the multimammate mouse *Mastomys natalensis*. *South African Journal of Wildlife Research* **28**: 1–3.

Monadjem, A. & Perrin, M. (2003) Population fluctuations and community structure in a Swaziland grassland over a three-year period. *African Zoology* **38**: 127–137.

Oguge, N. O. (1995) Diets, seasonal abundance and microhabitats of *Praomys* (*Mastomys*) *natalensis* and other small rodents in a Kenyan sub-humid grassland community. *African Journal of Ecology* **33**: 211–223.

Okullo, P. & Moe, S. R. (2012) Termite activity, not grazing, is the main determinant of spatial variation in savanna herbaceous vegetation. *Journal of Ecology* **100**: 232–241.

Okullo, P., Greve, P. M. K., & Moe, S. R. (2013) Termites, large herbivores, and herbaceous plant dominance structure small mammal communities in savannahs. *Ecosystems* **16**: 1002–1012.

Owen-Smith, N. (1973) The behavioural ecology of the white rhinoceros. PhD thesis, University of Wisconsin.

Owen-Smith, N. (1980) A quantitative assessment of the avifauna of the Umfolozi thorn savannah. *Lammergeyer* **30**: 49–60.

Parr, C. L., Robertson, H. G., Biggs, H. C., & Chown, S. L. (2004) Response of African savanna ants to long-term fire regimes. *Journal of Applied Ecology* **41**: 630–642.

Parr, C. L., Gray, E., & Bond, W. J. (2012) Cascading biodiversity and functional consequences of a global change-induced biome switch. *Diversity and Distributions* **18**: 493–503.

Ridsdill-Smith, T. J. & Edwards, P. B. (2011) Biological control: ecosystem functions provided by dung beetles. In: *Ecology and evolution of dung beetles* (eds L. W. Simmons & T. J. Ridsdill-Smith), pp. 254–266. Blackwell, Oxford.

Schweiger, E. W., Diffendorfer, J. E., Holt, R. D., Pierotti, R., & Gaines, M. S. (2000) The interaction of habitat fragmentation, plant, and small mammal succession in an old field. *Ecological Monographs* **70**: 383–400.

Sinclair, A. R. E., Metzger, K. L., Fryxell, J. M., *et al.* (2013) Asynchronous food web pathways could buffer the response of Serengeti predators to El Nino Southern Oscillation. *Ecology* **94**: 1123–1130.

Smit, R., Bokdam, J., den Ouden, J., *et al.* (2001) Effects of introduction and exclusion of large herbivores on small rodent communities. *Plant Ecology* **155**: 119–127.

Stutterheim, C. J. (1980) Symbiont selection of red-billed oxpecker in the Hluhluwe-Umfolozi Game Reserve Complex. *Lammergeyer* **30**: 21–25.

Uys, V. (2002) *A guide to the termite genera of Southern Africa.* Plant Protection Research Institute Handbook No. 15. Agricultural Research Council, Pretoria.

van den Bossche, P., Esterhuizen, J., Nkuna, R., *et al.* (2006) An update of the bovine trypanosomosis situation at the edge of Hluhluwe-iMfolozi Park, KwaZulu-Natal Province, South Africa. *Onderstepoort Journal of Veterinary Research* **73**: 77–79.

Van der Plas, F., Howison, R., Reinders, J., Fokkema, W., & Olff, H. (2013) Functional traits of trees on and off termite mounds: understanding the origin of biotically-driven heterogeneity in savannas. *Journal of Vegetation Science* **24**: 227–238.

Walker, B. H., Emslie, R. H., Owen-Smith, R. N., & Scholes, R. J. (1987) To cull or not to cull: lessons from a southern African drought. *Journal of Applied Ecology* **24**: 381–401.

Wright, E. J., Müller, P., & Kerr, J. D. (1989) Agents for biological control of novel hosts – assessing an aleocharine parasitoid of dung-breeding flies. *Journal of Applied Ecology* **26**: 453–461.

Yunger, J. A. (2004) Movement and spatial organization of small mammals following vertebrate predator exclusion. *Oecologia* 139: 647–654.

10 · *Interactions between Fire and Ecosystem Processes*

SALLY ARCHIBALD, HEATH BECKETT, WILLIAM J. BOND, CORLI COETSEE, DAVE J. DRUCE, AND A. CARLA STAVER

10.1 Introduction

Fire is an important characteristic of grassy savanna-woodland vegetation globally (Lehmann *et al.*, 2011; Archibald *et al.*, 2013). The combination of fine fuels and seasonal rainfall produces environments that are fire-prone to a degree possibly unrealized before C4 grassy vegetation spread across the globe from ~8 million years ago (Keeley and Rundel, 2005; Osborne, 2008). However, the type, frequency, extent, and intensity of fire vary. Because these features can have big impacts on savanna structure and function (see Chapters 3, 6, 7 and 8), understanding and controlling savanna fires has always been a priority for managers of these systems.

Savannas are classically described as being influenced by rainfall, soils, herbivory, and fire (Scholes and Archer, 1997). A distinctive feature of Hluhluwe-iMfolozi Park (HiP) is that it spans a rainfall range that crosses the divide between 'arid savanna' – where the main above-ground consumers are herbivores, and 'mesic savanna' – where fire takes over as the major consumer of vegetation (see Figure 10.1 and Chapter 2). Below ~400 mm mean annual rainfall (MAR), fire is rare because there is not enough moisture in most years to produce continuous fuel-beds. Annual burnt area reaches a maximum at ~1000 mm MAR, above which it is limited not by the biomass of grassy fuel, but by the extent of this fuel (savanna environments become rare above ~1200 mm in Africa: Lehmann *et al.*, 2011). In contrast, the biomass density of grazers peaks at about 700 mm MAR, after which it drops due to declining forage quality accompanied by leached soils (also demonstrated by East, 1984, and

Conserving Africa's Mega-Diversity in the Anthropocene, ed. Joris P. G. M. Cromsigt, Sally Archibald and Norman Owen-Smith. Published by Cambridge University Press.

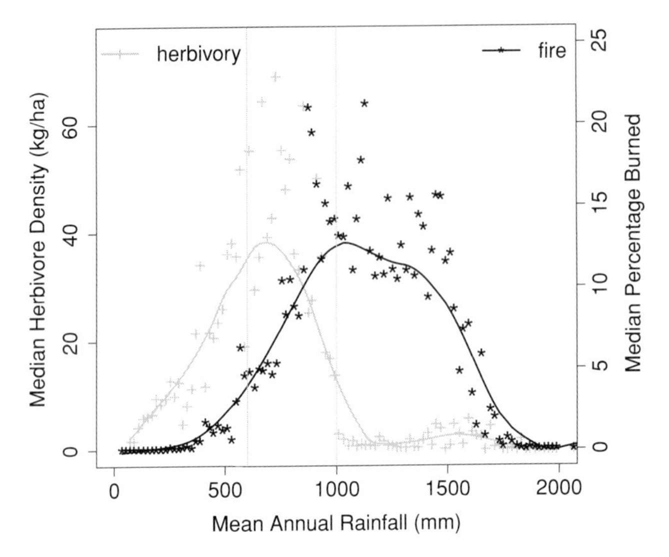

Figure 10.1 Demonstrating continental-scale patterns of fire and grazing in Africa across a rainfall gradient. The dominant consumer changes from herbivory in arid savannas to fire in mesic savannas, but most regions are affected by a combination of these agents (Scholes and Walker, 1993; Sankaran *et al.*, 2005). The rainfall domain of HiP is indicated by the box. Current herbivore densities in Hluhluwe and iMfolozi are ~90 kg/ha and ~120 kg/ha, respectively, if you exclude elephants (Chapter 4). About 36% and 26% of these two sections burn on average every year. (Reprinted with permission from Wiley; Hempson *et al.* (2015b) with data from Archibald *et al.* (2009).)

Hempson *et al.*, 2015a), although on high-nutrient sites herbivore density can increase up to 1000 mm of rainfall (Fritz and Duncan, 1994). The rainfall range within HiP from < 600 to ~1000 mm MAR spans part of this functional divide and thus provides a template to study both how the patterns of fire change in response to abiotic, biotic, and human drivers, and how these changing fire patterns impact the structure, function and composition of savanna ecosystems (see Chapter 8).

Fire has consequently been an active area of enquiry for scientists in HiP, and this has led to useful insights into interactions between herbivory and fire, the role of fire in enabling tree–grass coexistence, the variety of adaptations of plants to tolerate/survive fires, and factors controlling fire spread and therefore vegetation patterns. The scientific impact of this research has been considerable, but the extent to which research findings have influenced management of the park has varied over the years.

The history of fire management has been well described for other savanna conservation areas – especially the Kruger National Park (Trollope *et al.*, 1998; Biggs and Potgieter, 1999; Van Wilgen *et al.*, 2008), but not yet for HiP. Park staff have maintained a reliable spatial data set on fire occurrence dating back to the 1950s. This chapter documents interactions between fire and ecosystem processes in HiP and the history of fire management in the park. In the process we highlight some insights into savanna fire ecology that have been enabled by research at HiP. We will also look into how fire regimes might be impacted by global change, and the use of fire as a tool to prevent or mitigate unwanted change.

10.2 Fires in HiP and Other Savanna Ecosystems

Fires in savannas are generally surface fires fuelled by the grass layer. Globally, savanna fires are distinctive in their short return periods – on average every 2–4 years (Archibald *et al.*, 2010b). Tropical C4 grasses have highly combustible leaves (low bulk density) and regrow rapidly following defoliation. These features, combined with a predictable dry season lasting from 3 to 6 months, lead to a situation where every year there is a period when the fuel is both available and flammable (Bradstock, 2010). Grasses, with their fast regrowth rates and basal meristems, are especially adapted to this frequent fire regime. However, it represents a considerable evolutionary hurdle for other plants – particularly trees, which need to grow a fire-resistant trunk. Savanna trees and forbs are well known for their distinctive adaptations (Bond and Van Wilgen, 1996; Simon and Pennington, 2012; Maurin *et al.*, 2014) such as resprouting capacity, thick bark, and large underground storage organs which enable persistence under these extreme conditions (see Chapters 7 and 8). Fire-sensitive plants can also be present in savannas, however, and across the savanna biome there usually exist some closed-canopy formations that are resistant to grassy fires (Lehmann *et al.*, 2011). HiP, where patches of forest and thicket are common, is no exception here (see Chapter 8).

Fire characteristics such as the frequency, intensity, seasonality, extent of burn, and type of fuels describe the fire 'regime': the wildfire activity that prevails in a given area (Gill, 1975). In savannas, the intensity and frequency of fire are both important controllers of tree demographics (Chapter 7), and are generally what managers try to manipulate to maintain particular vegetation structure or composition. In contrast, the area burned and the season of burning can have implications for herbivore movements and nutrition, as well as soil resources, and are the characteristics considered when assessing management objectives related to herbivores. These characteristics are interlinked, with season and frequency affecting fire intensity, fire intensity affecting fire size, and fire size affecting burned area and, therefore, fire frequency. Fire managers need to confront this complexity (Bond and Archibald, 2003). Below, we describe the range of fire characteristics in HiP and savanna environments in general, their environmental controls, and the degree to which they can be manipulated.

10.2.1 Frequency

Savannas predictably have several dry months a year when weather and fuels are suitable for burning, so fire frequency is typically controlled by the rate at which grassy biomass accumulates – i.e. less-frequent fires in drier ecosystems, or during drier periods (Archibald *et al.*, 2010b). This is apparent in HiP both spatially and over time: the higher-rainfall regions have had more fires in the last ~60 years (Figure 10.2), and the area burned in any 1 year has depended on the accumulated rainfall over the previous 2 years (Balfour and Howison, 2001; Archibald *et al.*, 2010a; Figure 10.3B). A trend of increased fire frequency up to the 1980s occurred uniformly throughout the park, but in recent decades, fire frequencies have decreased in the south and increased or remained constant in the north (Figure 10.3C). Within this broad context, the type of vegetation and the extent of grazing also impact the area burned and frequency of fire: forests resist burning, closed-canopy thicket vegetation is less likely to burn, and heavily grazed areas are associated with reduced extent and frequency of fire (Archibald *et al.*, 2005; Chapter 6; Figure 10.3B). Moreover, given that up to 25% of fires in HiP can be human-ignited coming in from outside, proximity to the fence, combined with topographic features like roads that act as firebreaks, can affect the fire return period of particular sites (Figure 10.2).

10.2.2 Intensity

In savanna environments, fire intensity is controlled not only by fuel loads (amount of grass biomass), but also by the fuel moisture, and the weather conditions on the day, particularly wind speed, relative humidity, and air

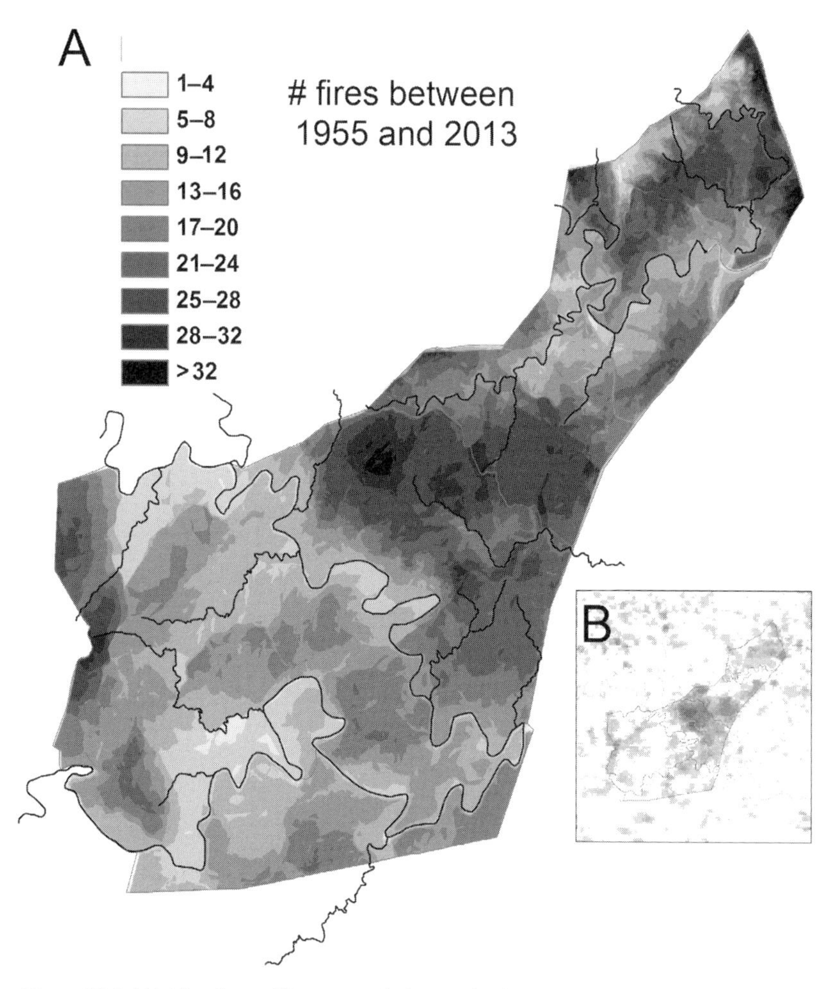

Figure 10.2 (A) Number of fires recorded over the last 59 years in HiP (data from Ezemvelo KZN Wildlife). The spatial pattern is a result of rainfall, mediated by the vegetation that is present, the extent of grazing, and the frequency of ignitions. Wetter areas have more fires, but even wet areas will not burn frequently if covered with thicket/forest vegetation. In contrast, dry sections of the park will burn when ignitions are frequent (e.g. close to the park fence) or when grazing pressure is low or after years of higher rainfall. A road break can clearly be seen in the western-most corner of the park, where fires burning from outside do not cross the road. (B) Fires in HiP and surrounding areas as recorded by MODIS satellite data (Archibald *et al.*, 2010a). HiP, like all conservation areas in Africa, burns far more than the outside landscape which is fragmented by human activities (see Figure 1.6). However, many of the fires in HiP come from outside the fence.

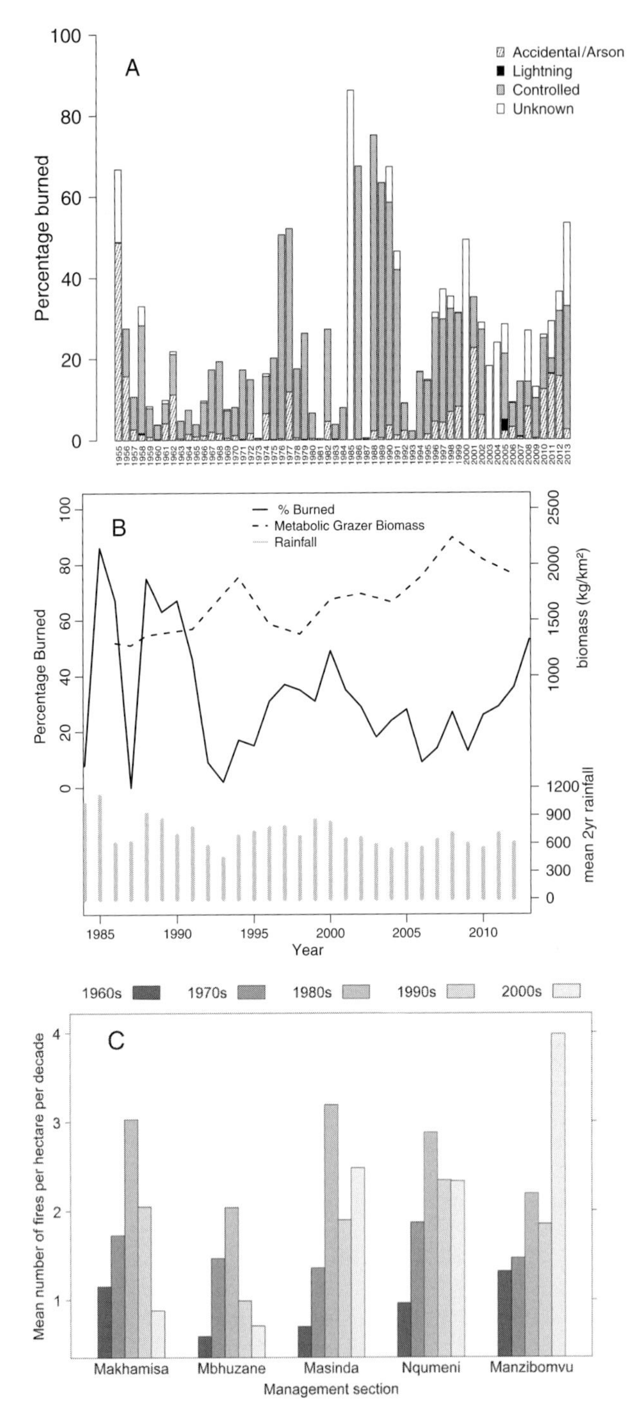

Figure 10.3 (A) Percentage of HiP burned over time, attributed to different causes. Prescribed burns became more common in the 1970s and lightning fires were

temperature (Govender *et al.*, 2006; Trollope and Tainton, 2007). Each of these factors can compensate for the other, but very intense and damaging fires only occur when all factors coincide. More intense fires are often larger, and result in 'clean' burns, with most of the above-ground vegetation combusted, whereas less-intense fires are 'patchy', and result in less top-kill of woody plants. Fire intensity also affects the type of carbon and other emissions that are released from a fire: lower-intensity 'smouldering' fires adding more of the greenhouse gases methane and carbon monoxide than high-intensity fires (Korontzi *et al.*, 2003). Very intense fires have the potential to be used to control invasive plants or shrub encroachment (see Chapters 7 and 15). In HiP the most intense fires occur in September/October just before the first rains when the weather conditions are most extreme and the fuel most dry, while fires occurring in summer or earlier in the dry season are less intense because most grass is still green. Because of this, managers can control fire intensity through prescribing the season, time of day, and weather conditions when fires are ignited. However, fire intensity and fire frequency are related because intense fires can only occur after sufficient build up of fuels (Archibald *et al.*, 2013). When manipulating fire return periods managers therefore inadvertently also affect fire intensity.

10.2.3 Season

Savanna fires show a distinct seasonal pattern where most of the burning occurs during the long (4–6-month) dry season when the grass is dry and flammable (Archibald *et al.*, 2010b). Within this window larger, more intense fires are associated with the later dry season, and smaller, more smouldering fires with the early dry season (due to changes in weather

Figure 10.3 (*caption continued*) always rare. After little fire in the 1950s and 1960s, large areas burned in the 1980s, but declined again in recent decades. (B) Relationships between rainfall, grazer density, and area burned for the time period for which we have all the data. Although periods of low fire coincided with periods of high grazer densities (linear model $p < 0.05$), annual percentage area burned is also strongly correlated with accumulated rainfall (Balfour and Howison, 2001), and a model which includes grazer densities explains little more of the variance in the data ($\Delta AIC < 1$, ANOVA $p = 0.1$). (C) Changes in the mean number of times a hectare burned per decade by management section over time. Sections are ordered from south to north (see park map in the prelim for exact location of sections).

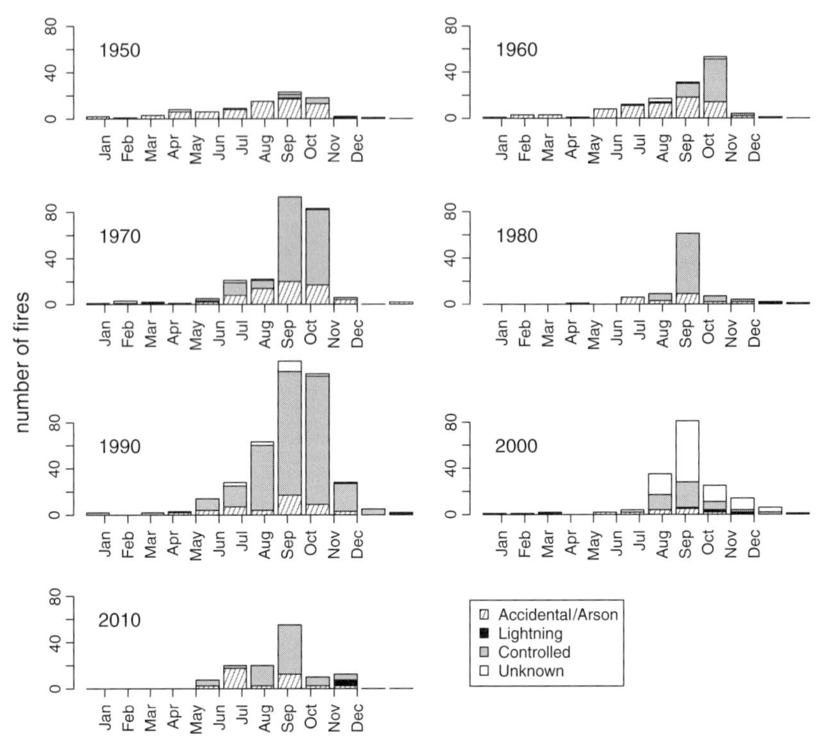

Figure 10.4 Cause of fires in HiP by decade. The extent of prescribed burns increased dramatically in the 1970s to become 70% of all fires, but this had the consequence of moving the fire season more strictly into the late dry season (August and September). Prescribed burning was at its height in the 1990s, but the increased number of fires did not result in a higher area burned (see Figure 10.3) as many of them were probably small and controlled. Lightning fires have not been a significant ignition source in HiP since the 1950s and probably not since before Iron Age settlement.

conditions and fuel continuity). Before humans were able to ignite fires, most of the burning probably occurred late in the dry season, associated with dry-lightning events (Archibald *et al.*, 2012). Once humans became the main ignition source, fires occurred throughout the dry season (supported by current data from HiP and elsewhere: Figure 10.4; Archibald *et al.*, 2010b). In HiP, prescribed burns almost always happen in August/September (Figure 10.4). Before prescribed burning became prevalent in the 1970s there was a much wider range of fire seasons – and consequently a much more varied fire regime – than currently (see management section).

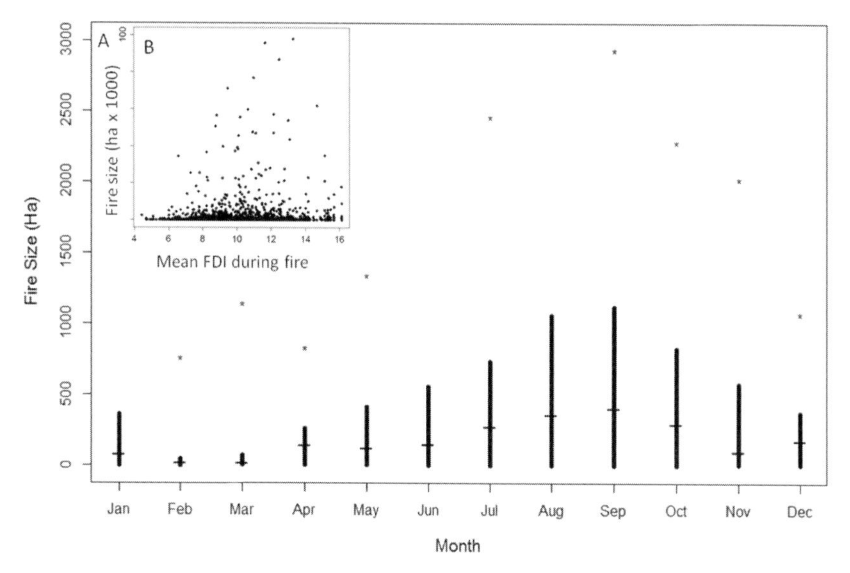

Figure 10.5 (A) Seasonal patterns of fire size in Hluhluwe-iMfolozi Park. Horizontal bars represent the median fire size, vertical bars the 75th quantile, and stars the 95th quantile. Distributions are highly skewed towards small fires. The largest fires occur from July to October. (B) Maximum fire size appears to be constrained by fire weather (indexed here by the McArthur Fire Danger Index).

10.2.4 Fire Size

Whether a fire continues to burn or not once ignited depends on the flammability and continuity of the fuel. Small fires are much more common than large ones (McKenzie *et al.*, 2011; Hantson *et al.*, 2015). In savannas, fire spread is affected by static or 'exogenous' factors like topography, roads, and rivers, as well as by factors that change from minute to minute (weather conditions), month to month (the degree of grazing, the dryness of the fuel, previous burning), and year to year (the extent of fire-resistant closed-canopy vegetation). Thus the size of any particular fire is a complex outcome of intrinsic and extrinsic factors. Understanding these processes becomes important both for appropriate application of prescribed fires (see later) and because of the impacts that extreme weather-driven mega-fires can have on ecosystems (Williams and Bradstock, 2009).

In HiP, weather conditions appear to set the maximum extent of burns (Figure 10.5B), but many other factors can keep fires below this upper limit. The heterogeneous landscape of HiP appears to affect fire patterns: fires in the Hluhluwe section are half the size on average that they are in the Corridor or Mfolozi regions (ANOVA, $p < 0.0001$), presumably due to more dissected topography (including high road density), with patches of forest and thicket that prevent the fuels from becoming connected. Moreover, in Hluhluwe fires get larger as the dry season progresses towards a maximum in August. In Mfolozi the maximum fire size occurs in June, presumably because grazing during the dry season reduces the available fuel. Area burned is the result of both the number and the size of fires. In HiP, total area burned is correlated with both number of fires ($r^2 = 0.48$) and maximum fire size ($r^2 = 0.32$), indicating that small fires do contribute to the total.

Concluding this section, fire regimes in savanna environments and elsewhere are the result of complex interactions between human activities (controlling the frequency and seasonal timing of ignitions), landscape, climate (controlling the fuel loads and the connectivity of the fuel bed), herbivory (again influencing fuel load), and weather conditions (controlling the fuel moisture and the intensity of fires). Moreover, feedbacks between fire characteristics mean that managing for one fire characteristic inadvertently results in changes in other aspects of the fire regime. Within the broad limits imposed by climate and vegetation, there is room to use fire to achieve particular management and ecological objectives. One of the main challenges with managing savanna conservation areas is determining what sort of fire regime is appropriate, and how to influence fire events to achieve this regime.

10.3 History of Fire Management at HiP

Little is known about patterns of fire before or during the late Iron Age period, but it is assumed that fires were frequent and ignited throughout the year by people (Berry and Macdonald, 1979). Historically, it was reported that 'Scarcely a night during winter but that a grass fire is seen somewhere on the veld, and, as the country is not divided into farms, a single fire may rage for three days before it dies out' (Aitken and Gale, 1921).

The early controlled burning applied by the game conservator from 1911 to 1930 was probably similar to the earlier regime, as the rangers simply set fires 'whenever it seemed necessary' (Vincent, 1970). From 1930 to 1940, fire was actively suppressed in the Mfolozi section to prevent burning the hundreds of Harris traps that were placed to combat tsetse fly (see Chapter 1). However, aerial photographs still show substantial numbers of fires in this region of HiP during this period

(Berry and Macdonald, 1979). Whether a substantial reduction in area burned was achieved is questionable in the light of later findings that fire application/suppression activities have very little impact on total area burned (Balfour and Howison, 2001; Van Wilgen *et al.*, 2004; Archibald *et al.*, 2010a). From 1955 onwards the fire programme in HiP was managed by the first ecologist appointed, Mr C. J. Ward, who initiated controlled burning for various management objectives and the detailed recording and mapping of fire events which still continues today (Berry and Macdonald, 1979). During the following decades, fire was applied for various reasons: to attract game, to promote visibility for tourists, to protect and promote fire-sensitive plants (by early-season fires), and as a method to counter bush increase (by late-season fires). Firebreaks were also applied to exclude fires from outside the reserve (Vincent, 1970) and to prevent fires from escaping the reserve. Later, during the 1960s and 1970s, rangers were encouraged to assess fuel loads and to apply burns after sufficient moribund material had accumulated. Concern over soil degradation in the 1970s resulted in systematic burning of upland grasslands to move grazers from heavily utilized bottomlands into the newly burned uplands (Roger Porter, personal comments). During this time, the practice of only burning after the spring rains and avoiding annual burns was pervasive in the region due to the influence of Professor J. D. Scott (1955) from the Natal University, and became prescribed by law in the mid-1970s. The reasoning was that early-season burns exposed soils to erosion, and reduced the productivity of the grass sward. Despite questions from ecologists and farmers about the wisdom of this approach (Tainton *et al.*, 1977), it became the general fire policy in HiP throughout the 1970s and 1980s (Figure 10.4). Berry and Macdonald (1979) speculate that it 'could possibly be a contributory factor in the tendency for woody plants to encroach what were formerly more open communities throughout the Complex since the time of proclamation' because these post-rain fires might not have been intense enough to knock back the sapling trees. Alternative viewpoints suggest that periods of low fire associated with high grazer numbers had been the source of the rampant bush encroachment (see Chapter 3).

Prescribed burns reached their peak in the 1990s (Figure 10.3A and Figure 10.4). These were widely spread between April and November and probably resulted in a more diverse fire regime. The concept of patch-mosaic burning – aiming for a variety of fire sizes, intensities, and frequencies – took off in the late 1990s and many conservation agencies have now embraced various versions of patch-mosaic burning (Brockett *et al.*,

2001). It is seen as a means to create more spatially and structurally diverse fire regimes and vegetation communities (Parr and Anderson, 2006). Despite the prevalence of this paradigm, the 2000s have seen a move back to mainly late-season fires in HiP for various reasons. Early-season burning in HiP is hampered because rangers need to wait for the harvesting of thatching grass by local communities to be completed before igniting fires. Moreover, while burning to control bush encroachment has always been a priority in HiP, in recent years this has dominated the fire management plans for each section (especially in the north) and has probably driven rangers to prioritize frequent, intense, late-season fires.

The latest HiP Integrated Management Plan (2011) states quite broadly that the intention of fire management in HiP is to 'actively manage for a shifting mosaic of differing fire impact and size, thereby creating a diversity of habitats that should ensure the conservation of the biodiversity representative of the area'. This policy is in line with process-based management and the 'pyrodiversity begets biodiversity' paradigm currently followed by other conservation areas. However, it contrasts with the use of fire as a tool to achieve particular objectives, and seems not to reflect the actual application of fires in much of HiP. The area burned in the park has been consistently increasing in the last 5 years (Figure 10.3A) with more than half the park burned in 2013 and 2014. However, this happened disproportionately in the most northern section (Figure 10.3C), and whether prompted by high rainfall years, changes in grazing, or control of bush encroachment is an open question.

In summary, there have been many different, and not necessarily complementary, objectives for applying prescribed fires in HiP (Bond and Archibald, 2003). The theme of bush encroachment has loomed large since at least the 1930s. Another theme has been burning to maintain a 'healthy' grass community, and thereby promote grazing, but exactly what sorts of fires achieve this has been hotly debated over the years. Over this same period of time, scientific research that has the potential to guide these management outcomes has been ongoing in the park. In the following section we summarize some of these advances, and the impact that they have had on fire management.

10.4 Savanna Fire Ecology: Insights from Work at HiP

10.4.1 Tree–Grass Interactions: Savanna Structure

The role of fire in maintaining an 'equilibrium' (Henkel, 1937 in Vincent, 1970) between trees and grasses was recognized early by HiP ecologists, and preventing shrub encroachment has long been one of the main motivations for applying controlled fires (Ward, 1962; Vincent, 1970; Berry and Macdonald, 1979). Part of HiP lies within a rainfall range where climate has the potential to support forest (Bond, 2005; Sankaran *et al.*, 2005) and frequent intense fires are perceived to be important for retaining open, park-like conditions. As savanna fires rarely result in tree death, the aim of applied fires is to maintain saplings in a short, juvenile state within the fire trap (Higgins *et al.*, 2000). Recent research in HiP has helped transform this broad understanding into a mechanistic, process-based model, which can be used (a) to determine appropriate fire regimes for preventing sapling escape, and (b) to predict consequences of changing climate, CO_2 concentrations, and fire regimes on tree demographics and savanna structure (Higgins *et al.*, 2000; Hoffmann and Solbrig, 2003; Bond and Midgley, 2012).

The details of this demographic bottleneck model are comprehensively covered in Chapter 7. The key components of a fire regime that control vegetation structure are the intensity and the frequency of fires, which interact with tree growth rates (Figure 10.6). Fire intensity determines the height at which a tree can be 'top-killed' by a fire (see Figure 7.4 in Chapter 7). Intense fires will be able to keep more of the woody individuals within the fire trap. Fire return period interacts with tree growth rates to determine how tall (i.e. fire-resistant) a tree is when the next fire occurs (Figure 7.5, Chapter 7), and therefore more-frequent fires will also be able to keep more of the woody individuals within the fire trap. The subtleties of this model relate to the fact that it is stem diameter, not tree height per se, that confers resistance to top-kill (Hoffmann and Solbrig, 2003), although these measures are related. Moreover, it is the maximum growth rates that are key, not average rates (Wakeling *et al.*, 2011; Chapter 7).

This conceptual model implies that anything that alters fire frequency, fire intensity, or tree growth rates has the potential to change the vegetation structure, but these can also compensate for each other. For example, fires that occur during the growing season, when stored reserves of woody plants are at a minimum, might reduce regrowth rates (Schutz, 2003 in Wakeling and Bond, 2007), as do browsers (Trollope, 1974; Staver *et al.*, 2009). Reduced growth rates in turn mean that shrub encroachment can be controlled by lower-intensity or less-frequent fires. In contrast, more frequent or more intense fires could potentially offset the positive effect of elevated CO_2 on tree growth rates (Figure 10.6).

The demographic bottleneck model has formalized previous thinking about the role of fire in preventing woody thickening. There are alternative views, however. Savanna trees and shrubs have short life spans

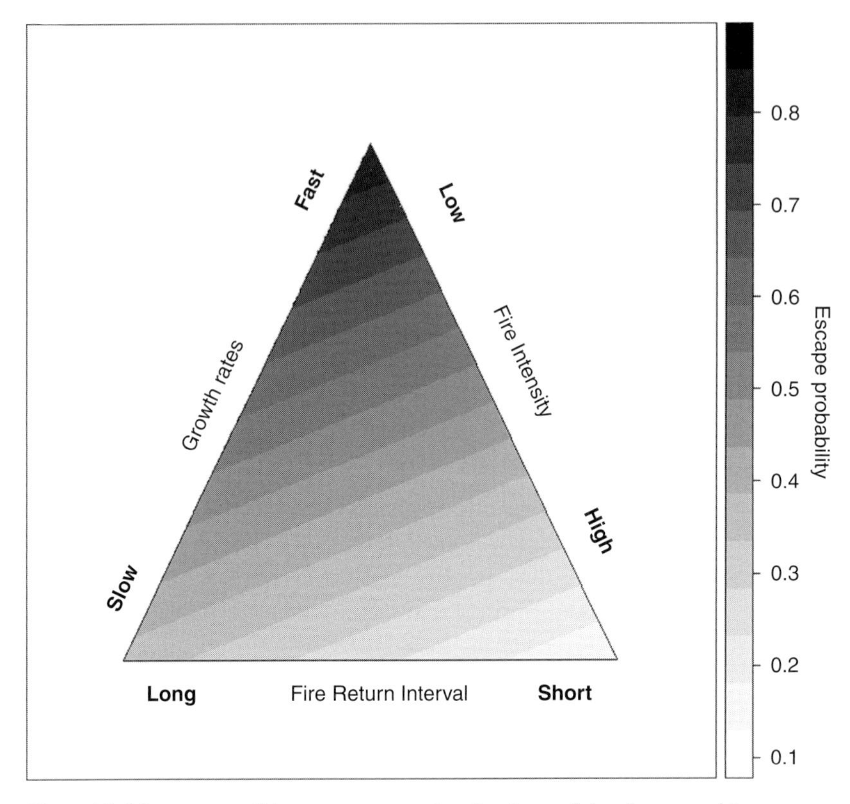

Figure 10.6 Summary of the management implications of the demographic bottleneck model for savanna vegetation structure. When a tree has grown tall enough not to be top-killed by a fire, it is considered to have escaped the fire trap (see Figure 7.4). Because escape probability can be altered by changes in growth rates, fire intensity, and fire return interval, a reduction in one variable can be compensated for by an increase in another to result in a similar outcome. This has implications for managers deciding on season of burn (affects fire intensity and tree regrowth rates) and frequency of burning, as well as providing a framework for dealing with unwanted environmental change (CO_2 increasing tree growth rates, for example, or increased temperatures reducing fire return period/fireline intensity).

and allowing saplings to escape the fire trap and become adult plants could facilitate long-term processes of succession from open grasslands to closed woodlands, and back to open grasslands (Susan J. van Rensburg, personal communication). Critiques of this thinking are that once mature trees or shrubs have established, (1) they are reproductively mature and can promote further woody plant expansion, and (2) they provide establishment sites for forest/thicket species that can transform a savanna into a more fire-resistant closed-canopy forest or thicket (Skowno *et al.*, 1999).

The frequent, intense fires proposed by the demographic bottleneck model can be counter to many other fire management objectives in a conservation area (protection of fire-sensitive species and habitats, protecting grazing resources, promoting pyrodiversity). Several park ecologists have tried to introduce a more informed approach to fire application – i.e. using the model to assess *when* and *where* high-intensity burns will be most important (e.g. locations in the park which are on the verge of switching from a desirable vegetation type to an undesirable one) but allowing for other fire types elsewhere in the system.

10.4.2 Firestorms and Grassland–Forest Transitions

Savanna fires are not generally able to penetrate adjacent forest or thicket vegetation because the understorey switches from flammable C4 grasses to forbs or largely bare ground (Hennenberg *et al.*, 2006). Forests and thickets also alter microclimates, reducing wind speeds and increasing plant moisture content (Little *et al.*, 2012), further reducing the penetration of fire. Within northern HiP the spatial distribution of savanna, forest, and thicket is quite complex (Chapter 2) and there is evidence that thicket has expanded at the expense of open grassland in the last century (Berry and Macdonald 1979; Wigley *et al.* 2010; Chapter 3). However, under certain conditions, fires can cross the boundary from grassland into thicket or forest. Such conditions have been studied during an intense fire that occurred in September 2008 in northern Hluhluwe. An analysis of the weather and fuel conditions at that time suggests that extreme temperature, wind, and humidity conditions produced a fire intensity that started to generate its own weather conditions by creating a convective column caused by warm rising air (Browne and Bond, 2011). These types of fires have become known as firestorms (Radke, 2013). Browne and Bond (2011) suggest that a firestorm will develop if the air temperature reaches above 30°C, the relative humidity falls below 30%, and wind speed reaches approximately 30 km/h (a rule of thumb borrowed from South African fire-fighters who assess the probability of grassy fires burning into plantations). However, some form of ladder fuel may assist the penetration of the surface fire into the woody canopy. From the 1980s, the invasive scrambler *Chromolaena odorata* spread widely across Hluhluwe, and created dense stands along the boundaries of savanna grasslands and thicket/forest (see Chapter 15). It may have facilitated the 2008 firestorm, because alien plant control teams cleared *Chromolaena* along forest and thicket boundaries, leaving behind highly flammable stacks of dry stems (Macdonald and Frame, 1988; te Beest *et al.*, 2012; Chapter 15). Its role in driving the firestorm of 2008 has not been resolved – but *Chromolaena* can cause fires to burn into thicket vegetation under normal conditions in HiP (te Beest *et al.*, 2012).

The ecological effects of firestorms depend on the type of vegetation. Although the 2008 fire caused 21% mortality of adult trees in thicket patches, many trees resprouted from the base (Browne and Bond, 2011) and there was rapid colonization by grasses. These sites accumulated enough grassy fuel to burn again in 2010, 2012, and 2014, and can now be considered functionally 'savanna'. Nevertheless, most of the resprouting trees in these sites are still thicket species such as *Euclea racemosa*, *Euclea divinorum*, *Berchemia zeyheri*, and *Sideroxylon inerme*, which are now kept in a fire trap. In contrast, there was complete species turnover in burnt forest sites after 2008 (Heath Beckett, unpublished data). These areas were colonized by rapidly growing pioneer forest species (e.g. *Trema orientalis*, *Croton sylvaticus*), which recovered from a seed bank rather than resprouting. Thus unlike thicket species, forest species appear not to have the ability to resprout and are severely impacted when fires do penetrate into the forest. If the pioneer species establish quickly enough to generate a microclimate/environment that excludes fire, then presumably climax forest species too will be able to re-establish. Little is known about this process in HiP or elsewhere. The savanna–thicket–forest mosaic of northern HiP is a prime example of alternate vegetation states that can occur under the same climatic conditions. Firestorms provide a destabilizing mechanism (a 'catastrophic regime shift' in the jargon of Scheffer and Carpenter, 2003) which can result in novel vegetation formations. They have been experimentally applied in the Kruger National Park, Eastern Cape, and even the south-western USA (Twidwell *et al.*, 2013).

However, applying firestorms requires a fire management policy that is flexible enough to allow for burning under the most extreme weather and fuel conditions, despite the risks involved. The difficulty in applying firestorms as a management tool lies in their association with extreme conditions. Over the period 2001–2008, fewer than three days of 30′ 30′ 30′ conditions were recorded, with the longest consecutive period of 11 hours occurring on the day of the firestorm in September 2008 (Browne, unpublished data). These conditions could become more common as temperatures increase, but high temperatures alone are not sufficient. Determining ways to achieve firestorms under more controlled situations is an active area of current research (Winston Trollope, personal comments). An important concern is preventing firestorms from threatening properties adjacent to conservation areas as well as infrastructure within the conservation area itself (e.g. tourist camps). Furthermore, concerns exist over the impact on fire-sensitive habitats such as indigenous forests (te Beest *et al.*, 2012). A meagre 0.1% of South Africa, of a potential 7%, is covered by indigenous forest (Mucina and Rutherford, 2006). The high biodiversity associated with scarp forests in HiP (six endemic genera and one endemic family) is reason for their priority conservation status. The data available so far demonstrate that firestorms have the potential to wreak havoc in fire-sensitive indigenous forests, so their routine application as a management tool is unlikely in the future.

10.4.3 Fire–Herbivore Interactions

As consumers of vegetation, fire and herbivory differ in their selectivity, temporal occurrence, geographic extent, and the impact they have on grass and woody vegetation. Grasses that tolerate heavy grazing hold their foliage close to the ground and spread laterally (Chapter 6), whereas those that benefit from fire tend to be tall in stature. Similar contrasts in species traits occur in the tree layer (Chapter 8). Hence, grasslands supporting frequent fires are structurally and compositionally very different from those subject to high herbivory. Moreover, these two consumers interact, and can affect the extent and prevalence of each other both positively and negatively. Work at HiP has been influential in elucidating these feedbacks, the mechanisms by which fire and herbivory affect each other, and the consequences of these interactions for savanna functioning.

Grazing affects fire in the short term (over a growing season) by removing above-ground fuel. A connected fuel bed is necessary for fire to spread and under heavy grazing short-grass patches can become so extensive that fires are no longer able to burn the fuel that remains. Waldram *et al.* (2008) described how the local removal of white rhinos increased fire size by an order of magnitude in HiP and reduced fire patchiness. In the long term (several years or decades), continued heavy grazing promotes the spread of short lawn-grass systems (Chapter 6), which are associated with more permanent reduction in the amount of fire in an ecosystem (Archibald *et al.*, 2005).

Fire impacts herbivory over a growing season by redistributing grazing impacts in the landscape. Grazers are attracted to the post-burn green flush (Wilsey, 1996; Tomor and Owen-Smith, 2002), both because of the

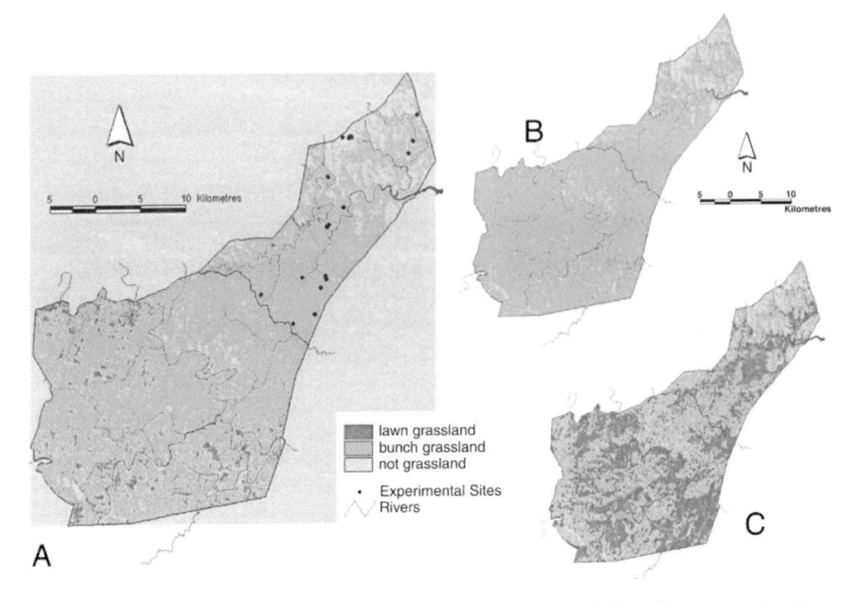

Figure 10.7 Current (A) and predicted (B, C) proportion of short lawn-grassland under two modelled uniform fire return period scenarios (3.2 years and 7 years, respectively), indicating the potentially large impacts that frequent fires can have on grass community state. (Reprinted with permission from the Ecological Society of America; Archibald *et al.*, 2005.)

high-quality forage that it provides, and because it represents a safer environment for predator avoidance (Burkepile *et al.*, 2013). Fires thus have a positive effect on grazer food availability in the short term. However, Archibald and Bond (2004) showed that large fires have the effect of dispersing animals and reducing the patchiness of grazing (Archibald *et al.*, 2005). As a result, large fires effectively relieve the grazing pressure on short-grazed patches, such as grazing lawns, each year (Yoganand and Owen-Smith, 2014) and may reduce the persistence of these short-grazed patches (Archibald *et al.*, 2005). The long-term impact of large fires on grazers may thus be negative (Archibald, 2008) because it ultimately leads to a reduction in short-grass habitat that is preferred by certain grazer species (Figure 10.7 – see also Chapter 6).

Both over time and across space, high grazing is associated with reduced fire and vice versa. In the Serengeti an increase in wildebeest from 1964 to 1974 resulted in a great decline in the area burned annually by fire (Norton-Griffiths, 1979). In HiP, grazer presence and fire frequency are also negatively correlated in space (Staver *et al.*, 2012). Hence, positive feedbacks with grazing, grazer-adapted grass communities and fire suppression on one hand; and fire, fire-adapted grass communities and grazer suppression on the other, can drive switches in ecosystem functioning. This implies that savanna landscapes are either herbivore-dominated (Brown World) or fire-dominated (Black World), depending on the past history of fire and herbivory and prevailing climatic conditions (Bond, 2005; Figure 10.7). In HiP there is a spatial transition from fire-dominated to herbivore-dominated landscapes moving southwards down the rainfall gradient – as evidenced by the adaptive traits of the main tree species (see Chapter 8). However, there is also evidence that parts of HiP have switched over time from dominance by browser-resistant *Acacia tortilis* to fire-resistant *Acacia karroo* (Bond *et al.*, 2001; Staver *et al.*, 2012).

That the vegetation structure and dominant consumer in a region are not set but can change over time depending on patterns and intensity of grazing and fire has important implications for our understanding and management of savanna ecosystems. In South Africa, rangeland managers generally apply fire to move animals off heavily grazed patches and promote tall tussock 'climax' grass communities (Tainton, 1985). However, the idea that the large frequent fires applied in many conservation areas throughout the latter half of last century might actually have been reinforcing a fire-feedback that has reduced the diversity and productivity of conservation areas is becoming more accepted (Hempson *et al.*, 2015b). Reducing fire extent could clash with other management goals, however, such as using fire to reduce shrub encroachment. It might also be difficult to achieve: grassy fuels tend to burn by accident when not burned by management fires (Van Wilgen *et al.*, 2004). It is likely that shifts from 'fire-dominated grass communities' to 'grazer dominated grass communities' are not easy to achieve through altering fire regimes, and that changes in grazer densities due to droughts, temporary release from predators, or the addition/removal of rhinos are more likely to have impacts.

10.4.4 Fire Spread, Threshold Behaviour, and System Switches

Fire spread is often modelled as an infection process (Sullivan, 2008), which means that one expects threshold behaviours: a small change in fuel continuity or fire weather could suddenly make a system 'connected' and result in large fire events (Cox and Durrett, 1988; Archibald *et al.*, 2012). In savannas, changes in fuel connectivity can happen very quickly as grasses cure in the dry season/after frost (increased connectivity) and as herbivores consume fuel (decreased connectivity). Evidence for threshold

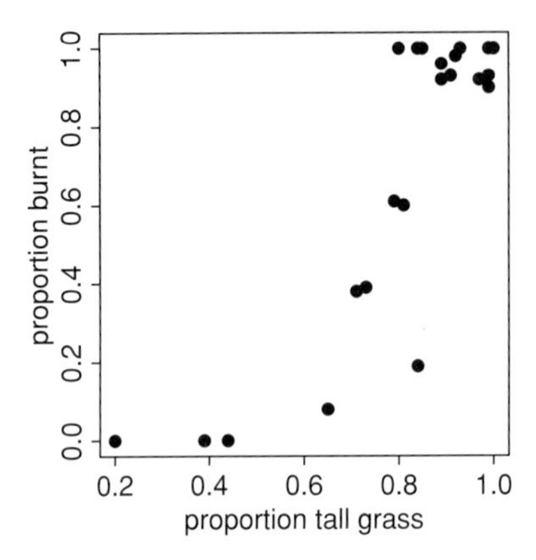

Figure 10.8 The proportion of a grassland that burns shows a clear threshold: unless ~60% of the area consists of tall (flammable) grass, fire will not spread. Above this threshold the proportion burnt quickly increases to cover the entire area. Therefore, as grazing reduces grass biomass over the season, it can drop below thresholds where fire can spread (unpublished data from William J. Bond).

behaviour in fire in savanna systems is rare: in Australia once the fuels and weather allow fires to burn through river channels there is a documented increase in fire size from the early to the late dry season (Yates *et al.*, 2008). Bond and Stock (unpublished data) demonstrated threshold behaviour in fire spread in HiP in relation to the proportion of short-grazed patches (Figure 10.8): fires stopped spreading through a plot when the proportion of non-flammable short grass was 40%, which fits precisely with theoretical percolation thresholds (Turner *et al.*, 1989).

That fire spread shows threshold behaviours has important implications for fire management. Below some threshold, fires fail to propagate and many ignitions would be needed to expand the area burned (Archibald *et al.*, 2012). For example, patch-mosaic burning requires that certain areas are burned early in the dry season when the grass is still green – i.e. when flammability is below the threshold that would allow fire to spread. This is achieved outside conservation areas through many ignitions (the combined efforts of very large human populations). Rangers asked to produce the same effect are often frustrated, and see the small, self-extinguishing fires as failed effort (Kruger Fire Management Workshop 2010). Because fire spread is controlled by both the spatial distribution of flammable material *and* the intrinsic flammability of the system (Turner *et al.*, 1989; Archibald *et al.*, 2012), these two drivers can compensate for each other; in cases where fire does not usually spread it may be possible to push systems beyond the fire spread threshold by burning under extreme weather conditions, or in particular spatial formations (see firestorm section).

Finally, this understanding is important when setting thresholds of potential concern (TPCs; Biggs and Rogers, 2003). Because savanna systems switch from being highly flammable to being much less flammable over a small range of environmental conditions, there are thresholds in tree cover or grassland composition where systems are likely to undergo rapid change that is potentially difficult to reverse. Identifying how these threshold dynamics play out in real landscapes would ensure that appropriate management actions are put into place before any thresholds are crossed.

10.5 Biogeochemistry and Fire

Studies of the effects of fire on biogeochemistry or, more specifically, nutrient cycling, for HiP are scant, and most of our knowledge is derived from research done in Kruger National Park (KNP) where experimental burn plots have been maintained over many decades. Findings suggest that frequent fires (annual and triennial burns) in KNP do not decrease carbon, plant available nitrogen and phosphorus pools, or nitrogen and phosphorus cycling rates (Coetsee *et al.*, 2008, 2010; Holdo *et al.*, 2012). However, Hartshorn *et al.* (2009) showed that triennial fire treatments burned under medium intensity had increased labile P when compared with plots that were not burned for about 50 years. Mills and Fey (2004) showed that annual fire in KNP caused increased crusting and decreased water infiltration in the short term. On basalt, frequent burning caused declines in organic matter and consequently in water retention capacity and increased compaction (Strydom *et al.*, 2014). These effects were manifested immediately post-fire but disappeared within about 2 years. Furthermore, less herbaceous cover on the annual burns led to lower evapotranspiration which left more water in the soil (Strydom *et al.*, 2014). Hence it appears that fire has negligible direct effects on soil nutrient and water availability in savannas.

Fire can indirectly affect soil nutrients by decreasing canopy cover (Coetsee *et al.*, 2010; Holdo *et al.*, 2012). Pellegrini *et al.* (2015) showed that soil carbon and nitrogen pools were decreased with frequent fire in KNP at a landscape scale as a combined result of decreased tree cover and increased combustion of grass. The stoichiometry (C:N:P) of plants did not change with fire return period. The resilience of African savannas to frequent fire is surprising in light of findings from savannas elsewhere (Ojima *et al.*, 1994; Reich *et al.*, 2001). African savannas are distinguished by retaining a full suite of extant herbivores, which remove some of the fuel for fires and in the process recycle some of the above-ground nutrients below ground. Resorption of nutrients by vegetation in the fire-prone season might also restrict losses (Bond and Van Wilgen, 1996; Hoffmann *et al.*, 2003). It appears that fire has limited direct effects on the availability of nutrients to plants and ultimately to animals. However, indirect impacts of fire on carbon and nitrogen dynamics via the impacts on woody cover may be substantial.

10.6 Savanna Fire Management in a Changing World

The future is projected to be a world of higher temperatures and higher atmospheric CO_2, with increasing human population densities outside protected areas, and continued spread of alien invasive plants into them. Consequences for fire regimes are varied. There is no evidence that increased temperatures affect the total area burned in savanna ecosystems (Scholes *et al.*, 2010) because fire spread is limited by fuel amount, not fuel moisture. However, higher temperatures might increase the likelihood of extremely large, and more intense, fires associated with extreme weather conditions (Figure 10.5). There is evidence that increased CO_2 has exacerbated woody thickening in HiP (Bond and Midgley, 2012; Chapter 7), which would reduce the extent of the area burned if grassy landscapes become fragmented below fire spread thresholds. There is much less information on the impacts of increasing CO_2 on the grass layer. High CO_2 can increase water-use efficiency (and reduce nutritional quality) in grasses (Polley *et al.*, 1994). This could increase fuel loads in dry areas, and negatively affect herbivore nutrition. *Chromolaena odorata*, the alien invader of greatest concern in HiP (Chapter 15), increases fire intensities by increasing fuel loads and flammability. It can also burn when it is green and thereby change the seasonality of fire. Most importantly, the concentration of *Chromolaena* infestations on boundaries between thicket or forest and grassland might increase the penetration of fires into these fire-sensitive communities.

Like many protected areas in Africa, HiP is surrounded by densifying human settlements (Chapter 1). Its fire regimes are increasingly disconnected from the surrounding landscape where human activities largely prevent extensive fires (Figure 10.2B). Nevertheless, many fires enter the park from outside and interact with prescribed burning activities.

Moreover, people allowed to harvest grass in HiP for thatching affect the times when managers can apply fires. Local Fire Protection Associations have the power to impose restrictions on when rangers can apply fires, but this is not currently a concern for HiP. However, the spread of fire into communities adjacent to the park is a major concern that influences the use of fire in HiP. Such escape fires lead to financial claims from surrounding communities, which makes it less likely for park managers to want to set intense fires.

Key issues for fire management in HiP in the future include the following.

1. Balancing the contrasting needs of a fire regime that controls woody thickening and alien invasive plants, but conserves indigenous forest, and does not constrain the heterogeneous environments created by grazing animals.
2. Understanding the constraints imposed by fuels, weather, and human ignitions, and the threshold dynamics of fire spread.
3. Addressing the needs of surrounding communities, the role of fire protection associations, and the expertise and time available for fire application by section rangers.

The mixture of landscapes in HiP, the wide rainfall range, and the largely intact mammal fauna, produce a diversity of fire types, sizes, intensities, and frequencies. Hence the park fulfils the goals of patch-mosaic burning in the main. Is this sufficient as a fire management goal? HiP is small enough for intensive fire management and in the past fire has often been directed towards achieving desired ecological outcomes. Management interventions with fire regimes (manipulating season, frequency, and intensity) are probably easier to achieve in the higher-rainfall Hluhluwe region where fire is the dominant consumer. Research in HiP and elsewhere has demonstrated that woody thickening can be combated through the application of particular types of fires (with the assistance of browsing animals). We understand the impacts of fire on thicket and forest vegetation, and the types of fires that could lead to state switches. Managing fire and herbivory in concert, rather than independently, has the potential to improve our ability to manipulate vegetation composition and structure. Finally, our understanding of fire behaviour should enable management actions to be initiated before thresholds are crossed. Thus it appears that we know how to use fire to achieve particular goals. Future challenges will be defining these goals – something that requires extending these discussions beyond the scientific community.

10.7 References

Aitken, R. D. & Gale, G. W. (1921) *Botanical survey of Natal and Zululand*. Government Printing and Stationery Office, Pretoria.

Archibald, S. (2008) African grazing lawns – how fire, rainfall, and grazer numbers interact to affect grass community states. *Journal of Wildlife Management* **72**: 492–501.

Archibald, S. & Bond, W. J. (2004) Grazer movements: spatial and temporal responses to burning in a tall-grass African savanna. *International Journal of Wildland Fire* **13**: 377–385.

Archibald, S., Bond, W. J., Stock, W. D., & Fairbanks, D. H. K. (2005) Shaping the landscape: fire–grazer interactions in an African savanna. *Ecological Applications* **15**: 96–109.

Archibald, S., Roy, D. P., van Wilgen, B. W., & Scholes, R. J. (2009) What limits fire? An examination of drivers of burnt area in Southern Africa. *Global Change Biology* **15**: 613–630.

Archibald, S., Nickless, A., Govender, N., Scholes, R. J., & Lehsten, V. (2010a) Climate and the inter-annual variability of fire in southern Africa. *Global Ecology and Biogeography* **19**: 794–809.

Archibald, S., Scholes, R. J., Roy, D. P., Roberts, G., & Boschetti, L. (2010b) Southern African fire regimes as revealed by remote sensing. *International Journal of Wildland Fire* **19**: 861–878.

Archibald, S., Staver, A. C., & Levin, S. A. (2012) Evolution of human-driven fire regimes in Africa. *Proceedings of the National Academy of Sciences* **109**: 847–852.

Archibald, S., Lehmann, C. E., Gómez-Dans, J. L., & Bradstock, R. A. (2013) Defining pyromes and global syndromes of fire regimes. *Proceedings of the National Academy of Sciences* **110**: 6442–6447.

Balfour, D. A. & Howison, O. E. (2001) Spatial and temporal variation in a mesic savanna fire regime: responses to variation in annual rainfall. *African Journal of Range and Forage Science* **19**: 45–53.

Berry, A. & Macdonald, I. A. W. (1979) Fire regime characteristics in the Hluhluwe–Corridor–Umfolozi Game Reserve Complex in Zululand. Area description and an analysis of causal factors and seasonal incidence of fire in the central complex with particular reference to the period 1955 to 1978. Unpublished report, Natal Parks Board, Pietermaritzburg.

Biggs, H. C. & Potgieter, A. L. F. (1999) Overview of the fire management policy of the Kruger National Park. *Koedoe* **42**: 101–111.

Biggs, H. C. & Rogers, K. H. (2003) An adaptive system to link science, monitoring and management in practice. In *The Kruger experience: ecology and management of savanna heterogeneity* (eds. J. T. du Toit, K. H. Rogers, & H. C. Biggs), pp. 59–80. Island Press, Washington, DC.

Bond, W. J. (2005) Large parts of the world are brown or black: a different view on the 'Green World' hypothesis. *Journal of Vegetation Science* **16**: 261–266.

Bond, W. J. & Archibald, S. (2003) Confronting complexity: fire policy choices in South African savanna parks. *International Journal of Wildland Fire* **12**: 381–389.

Bond, W. J. & Midgley, G. F. (2012) Carbon dioxide and the uneasy interactions of trees and savannah grasses. *Philosophical Transactions of the Royal Society B: Biological Sciences* **367**: 601–612.

Bond, W. J. & Van Wilgen, B. W. (1996) *Fire and plants*. Population and Community Biology series Vol. 14, Chapman and Hall, London.

Bond, W. J., Smythe, K. A. & Balfour, D. A. (2001) Acacia species turnover in space and time in an African savanna. *Journal of Biogeography* **28**: 117–128.

Bradstock, R. A. (2010) A biogeographic model of fire regimes in Australia: contemporary and future implications. *Global Ecology and Biogeography* **19**: 145–158.

Brockett, B. H., Biggs, H. C., & Van Wilgen, B. W. (2001) A patch mosaic burning system for conservation areas in southern African savannas. *International Journal of Wildland Fire* **10**: 169–183.

Browne, C. & Bond, W. (2011) Firestorms in savanna and forest ecosystems: curse or cure? *Veld & Flora* **97**: 62–63.

Burkepile, D. E., Burns, C. E., Tambling, C. J., *et al.* (2013) Habitat selection by large herbivores in a southern African savanna: the relative roles of bottom-up and top-down forces. *Ecosphere* **4**: 1–19.

Coetsee, C., February, E. C., & Bond, W. J. (2008) Nitrogen availability is not affected by frequent fire in a South African savanna. *Journal of Tropical Ecology* **24**: 647–654.

Coetsee, C., Bond, W. J., & February, E. C. (2010) Frequent fire affects soil nitrogen and carbon in an African savanna by changing woody cover. *Oecologia* **162**: 1027–1034.

Cox, J. T. & Durrett, R. (1988) Limit theorems for the spread of epidemics and forest fires. *Stochastic Processes and their Applications* **30**: 171–191.

East, R. (1984) Rainfall, soil nutrient status and biomass of large African savanna mammals. *African Journal of Ecology* **22**: 245–270.

Fritz, H. & Duncan, P. (1994) On the carrying capacity for large ungulates of African savanna ecosystems. *Proceedings of the Royal Society of London B: Biological Sciences* **256**: 77–82.

Gill, A. M. (1975) Fire and the Australian flora: a review. *Australian Forestry* **38**: 4–25.

Govender, N., Trollope, W. S., & Van Wilgen, B. W. (2006) The effect of fire season, fire frequency, rainfall and management on fire intensity in savanna vegetation in South Africa. *Journal of Applied Ecology* **43**: 748–758.

Hantson, S., Pueyo, S., & Chuvieco, E. (2015) Global fire size distribution is driven by human impact and climate. *Global Ecology and Biogeography* **24**: 77–86.

Hartshorn, A. S., Coetsee, C., & Chadwick, O. A. (2009) Pyromineralization of soil phosphorus in a South African savanna. *Chemical Geology* **267**: 24–31.

Hempson, G. P., Archibald, S., & Bond, W. J. (2015a) A continent-wide assessment of the form and intensity of large mammal herbivory in Africa. *Science* **350**: 1056–1061.

Hempson, G. P., Archibald, S., Bond, W. J., *et al.* (2015b) Ecology of grazing lawns in Africa. *Biological Reviews* **90**: 979–994.

Henkel, J. S. (1937) *Report on the plant and animal ecology of the Hluhluwe Game Reserve, with special reference to tsetse flies*. The Natal Witness, Pietermaritzburg, South Africa.

Hennenberg, K. J., Fischer, F., Kouadio, K., *et al.* (2006) Phytomass and fire occurrence along forest savanna transects in the Comoe National Park, Ivory Coast. *Journal of Tropical Ecology* **22**: 303–311.

Higgins, S. I., Bond, W. J., & Trollope, W. S. (2000) Fire, resprouting and variability: a recipe for grass–tree coexistence in savanna. *Journal of Ecology* **88**: 213–229.

Hoffmann, W. A. & Solbrig, O. T. (2003) The role of topkill in the differential response of savanna woody species to fire. *Forest Ecology and Management* **180**: 273–286.

Hoffmann, W. A., Orthen, B., & Nascimento, P. K. V. D. (2003) Comparative fire ecology of tropical savanna and forest trees. *Functional Ecology* **17**: 720–726.

Holdo, R. M., Mack, M. C., & Arnold, S. G. (2012) Tree canopies explain fire effects on soil nitrogen, phosphorus and carbon in a savanna ecosystem. *Journal of Vegetation Science* **23**: 352–360.

Keeley, J. E. & Rundel, P. W. (2005) Fire and the Miocene expansion of C4 grasslands. *Ecology Letters* **8**: 683–690.

Korontzi, S., Justice, C. O., & Scholes, R. J. (2003) Influence of timing and spatial extent of savanna fires in southern Africa on atmospheric emissions. *Journal of Arid Environments* **54**: 395–404.

Lehmann, C. E. R., Archibald, S. A., Hoffmann, W. A., & Bond, W. J. (2011) Deciphering the distribution of the savanna biome. *New Phytologist* **191**: 197–209.

Little, J. K., Prior, L. D., Williamson, G. J., Williams, S. E., & Bowman, D. M. (2012) Fire weather risk differs across rain forest–savanna boundaries in the humid tropics of north-eastern Australia. *Austral Ecology* **37**: 915–925.

Macdonald, I. A. W. & Frame, G. W. (1988) The invasion of introduced species into nature reserves in tropical savannas and dry woodlands. *Biological Conservation* **44**: 67–93.

Maurin, O., Davies, T. J., Burrows, J. E., *et al.* (2014) Savanna fire and the origins of the 'underground forests' of Africa. *New Phytologist* **204**: 201–214.

McKenzie, D., Miller, C. & Falk, D. A. (2011) *The landscape ecology of fire*. Springer, Berlin.

Mills, A. J. & Fey, M. V. (2004) Frequent fires intensify soil crusting: physicochemical feedback in the pedoderm of long-term burn experiments in South Africa. *Geoderma* **121**: 45–64.

Mucina, L. & Rutherford, M. C. (2006) *The vegetation of South Africa, Lesotho and Swaziland*. South African National Biodiversity Institute, Pretoria.

Norton-Griffiths, M. (1979) The influence of grazing, browsing, and fire on the vegetation dynamics of the Serengeti. In: *Serengeti: dynamics of an ecosystem* (eds A. R. E. Sinclair & M. Norton-Griffiths), pp. 310–352. University of Chicago Press, Chicago.

Ojima, D. S., Schimel, D. S., Parton, W. J., & Owensby, C. E. (1994) Long- and short-term effects of fire on nitrogen cycling in tallgrass prairie. *Biogeochemistry* **24**: 67–84.

Osborne, C. P. (2008) Atmosphere, ecology and evolution: what drove the Miocene expansion of C(4) grasslands? *Journal of Ecology* **96**: 35–45.

Parr, C. L. & Anderson, A. N. (2006) Patch mosaic burn for biodiversity conservation: a critique of the pyrodiversity paradigm. *Conservation Biology* **20**: 1610–1619.

Pellegrini, A. F. A., Hedin, L. O., Staver, A. C., & Govender, N. (2015) Fire alters ecosystem carbon and nutrients but not plant nutrient stoichiometry or composition in tropical savanna. *Ecology* **96**: 1275–1285.

Polley, H. W., Johnson, H. B., & Mayeux, H. S. (1994) Increasing CO_2: comparative responses of the C4 grass *Schizachyrium* and grassland invader *Prosopis*. *Ecology* **75**: 976–988.

Radke, J. (2013) Fire and firestorms. In: *Encyclopedia of natural hazards SE – 134* (ed. P. T. Bobrowsky), pp. 323–324. Encyclopedia of Earth Sciences series. Springer Dordrecht.

Reich, P. B., Peterson, D. W., Wedin, D. A., & Wrage, K. (2001) Fire and vegetation effects on productivity and nitrogen cycling across a forest–grassland continuum. *Ecology* **82**: 1703–1719.

Sankaran, M., Hanan, N. P., Scholes, R. J., *et al.* (2005) Determinants of woody cover in African savannas. *Nature* **438**: 846–849.

Scheffer, M. & Carpenter, S. R. (2003) Catastrophic regime shifts in ecosystems: linking theory to observation. *Trends in Ecology & Evolution* **18**: 648–656.

Scholes, R. J. & Archer, S. R. (1997) Tree–grass interactions in savannas. *Annual Review of Ecological Systematics* **28**: 517–544.

Scholes, R. J. & Walker, B. H. (1993) *An African savanna: synthesis of the Nylsvley study*. Cambridge University Press, Cambridge.

Scholes, R. J., Archibald, S., Colvin, C., *et al.* (2010) *Global change risk analysis: understanding and reducing key risks to ecosystem services associated with climate change in South Africa*. p. 21.

Scott, J. D. (1955) Principles of pasture management. In: *The grasses and pastures of South Africa* (ed. D. Meredith), pp. 601–623. Central News Agency, Johannesburg.

Simon, M. F. & Pennington, T. (2012) Evidence for adaptation to fire regimes in the tropical savannas of the Brazilian cerrado. *International Journal of Plant Sciences* **173**: 711–723.

Skowno, A. L., Midgley, J. J., Bond, W. J., & Balfour, D. (1999) Secondary succession in *Acacia nilotica* (L.) savanna in the Hluhluwe Game Reserve, South Africa. *Plant Ecology* **145**: 1–9.

Staver, A. C., Bond, W. J., Stock, W. D., van Rensburg, S. J., & Waldram, M. S. (2009) Browsing and fire interact to suppress tree density in an African savanna. *Ecological Applications* **19**: 1909–1919.

Staver, A. C., Bond, W. J., Cramer, M. D., & Wakeling, J. L. (2012) Top-down determinants of niche structure and adaptation among African Acacias. *Ecology Letters* **15**: 673–679.

Strydom, T., Rowe, T., Riddell, E., Govender, N., & Lorentz, S. (2014) *Pyrohydrology in African savannas*. Report 2146/1/14. Water Research Commission, Pretoria.

Sullivan, A. L. (2008) Wildland surface fire spread modelling, 1990–2007. 3: Simulation and mathematical analogue models. *International Journal of Wildland Fire* **18**: 387–403.

Tainton, N. M. (1985) Recent trends in grazing management philosophy in South Africa. *Journal of the Grassland Society of Southern Africa* **2**: 4–6.

Tainton, N. M., Groves, R. H., & Nash, R. (1977) Time of mowing and burning veld: short term effects on production and tiller development. *Proceedings of the Annual Congresses of the Grassland Society of Southern Africa* **12**: 59–64.

te Beest, M., Cromsigt, J. P. G. M., Ngobese, J., & Olff, H. (2012) Managing invasions at the cost of native habitat? An experimental test of the impact of fire on the invasion of *Chromolaena odorata* in a South African savanna. *Biological Invasions* **14**: 607–618.

Tomor, B. M. & Owen-Smith, N. (2002) Comparative use of burnt and unburnt grassland by grazing ungulates in the Nylsvley nature reserve, South Africa. *African Journal of Ecology* **40**: 201–204.

Trollope, W. S. W. (1974) Role of fire in preventing bush encroachment in the Eastern Cape. *Proceedings of the Annual Congresses of the Grassland Society of Southern Africa* **9**: 67–72.

Trollope, W. S. W. & Tainton, N. M. (2007) Effect of fire intensity on the grass and bush components of the Eastern Cape thornveld. *African Journal of Range and Forage Science* **3**: 37–42.

Trollope, W. S. W., Trollope, L. A., Biggs, H. C., Pienaar, D., & Potgieter, A. L. F. (1998) Long-term changes in the woody vegetation of the Kruger National Park, with special reference to the effects of elephants and fire. *Koedoe* **41**: 103–112.

Turner, M. G., Gardner, R. H., Dale, V. H., & O'Neill, R. V. (1989) Predicting the spread of disturbance across heterogeneous landscapes. *Oikos* **55**: 121–129.

Twidwell, D., Rogers, W. E., Fuhlendorf, S. D., *et al.* (2013) The rising Great Plains fire campaign: citizens' response to woody plant encroachment. *Frontiers in Ecology and the Environment* **11**: 64–71.

Van Wilgen, B. W., Govender, N., Biggs, H. C., Ntsala, D., & Funda, X. N. (2004) Response of savanna fire regimes to changing fire-management policies in a large African national park. *Conservation Biology* **18**: 1533–1540.

Van Wilgen, B. W., Govender, N. & MacFadyen, S. (2008) An assessment of the implementation and outcomes of recent changes to fire management in the Kruger National Park. *Koedoe* **50**: 22–31.

Vincent, J. (1970) The history of Umfolozi Game Reserve, Zululand, as it relates to management. *Lammergeyer* **11**: 7–48.

Wakeling, J. L. & Bond, W. J. (2007) Disturbance and the frequency of root suckering in an invasive savanna shrub, *Dichrostachys cinerea*. *African Journal of Range and Forage Science* **24**: 73–76.

Wakeling, J. L., Staver, A. C., & Bond, W. J. (2011) Simply the best: the transition of savanna saplings to trees. *Oikos* **120**: 1448–1451.

Waldram, M. S., Bond, W. J., & Stock, W. D. (2008) Ecological engineering by a mega-grazer: white rhino impacts on a South African savanna. *Ecosystems* **11**: 101–112.

Ward, C. J. (1962) Report on scrub control in the Hluhluwe Game Reserve. *Lammergeyer* **2**: 57–62.

Wigley, B. J., Bond, W. J., & Hoffman, M. (2010) Thicket expansion in a South African savanna under divergent land use: local vs. global drivers? *Global Change Biology* **16**: 964–976.

Williams, R. J. & Bradstock, R. A. (2009) Large fires and their ecological consequences: introduction to the special issue. *International Journal of Wildland Fire* **17**: 685–687.

Wilsey, B. J. (1996) Variation in use of green flushes following burns among African ungulate species: the importance of body size. *African Journal of Ecology* **34**: 32–38.

Yates, C. P., Edwards, A. C., & Russell-Smith, J. (2008) Big fires and their ecological impacts in Australian savannas: size and frequency matters. *International Journal of Wildland Fire* **17**: 768–781.

Yoganand, K. & Owen-Smith, N. (2014) Restricted habitat use by an African savanna herbivore through the seasonal cycle: key resources concept expanded. *Ecography* **37**: 969–982.

Part III

Where Science and Conservation Management Meet

11 · *Rhino Management Challenges: Spatial and Social Ecology for Habitat and Population Management*

WAYNE L. LINKLATER AND ADRIAN M. SHRADER

11.1 Introduction

No other protected area has contributed as much to the conservation of Africa's white (*Ceratotherium simum* var. *simum*) and black (*Diceros bicornis* var. *minor*) rhinos as the Hluhluwe-iMfolozi Park (HiP). In 1895, when the Hluhluwe and Umfolozi Game Reserves (GR) were proclaimed, southern Africa's only remaining white rhinos were restricted to the Umfolozi GR and probably numbered fewer than 100 individuals (Owen-Smith, 1981). Black rhinos survived more widely in Zululand, but estimates of their numbers were not reported. By 1970, the white rhino population in HiP had grown to ~2000 animals (Owen-Smith, 1981; Figure 11.1A), while black rhino numbers exceeded 300 (Brooks and Macdonald, 1983; Figure 11.1B).

During the 1970s and 1980s, illegal hunting escalated across Africa due to growing demand for rhino horn for Eastern medicine and handles for Yemeni jambiya daggers (Western and Vigne, 1985). In 1976, both black and white rhinos (along with the three Asian species) were listed in Appendix I of the Convention on International Trade in Endangered Species, in an attempt to eliminate international trade in rhino products. Nevertheless, the continental black rhino population was reduced from ~65,000 individuals in 1970 to only ~2480 by 1992 (African Rhino Specialist Group, 1991, 1992). Within southern Africa, black rhinos were eliminated from Botswana and severely reduced in Zimbabwe (Milliken,

Conserving Africa's Mega-Diversity in the Anthropocene, ed. Joris P. G. M. Cromsigt, Sally Archibald and Norman Owen-Smith. Published by Cambridge University Press.

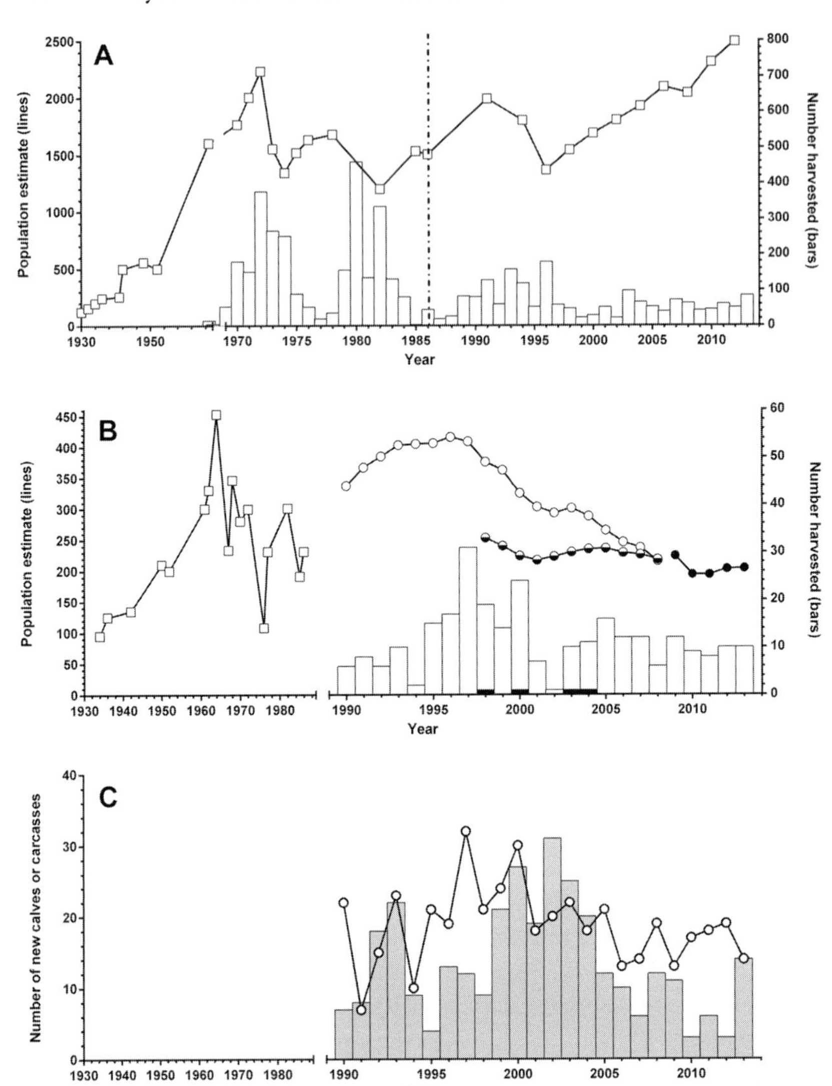

Figure 11.1 (A) White rhino population estimates (□, 1930–) and live harvest (white bar, 1970–). The vertical dashed line indicates when the Sink Management Policy was initiated. (B) Black rhino population estimates (i) prior to 1990 (□, running totals of sporadic separate estimates from Hluhluwe GR, Umfolozi GR and the Corridor), (ii) 1990–2008 (O) using ear-notched individuals and mark–resight techniques, (iii) 1998–2008 (◒) from revised mark–recapture estimates (Clinning *et al.*, 2009), and (iv) the known entirely marked population 2009–2013 (●), plus live harvest (white bar) and introduction numbers (black bar) (1990–2013). (C) Black rhinoceros calves (O) and carcasses (grey bar) reported each year (1990–2013).

1993). In contrast, rhinos within South Africa were less affected. From 1970 to 1994, poaching of both rhino species within HiP was restricted to fewer than five rhinos per year. Hence this park became the source of rhinos for establishing numerous small populations in captivity and for re-introduction and restocking across the continent.

Growing populations of both rhino species in HiP initially represented an international conservation success (Player, 1972; Emslie and Brooks, 1999). However, the strategic importance of HiP for rhino conservation brought more complex challenges. In this chapter, we contrast the population and management histories of the two rhino species in HiP, and consider the uncertainties for future management. As a framework, we provide hypotheses and predictions for adaptive management so that the role of density-dependent and density-independent influences on rhino vital rates and dispersal in HiP can be tested.

11.2 Population Success, Ecological Challenges

By the 1970s, concerns were being raised about the impact of the growing white rhino population on the grass cover and soils in HiP (Owen-Smith, 1973). White rhino grazing was responsible for a reduction in grass cover which exposed soils and stream banks to increased erosion. The completion of a rhino-proof fence around the park by 1966 threatened to compound this perceived problem (Owen-Smith, 1981; Chapter 5). The spectre grew of an overpopulation situation with adverse consequences for the environment and other animal species.

Black rhinos posed the opposite problem. After the discovery of 46 fresh black rhino carcasses in Hluhluwe GR in 1961 (Emslie, 1999) – a death-event for which reasons remain unknown – estimates of black rhino numbers continued to decline into the early 1970s (Figure 11.1B). Despite subsequent recovery, mainly through expansion in numbers in Umfolozi GR, an apparent further decline ensued during the late 1990s. This trend is now understood to have been an artefact of mark–resight estimates with certain assumptions violated (Clinning *et al.*, 2009). Nevertheless, concerns about the health of the black rhino population and the quality of its environment, especially its browse food resource, came to the fore.

Hence white and black rhinos pose different challenges requiring divergent management approaches. New strategies for both species were facilitated by technical advances in large animal capture.

11.3 Advances in Large Animal Capture

Collaboration between park managers and wildlife veterinarians during the late 1950s pioneered the use of chemical immobilization to capture animals as large as rhinos (Harthoorn, 1962a, 1962b; Player, 1972). This made it possible to redistribute rhinos not only in Africa but also internationally (King and Carter, 1965). Initially, rhinos were sent to zoos as an insurance against extinction in the wild (Player, 1972). Furthermore, the routine capture of large numbers of rhinos enabled HiP to become the source for the re-introduction and restocking of rhinos elsewhere.

Since the 1960s, > 4500 white rhinos and > 250 black rhinos have been transferred from HiP to various zoos, wildlife reserves, and private game ranches. For example, the white rhino population within Kruger National Park, which reached around 10,000 animals (Ferreira *et al.*, 2012), was derived from 336 white rhinos relocated from HiP between 1961 and 1972. The black rhino population there exceeds 600 animals (Ferreira *et al.*, 2011), founded when 20 black rhinos were relocated from HiP in 1971 (later augmented with others from Zimbabwe; Pienaar, 1970; Pienaar *et al.*, 1992; Emslie *et al.*, 2009). Twenty-eight black rhinos were brought from HiP to Malilangwe, Zimbabwe in 1998 (Emslie *et al.*, 2009). White rhinos have been re-introduced into protected areas in Botswana, Mozambique, Zimbabwe, and Namibia, and moved even as far as Kenya, although the species was historically absent there.

Currently, HiP is home to ~2500 white rhinos and ~200 black rhinos (Figure 11.1A,B), with populations of both species managed by live capture and translocation elsewhere. However, this approach has posed new challenges requiring a deeper understanding of rhino ecology and behaviour.

11.4 Density Dependence and Compensatory Population Growth

The recovery of both African rhinos globally depends on remaining populations functioning as sources of animals for re-introduction and restocking. According to the theory underlying sustainable harvesting, reductions in population density should promote compensatory increases in reproduction and survival because food or other resource limitations are alleviated (Rosenberg *et al.*, 1993). Hence, source populations should be able to provide a sustained supply of 'surplus' animals for relocation elsewhere. Being initially stocked at low density, the relocated animals should show higher rates of reproduction and better survival, particularly of young animals, than the more crowded source population.

Logistic harvest models assume immediate demographic responses as soon as the population density is reduced. However, the spatial and social processes that underlie the population trend are more complex. Individual animals differ not only in age and sex, but also in their social relationships and hence ability to move freely in response to density reductions. The effects of removals are not experienced by individuals living remote from the place from which animals were taken. There may also be delays before the effects of reduced competition for food become manifested in survival and reproductive rates. Over some range in density, vital rates may remain uninfluenced by density reductions (Fowler, 1981). Limitations of current behavioural and ecological knowledge are recognized impediments to the sustainable harvest of exploited species (Reynolds *et al.*, 2001).

Following animal removals, density-dependent responses in population growth rate might not take place, or be delayed, if any of the following circumstances apply.

1. More productive individuals are preferentially removed.
2. The availability of food resources is not improved.
3. The space made vacant is not immediately utilized by the remaining animals.
4. Remaining animals experience greater pressure from predators responding to higher densities of other prey species (Courchamp *et al.*, 2000).

Rates of recolonization following harvesting are rarely documented. Recolonization can be rapid for highly mobile species, particularly where the population surrounding the harvested area includes many young, non-breeding animals. However, species with slow life histories, like rhinos (Owen-Smith, 1988), are generally slow to disperse into vacant habitat. Population redistribution may be delayed by spatially loyal behaviours (territoriality or home range fidelity; Stamps and Swaisgood, 2007). Where recolonization is slow, the size and spatial arrangement of harvested areas become important considerations (Novaro *et al.*, 2005). Larger harvested areas are recolonized more slowly and recolonization rates are reduced if they are in close proximity to other harvested areas. Moreover, higher-quality habitats support higher animal densities, and if these localities are repeatedly harvested (e.g. density-dependent capture bias) a greater proportion of the population will be left in poorer habitat. Thus, for species that do not respond spatially to harvest by rapid recolonization, a poor recruitment response overall might result.

The effects of consumption on food quality have been demonstrated for large grazers, like white rhinos, which can maintain 'grazing lawns' providing low-fibre and hence relatively nutritious forage (Waldram, 2005; Chapter 6). The concept has been extended to browsers that maintain 'hedges' of low-growing shrubs (Makhabu and Skarpe, 2006; Fornara and du Toit, 2007; Cromsigt and Kuijper, 2011). Grasses and shrubs can grow taller and hence more fibrous, or beyond the reach of herbivores, if grazing or browsing pressure is reduced by removals.

Lastly, smaller species able to respond more rapidly might benefit from the food made available by rhino removals, and pre-empt gains by the larger competitor. For black rhinos, these include kudu (*Tragelaphus strepsiceros*), nyala (*T. angasi*), and impala (*Aepyceros melampus*), which also browse the low-growing shrubs favoured by black rhinos.

11.5 Live Harvest of Rhinos

The initial motivation for live removals of white rhinos was to alleviate a threatened overpopulation situation associated with the risk of a population crash and adverse consequences for other species. However, an important aim was also to distribute the species more widely as insurance against a possible disaster in HiP, such as a lethal disease outbreak. For black rhinos, population management was more narrowly aimed at establishing viable populations outside HiP. Hence the contexts for managing the two rhino species within HiP were quite different – white rhinos were numerous and growing rapidly while black rhino numbers were much lower and the population seemed to be shrinking.

11.5.1 Dispersal Sink Management of White Rhinos

The growth rate of the white rhino population up until 1971, incorporating animals removed after 1962, was a constant 9.5% per year (Owen-Smith, 1981). Assuming logistic growth, maximum sustainable yields are obtained by reducing the population towards half of the ultimate carrying capacity and harvesting at half of the maximum growth rate. However, judging from the habitat changes occurring, the total of 2000 white rhinos present in HiP in 1970 was not far below the maximum number that could be supported despite the lack of any density-dependent reduction in the population growth rate. This is consistent with Fowler's

(1981) finding that large mammal populations do not show demographic responses to food shortfalls until their abundance is close to carrying capacity (see Chapter 5). Simply counteracting the annual population growth in this situation would require the removal of almost 200 rhinos every year. Removals could be reduced by about one-third if concentrated selectively on subadults approaching maturity, which have a high future reproductive value. However, this high rate of harvest would be difficult to maintain and disruptive.

It would be more effective to allow the population to approach closer to carrying capacity so that density-related influences on birth intervals, calf survival, and age at first reproduction would come into play, but the ultimate carrying capacity was not known, and changes in these vital rates might be too slow to avoid an over-shoot of the carrying capacity. The only regulatory mechanism that could be sufficiently quick-acting is dispersal, i.e. movements of animals out of densely settled areas. Estimates of dispersal rates could be obtained from changes in the regional distribution of white rhinos recorded in successive aerial censuses, particularly in the numbers of white rhinos moving beyond the protected area before the fence was completed. Between 1953 and 1970, the difference between the overall population growth rate and the rate of increase within the highest-density region between the two Mfolozi rivers amounted to about 3% per year (Owen-Smith, 1988). Dispersal movements were mostly by subadult animals of both sexes, plus some adult males. For the subadult segment, the specific dispersal rate was estimated to be 7.5% per year. This dispersal rate could potentially stabilize the population if coupled ultimately with modest density-dependent reductions in fecundity and offspring survival as well as delayed maturity. However, changes in rates of reproduction and survival would need to be more drastic to halt population growth in the absence of dispersal.

The biggest question, however, was how dispersal could take place after the park became completely fenced. The proposed solution was to establish sink zones within the fenced area from which most rhinos settling would be removed (Owen-Smith, 1973, 1981, 1988). No removals should take place in the remainder of HiP, where white rhinos would be allowed to establish the carrying capacity naturally through dispersing into the sinks when food ran short. Positive intrinsic growth within the core area would be counterbalanced by negative growth within the sink zones (Pulliam, 1988), maintained by live removals. Other species threatened by the grassland changes induced by high densities of white rhinos in the core area could persist in the sink zones.

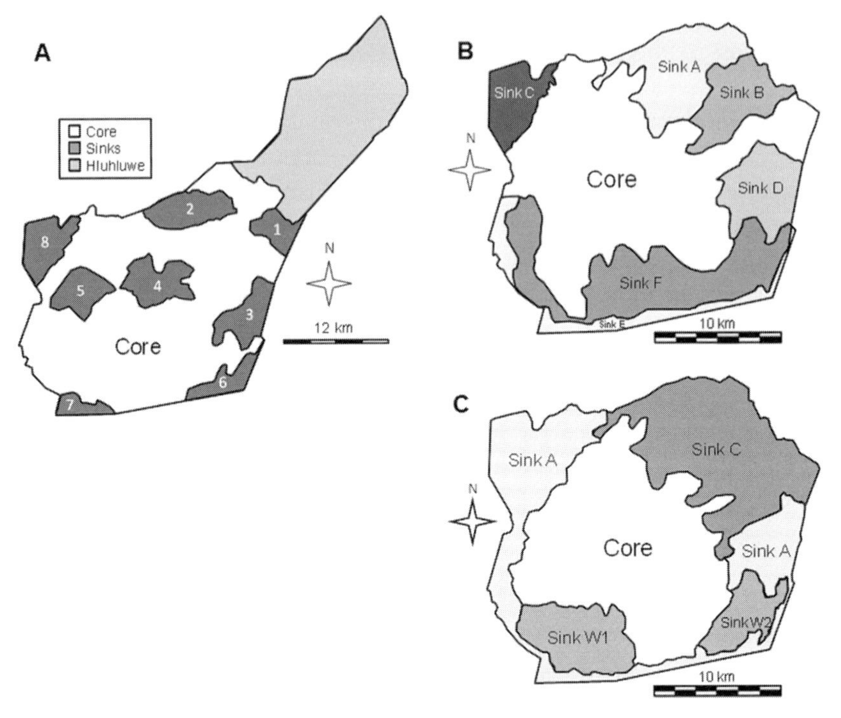

Figure 11.2 Changing locations of designated dispersal sinks for white rhinos within the Hluhluwe-iMfolozi Park. (A) Sinks initially established in 1986. (B) Sinks maintained from 1992 to 1996. (C) Current location of sinks established in 1997.

Dispersal sink management of the white rhino population was eventually initiated in 1986 (Conway *et al.*, 2001). The aim was to maintain white rhino density within the sinks at around one rhino/km^2, with the number of animals exceeding this density being removed annually. However, some of the sink zones seemed too small to be effective and two of the sink zones were inappropriately located centrally within Mfolozi (Figure 11.2A). Therefore, in 1992 the designated sinks were enlarged and shifted so that all lay on the periphery of Mfolozi (Figure 11.2B; Maddock, 1992). In Mfolozi, the resulting central core area covered 302 km^2 and was surrounded by sink zones totalling 433 km^2. Within Hluhluwe, merely a fixed upper limit of 500 white rhinos was set because of its narrow confines. A benefit of the peripheral arrangement was that white rhino densities were lowered near park boundaries where poaching risks were highest (Maddock, 1992; Balfour, 1999). In 1997, minor adjustments were made to the sink boundaries so that they ran around the entire periphery of the park (Figure 11.2C). For Hluhluwe, the fixed upper limit was raised to 700 animals because the white rhino population there had remained relatively stable at ~500 individuals since the late 1980s despite removals amounting to only 2% of the local population per year. The limit was then removed in 2005 as the white rhino density within Hluhluwe was thought to be too low to maintain grazing lawns in that high-rainfall section of the park. As shown in Figure 11.1A, removals overall have generally remained at fewer than 100 white rhinos per year over the past two decades. Nevertheless, the white rhino population has continued to grow.

Sinks within the current configuration (Figure 11.2C) are managed differently. In the two sinks labelled A (144 km^2), all animals found within them, except for females with calves <1 year old, are removed annually. In the Corridor Sink C (114 km^2), white rhinos are removed only once the density in the core reaches 2.5 rhinos/km^2, as a safety valve to prevent the white rhinos in the core from potentially exceeding what available resources can support. Sinks W1 and W2 (30 and 20 km^2, respectively) lie within the wilderness section of Mfolozi where roads are excluded. Hence, capture vehicles are not allowed to enter, making removing rhinos from these areas logistically difficult. To overcome this restriction, helicopters have been used to transport rhinos out of the sink (Cooke, 1998), meaning that capture is more costly. Accordingly, white rhinos are removed from the wilderness sinks only after the white rhino density in the core area of the park exceeds three rhinos/km^2.

11.5.2 Metapopulation Management of Black Rhinos

In contrast to the sink management policy for white rhinos, the conservation strategy for black rhinos in HiP is outwardly focused towards metapopulation expansion to overcome the isolation and relatively small size of the HiP population. However, much dissension exists about the magnitude of the removals of black rhinos that the park can sustain. Between 1990 and 2013, 2–31 black rhinos were harvested from HiP each year (Figure 11.1B), totalling 276 animals. The failure of the population to compensate has been interpreted as evidence for underharvesting (Emslie, 2001). Based on the logistic model, reducing the population density to 75% or less of ecological carrying capacity would be required to generate compensatory increases in rates of survival and reproduction and hence much higher population growth. Others have suggested that the population has been reduced beyond its capacity to compensate given current habitat conditions (Balfour, 2001). More recently, it has been recommended that a constant proportional harvest of 5% should be removed each year (Goodman, 2001; Cromsigt *et al.*, 2002).

Uncertainties in estimates of the black rhino population by different methods may have contributed to overharvesting (Clinning *et al.*, 2009). Historical highs and the subsequent declines were not as large as formerly believed (Figure 11.1B). Taking into account animals removed, the black rhino population is estimated to have increased intrinsically at 3.4% per annum between 1998 and 2008. However, removals during this period averaged 5.2 ± 0.8% of the population per year, and have been opportunistically and patchily distributed over the park (Figure 11.3). Whether the population growth rate would increase if the density were to be reduced further is debatable, particularly following Fowler's (1981) suggestion that density dependence only kicks in close to carrying capacity.

Choices of the black rhinos to be removed were influenced both by HiP section rangers and the expectations or requirements of those receiving the rhinos. Only females without calves were captured, potentially including a high proportion of young animals contributing most to future population growth (Clinning *et al.*, 2009). Rangers tended to favour removing old males that might already have contributed genes to the population, or subadults yet to reproduce, rather than prime-aged animals. Buyers wanted young or prime-aged adults with high reproductive potential.

11.6 Evaluating Rhino Population Responses

11.6.1 White Rhino

Between 1962 and 1974, white rhino removals were aimed at reducing the population to alleviate 'overgrazing'. Thereafter until 1985 the number of white rhinos harvested was adjusted annually to maintain numbers within some crudely estimated carrying capacity, with more individuals being removed during dry years (Brooks and Macdonald, 1983). The annual off-take between 1974 and 1985 averaged 146 animals (range 3–460) and amounted to 10.5% of the population. This effectively suppressed population increase. When surprisingly few herbivores died during the severe 1982/3 drought (Walker *et al.*, 1987), it became apparent that removals of white rhinos and other grazers had been excessive. From 1986 onwards the sink management policy was applied and the number

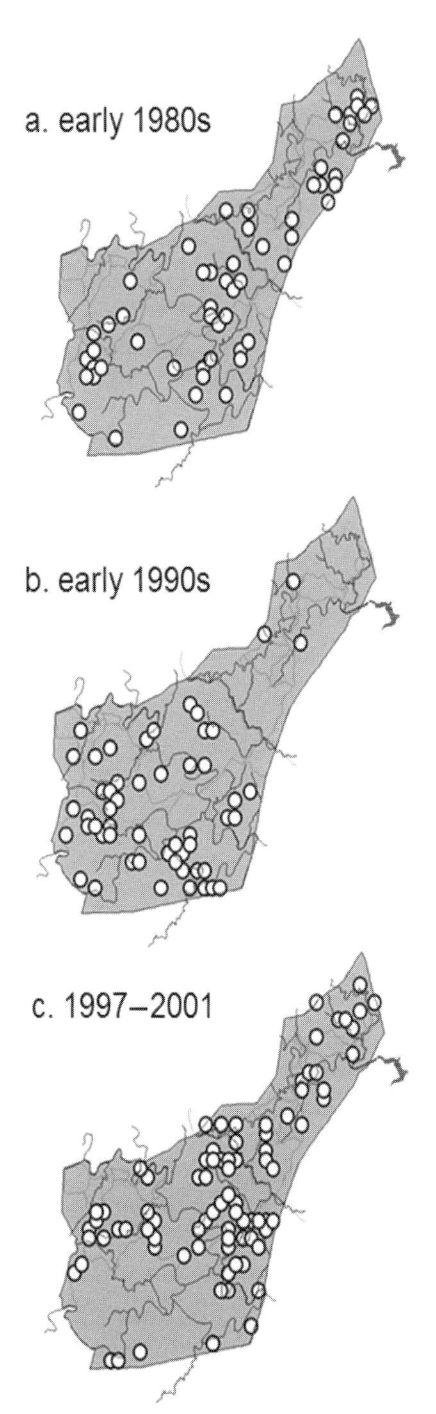

Figure 11.3 Changing locations of the places where black rhino removals took place from 1980 to 2001.

of white rhinos harvested annually was reduced to a mean of 66 animals (range 19–179), representing on average 3.6% of the population. This slowed the net growth of the white rhino population between 1986 and 2012 to an annual average of 1.8%, indicating an intrinsic growth rate of 5.4% per year. This is substantially lower than the annual rate of increase of 9.5% prior to 1971 (Owen-Smith, 1981). Hence, density-dependent feedbacks on reproduction and survival evidently became increasingly effective. However, the number of rhinos dispersing into the sinks and consequently removed has shown no change since the policy was implemented, despite a strong increase in rhino numbers. Nevertheless, overgrazing within the core region has no longer been raised as a concern.

There are practical problems in applying the sink removal strategy, apart from the barrier to motor vehicles entering into the wilderness sinks. The number of rhinos counted in the sinks, which is used to set the annual removal quotas, can fluctuate unpredictably from year to year, affecting the number that can be offered to potential buyers. This complicates planning and logistics for both management and game capture staff, and may not match the demand (market) for white rhinos. Despite such concerns, dispersal sink management of white rhinos has been retained over the past 28 years, and is consistent with the policy of process-based management adopted by the conservation agency (Chapter 1).

11.6.2 Black Rhinos

The maximum growth rate of a black rhino population should be similar to that of white rhinos, i.e. around 9% per year (Knight, 2001). However, with allowance for removals, the black rhino population in HiP has grown by only 3.6% per year over 1998–2013 (Figure 11.1B). The population total within the park has actually declined because live harvests during this period averaged 5.0% per year during this period. Surprisingly, there has been no compensation for the progressively reduced density. Records supplied by field rangers show no changes in numbers of new calves and found carcasses recorded annually.

Of 22 fully adult female black rhinos captured for translocation between 2004 and 2006, just seven (32%) were pregnant (EKZN Wildlife, unpublished data). Estimates of reproductive performance based on captured animals are probably larger than the overall population-wide rate because cows with young calves are less likely to be pregnant and are avoided by the capture team. Hence the average birth interval probably exceeds 3 years, which is somewhat longer than the mean calving interval of 2.5 years recorded for white rhinos in HiP (Owen-Smith, 1981). Predation on young calves may further reduce the population performance of black rhinos in HiP (Plotz and Linklater, 2009). The filarial nematodes and associated lesions typical of black rhinos in HiP may also debilitate animals there (Plotz, 2014). All in all, however, drivers of the poor growth rate of HiP's black rhino population remain poorly understood.

11.7 Social Ecology of Dispersal Movements

For both rhino species, the challenge to improving harvest strategies for population and habitat management is the same – a better understanding of the behaviour and ecology of individuals and resources under a harvest regime. In particular, understanding dispersal movements and habitat colonization is crucial to testing hypotheses about how their populations might respond to removals. Simple harvest models assume that resources released by the population reduction will be taken up by the remaining animals. Vacant home ranges could become reoccupied by the expansion of neighbouring home ranges or by recolonization from elsewhere. The source for recolonization could be either socially or spatially displaced adults, or settlement by dispersing subadults. Larger body size is generally associated with slower life histories and, therefore, slow dispersal rates (Owen-Smith, 1988). Rhinos and equids (Linklater and Cameron, 2009) are unusual among ungulates because natal dispersal movements are made by both sexes. More is known about such movements for white rhinos than for black rhinos.

For white rhinos, natal dispersal following the breaking of maternal bonds is a slow process extending over several years. Mothers drive away their previous offspring following the birth of a new infant typically when the older calf is ~2.5 years of age. Former calves, now termed subadults, form associations (i.e. become 'buddies') with other subadults and/or with adult females that have calves > 3months of age (Owen-Smith, 1975). Subadult-only groups may explore and travel extensively, but at some stage they may join and move with adult females lacking a small calf (Owen-Smith, 1975; Shrader and Owen-Smith, 2002). By doing this, subadults gain experience of the locations of grazing areas, distribution of water, and of presence of other white rhinos in a wide area (Shrader and Owen-Smith, 2002). These associations last from a single day to several years. As a result, individuals can move between a number of different 'buddies' before settling into a home range (females) or territory (males). As females only settle into home ranges at around 7 years of age after giving birth (Owen-Smith, 1975), their dispersal phase lasts about 4.5 years. Males first occupy territories when they are around 10 years of age, so dispersal can last for around 8 years.

The physical features that defined the boundaries of dispersal sinks (rivers, roads, streams) do not inhibit the movements of white rhinos. The sink zones are large enough to enclose the home ranges and territories of resident adults. However, some of the animals removed from the sinks may have been neighbouring residents with territories or home ranges extending across the core–sink boundary. The removal of these adult females could reduce the availability of dispersal opportunities for subadults into the sinks. Among adults dispersal movements are mainly made by males, but adult females can also shift their home ranges. This is evident from the continuing harvests of pregnant females as well as adult males from the sink zones. Nevertheless, the origins of these colonists are unknown.

Natal dispersal by black rhinos has not been studied. We expect longer times and distances of dispersal for males, as among white rhinos, but this has not yet been documented. Nevertheless, we expect dispersal to be a slow process as in white rhinos. Observations on black rhinos following their release into new reserves (Linklater *et al.*, 2006) suggest that their dispersal and settlement may likewise be socially mediated. Sighting records of individual black rhinos within HiP before and after the removal of neighbouring rhinos indicate that black rhinos do not shift their ranges into unoccupied habitat during the first 2 years after the harvest. Black rhinos of opposite sex to the animal removed actually shifted their ranges away from the vacated habitat (Linklater and Hutcheson, 2010). Reproductively mature individuals seem slow to colonize uninhabited or under-utilized space (Lent and Fike, 2003).

11.8 Uncertainties as Hypotheses for Rhino Responses to Harvest

Based on the above findings, we formulated four hypotheses to explain why rhinos of both species have not responded to live harvests as expected. These hypotheses are distinguished by at least one of their predictions about (1) the abundance and feeding pressure of rhino, (2) rhino competitors, and (3) rhino fecundity and recruitment after harvest (Table 11.1). Also pertinent are dispersal rates by rhino and competing species, the responses of key food species, and predator diets after a

Table 11.1. *The predictions of increases (🡅), decreases (🡇) or 'no change' that four hypotheses for a poor response by rhino to harvest make about abundance and feeding pressure of (1) rhino and (2) rhino competitors, and (3) rhino fecundity and recruitment after its harvest in the area harvested*

Hypothesis	Rhino feeding and density (1)	Competitor density and feeding (2)	Rhino fecundity and recruitment (3)
Social constraints	🡇	No change	No change
Depredation constraints	No change	No change	🡇
Competitor substitution	🡇	🡅	No change
Ecological constraints	🡅	No change	🡇

harvest, but these are more difficult to measure without recourse to individual animal studies or intensive and extensive vegetation sampling. The three predictions of our framework, however, are comparatively easy to measure, with most rhino monitoring programmes and supporting infrastructures, and are the minimum necessary for a differential diagnosis of the problem.

The *social constraint hypothesis* suggests that when harvests depress local rhino densities, recolonization is slow such that habitat and resources remain under-utilized. Thus, distributed harvests result in a patchwork of low- and high-density areas rather than freeing resources more generally through the population. Accordingly, resource recovery in depressed-density areas does not influence the wider population performance, meaning that in high-density areas resource limitations remain effective. In particular, the social constraints hypothesis predicts that the density of animals in the harvested area will remain depressed for a prolonged period.

The *depredation constraint hypothesis* suggests that the compensatory recruitment response of rhinos to reduced density is prevented or limited by predation. Where densities are reduced, but predator densities remain the same, the proportional predation rate on the harvested species might increase. This hypothesis predicts that there is reduced recruitment into the breeding population, which slows the population increase despite occupation of the vacated home range by other rhinos. It is relevant only for black rhinos where predation on calves could be great enough to be demographically important (Brain *et al.*, 1999; Plotz and Linklater, 2009). Predation on white rhino calves appears to be rare (Owen-Smith, 1973, 1988).

The *competitor substitution hypothesis* suggests that rhinos are slower than their competitors at recolonizing areas from which rhinos have been harvested. This hypothesis predicts that the removal of rhinos is followed by an increase in the local abundance and feeding activity of other browsers (e.g. kudu for black rhino and wildebeest for white rhino). The ongoing and increased impact of competitors prevents the recovery of food resources and thus may inhibit recolonization of vacant home ranges by rhinos.

The *ecological constraint hypothesis* suggests that historical harvesting has not reduced densities sufficiently to free food resources, particularly for black rhinos. This hypothesis predicts that the feeding pressure of rhinos will be only briefly depressed in areas from which rhinos have been removed because recolonization is rapid. This hypothesis best matches the expectations of simple harvest models.

We do know that neighbouring black rhinos appear reluctant to colonize the ranges of harvested rhino, partly because reproductive relationships are disrupted (Linklater and Hutcheson, 2010). We also know that home range sizes of black rhinos have not changed appreciably over the past 50 years (Linklater *et al.*, 2010; Plotz *et al.*, 2016) since first measured (Hitchins, 1969, 1971). Thus, there is no firm evidence that black rhino in HiP are expanding their ranges in response to deteriorating habitat conditions (contrary to the conclusions of Reid *et al.*, 2007; Slotow *et al.*, 2010). We know that lions and spotted hyenas are responsible for some level of predation on black rhino calves (Plotz and Linklater, 2009), although the impact on population performance needs to be assessed. Taken together, observations of slow recolonization, range size stasis, and predation on calves could indicate that social and depredation constraints are playing a role in reducing the response of the black rhino population to removals. How potential competitors respond to the resources released has yet to be investigated.

Harvesting models based on simple density-dependent functions continue to be used as a guide to optimal sustainable harvests in fisheries management (Punt and Smith, 2001). However, recent experience has emphasized the importance of animal behaviour and spatial ecology for the reliability of harvest quotas (Milner-Gulland, 2001; Sutherland and

Gill, 2001). We have presented a realistic hypothetical framework for the role of density-dependent and density-independent influences on rhino vital rates and dispersal in HiP, which is testable even where detailed, individual-based data are lacking. Our hope is that this framework will help design variations in rhino harvest regimes in order to evaluate these hypotheses.

11.9 Final Considerations

HiP has been a laboratory of innovation and enlightened conservation management for both white and black rhinos. Collaborative relationships and an outward focus have meant that what has happened in HiP has been influential for international rhino conservation over half a century. The overall success in rhino conservation will ultimately be judged by the diminishing importance of HiP as a source for metapopulation management. However, as we write, escalating illegal hunting is once more threatening the gains achieved in rhino conservation in South Africa. During 2008–2015, official figures show that 5048 rhinos were killed, with the largest proportion being white rhinos from Kruger Park (www.stoprhinopoaching.com/, accessed 16 February 2016). Since 2013, the poaching rate has escalated to over 1000 rhino killed nationally per year. Fortress protection is again becoming the primary conservation activity.

Appropriately, a small museum in HiP emphasizes the critical role of this protected area in the conservation of all of the world's rhinos through innovative advances in technology, science, and adaptive management. In the spirit of that tradition, we hope that our framework provides the foundation for the next advance in the management of rhinos.

11.10 References

African Rhino Specialist Group (ARSG) (1991) *Population estimates for black rhinoceros (Diceros bicornis) and white rhinoceros (Ceratotherium simum) in Africa in 1991, and trends since 1987*. IUCN/SSC, Gland, Switzerland.

African Rhino Specialist Group (ARSG) (1992) *Proceedings of the African Rhino Specialist Group Held at Victoria Falls, Zimbabwe from 17–22 November 1992*. ARSG, Pietermaritzburg.

Balfour, D. (1999) Managing the white rhino population in Hluhluwe-Umfolozi Park. Unpublished report, Natal Parks Board, Pietermaritzburg.

Balfour, D. (2001) Managing black rhino for productivity: some questions about current RMG assumptions and guidelines and some ideas about data use. In: *Proceedings of a SADC rhino management group (RMG) workshop on biological management to meet continental and national black rhino conservation goals* (ed. R. Emslie), pp. 35–36. SADC Regional Programme for Rhino Conservation, Giants Castle, South Africa.

Brain, C., Forge, O. & Erb, P. (1999) Lion predation on black rhinoceros (*Diceros bicornis*) in Etosha National Park. *African Journal of Ecology* **37**: 107–109.

Brooks, P. M. & Macdonald, I. A. W. (1983) The Hluhluwe-Umfolozi Reserve: an ecological case history. In: *Management of large mammals in African conservation areas* (ed. N. Owen-Smith), pp. 51–77. Haum Educational Publishers, Pretoria.

Clinning, G., Druce, D., Robertson, D., Bird, J., & Nxele, B. (2009) Black rhino in Hluhluwe-iMfolozi Park: historical records, status of current population and monitoring and future management recommendations. Unpublished report, Ezemvelo KwaZulu-Natal Wildlife, Pietermaritzburg.

Conway, A., Balfour, D., Dale, T., *et al.* (2001) Hluhluwe-Umfolozi Management Plan 2001. Unpublished report, Ezemvelo KZN Wildlife, Pietermaritzburg.

Cooke, M. (1998) Air lifting immobilized rhinoceros. Unpublished report, Natal Parks Board document, Pietermaritzburg.

Courchamp, F., Langlais, M., & Sugihara, G. (2000) Rabbits killing birds: modelling the hyperpredation process. *Journal of Animal Ecology* **69**: 154–164.

Cromsigt, J. P. G. M. & Kuijper, D. P. J. (2011) Revisiting the browsing lawn concept: evolutionary interactions or pruning herbivores? *Perspectives in Plant Ecology Evolution and Systematics* **13**: 207–215.

Cromsigt, J. P. G. M., Hearne, J., Heitkönig, I. M. A., & Prins, H. H. T. (2002) Using models in the management of black rhino populations. *Ecological Modelling* **149**: 203–211.

Emslie, R. H. (1999) The feeding ecology of the black rhinoceros (*Diceros bicornis minor*) in Hluhluwe-Umfolozi Park, with special reference to the probable causes of the Hluhluwe population crash. PhD thesis, University of Stellenbosch.

Emslie, R. (2001) Black rhino in Hluhluwe-Umfolozi Park. In: *Proceedings of a SADC rhino management group (RMG) workshop on biological management to meet continental and national black rhino conservation goals* (ed. R. Emslie), pp. 86–91. SADC Regional Programme for Rhino Conservation, Giants Castle, South Africa.

Emslie, R. & Brooks, M. (1999) *African rhino: status survey and conservation action plan.* IUCN/SSC African Rhino Specialist Group, Gland, Switzerland.

Emslie, R., Amin, R. & Kock, R. (2009) Guidelines for the in situ re-introduction and translocation of African and Asian rhinoceros. Occasional Paper of the IUCN Species Survival Commission No. 39.

Ferreira, S. M., Greaver, C. C., & Knight, M. H. (2011) Assessing the population performance of the black rhinoceros in Kruger National Park. *South African Journal of Wildlife Research* **41**: 192–204.

Ferreira, S. M., Botha, J. M., & Emmett, M. (2012) Anthropogenic influences on conservation values of white rhinos. *PLoS ONE* **7**: e45989.

Fornara, D. A. & du Toit, J. T. (2007) Browsing lawns? Responses of *Acacia nigrescens* to ungulate browsing in an African savanna. *Ecology* **88**: 200–209.

Fowler, C. W. (1981) Density dependence as related to life history strategy. *Ecology* **62**: 602–610.

Goodman, P. (2001) Black rhino harvesting strategies to improve and maintain productivity and minimise risk. In: *Proceedings of a SADC rhino management group (RMG) workshop on biological management to meet continental and national black rhino conservation goals* (ed. R. Emslie), pp. 57–63. SADC Regional Programme for Rhino Conservation, Giants Castle, South Africa.

Harthoorn, A. (1962a) The capture and relocation of the white (square-lipped) rhinoceros, *Ceratotherium simum simum*, using drug-immobilising techniques, at the Umfolozi Game Reserve, Natal. *Lammergeyer* **2**: 1–9.

Harthoorn, A. (1962b) Capture of white (square-lipped) rhinoceros, *Ceratotherium simum simum* (Burchell), with the use of the drug immobilization technique. *Canadian Journal of Comparative Medicine* **26**: 203–208.

Hitchins, P. M. (1969) Influence of vegetation types on sizes of home ranges of black rhinoceros, Hluhluwe Game Reserve, Zululand. *Lammergeyer* **12**: 48–55.

Hitchins, P. (1971) Preliminary findings in a telemetric study of the black rhinoceros in Hluhluwe Game Reserve, Zululand. In: *Proceedings of a Symposium on Biotelemetry*. CSIR, Pretoria.

King, J. & Carter, B. (1965) The use of the oripavine derivative M-99 for the immobilization of the black rhinoceros (*Diceros bicornis*) and its antagonism with the related compound M-285. *East African Wildlife Journal* **3**: 19–27.

Knight, M. (2001) Current and possible population performance indicators for black rhinos. In: *Proceedings of a SADC rhino management group (RMG) workshop on biological management to meet continental and national black rhino conservation goals* (ed. R. Emslie), pp. 49–56. SADC Regional Programme for Rhino Conservation, Giants Castle, South Africa.

Lent, P. C. and Fike, B. (2003). Home ranges, movements and spatial relationships in an expanding population of black rhinoceros in the Great Fish River Reserve, South Africa. *South African Journal of Wildlife Research* **33**: 109–118.

Linklater, W. & Cameron, E. (2009) Social dispersal but with philopatry reveals incest avoidance in a polygynous ungulate. *Animal Behaviour* **77**: 1085–1093.

Linklater, W. L. & Hutcheson, I. (2010) Black rhinoceros are slow to colonize a harvested neighbour's range. *South African Journal of Wildlife Research* **40**: 58–63.

Linklater, W. L., Flammand, J., Rochet, Q., *et al.* (2006) Preliminary analyses of the free-release and scent-broadcasting strategies for black rhinoceros reintroduction. *Ecological Journal* **7**: 26–34.

Linklater, W. L., Plotz, R., Kerley, G. I. H., *et al.* (2010) Dissimilar home range estimates for black rhinoceros (*Diceros bicornis*) can not be used to infer habitat change. *Oryx* **44**: 16–19.

Maddock, A. (1992) White rhino sink strategy. Unpublished report, Natal Parks Board, Pietermaritzburg.

Makhabu, S. W. & Skarpe, C. (2006) Rebrowsing by elephants three years after simulated browsing on five woody plant species in northern Botswana. *South African Journal of Wildlife Research* **36**: 99–102.

Milliken, T. (1993) *The decline of the black rhino in Zimbabwe: implications for future rhino conservation*. Traffic International, Cambridge.

Milner-Gulland, E. (2001) The exploitation of spatially structured populations. In: *Conservation of exploited species* (eds J. Reynolds, G. Mace, K. Redford, & J. Robinson), pp. 41–66. Cambridge University Press, Cambridge.

Novaro, A. J., Funes, M. C., & Walker, R. S. (2005) An empirical test of source–sink dynamics induced by hunting. *Journal of Applied Ecology* **42**: 910–920.

Owen-Smith, R. N. (1973) The behavioral ecology of the white rhinoceros. PhD thesis, University of Wisconsin.

Owen-Smith, R. N. (1975) The social ethology of the white rhinoceros *Ceratotherium simum* (Burchell 1817). *Zeitschrift fur Tierpsychologie* **38**: 337–384.

Owen-Smith, N. (1981) The white rhino over-population problem, and a proposed solution. In: *Problems in management of locally abundant wild animals* (eds J. Jewell, S. Holt, & D. Hart), pp. 129–150. Academic Press, New York.

Owen-Smith, R. N. (1988) *Megaherbivores: the influence of large body size on ecology*. Cambridge University Press, Cambridge.

Pienaar, D. J., Bothma, J. d. P., & Theron, G. K. (1992) Landscape preference of the white rhinoceros in the southern Kruger National Park. *Koedoe* **35**: 1–7.

Pienaar, U. d. V. (1970) The recolonisation history of the square-lipped (white) rhinoceros in the Kruger National Park (October 1961–November 1969). *Koedoe* **13**: 157–169.

Player, I. (1972) *The white rhino saga*. William Collins Sons & Co. Ltd, London.

Plotz, R. (2014) The inter-specific relationships of black rhinoceros (*Diceros bicornis*) in Hluhluwe-iMfolozi Park. PhD thesis, Victoria University of Wellington.

Plotz, R. & Linklater, W. (2009) Black rhinoceros (*Diceros bicornis*: Rhinocerotidae) calf succumbs after lion predation attempt: implications for conservation management. *African Zoology* **42**: 283–287.

Plotz, R. D., Grecian, W. J., Kerley, G. I. H., & Linklater, W. L. (2016) Standardising home range studies for improved management of the critically endangered black rhinoceros. *PLoS ONE* **11**(3): e0150571.

Pulliam, H. R. (1988) Sources, sinks, and population regulation. *American Naturalist* **132**: 652–661.

Punt, A. & Smith, A. (2001) The gospel of maximum sustainable yield in fisheries management: birth, crucifixion and reincarnation. In: *Conservation of exploited species* (eds J. Reynolds, G. Mace, K. Redford, & J. Robinson), pp. 41–66. Cambridge University Press, Cambridge.

Reid, C., Slotow, R., Howison, O., & Balfour, D. (2007) Habitat changes reduce the carrying capacity of Hluhluwe-Umfolozi Park, South Africa, for critically endangered black rhinoceros *Diceros bicornis.Oryx* **41**: 247–254.

Reynolds, J., Mace, G., Redford, K., & Robinson, J. (eds) (2001) *Conservation of exploited species*. Cambridge University Press, Cambridge.

Rosenberg, A. A., Fogarty, M. J., Sissenwine, M. P., Beddington, J. R., & Shepherd, J. G. (1993) Achieving sustainable use of renewable resources. *Science* **262**: 828–829.

Shrader, A. M. & Owen-Smith, N. (2002) The role of companionship in the dispersal of white rhinoceros (*Ceratotherium simum*). *Behavioral Ecology and Sociobiology* **52**: 255–261.

Slotow, R., Reid, C., Balfour, D., & Howison, O. (2010) Use of black rhino range estimates for conservation decisions: a response to Linklater *et al. Oryx* **44**: 18–19.

Stamps, J. & Swaisgood, R. (2007) Someplace like home: experience, habitat selection and conservation biology. *Applied Animal Behaviour Science* **102**: 392–409.

Sutherland, W. & Gill, J. (2001) The role of behaviour in studying sustainable exploitation. In: *Conservation of exploited species* (eds J. Reynolds, G. Mace, K. Redford, & J. Robinson, pp. 259–280. Cambridge University Press, Cambridge.

Waldram, M. (2005) The ecological effects of grazing by the white rhino (*Ceratotherium simum simum*) at a landscape scale. MSc thesis, University of Cape Town.

Walker, B. H., Emslie, R. H., Owen-Smith, R. N., & Scholes, R. J. (1987). To cull or not to cull: lessons from a southern African drought. *Journal of Applied Ecology*, **24**: 381-401.

Western, D. & Vigne, L. (1985) The deteriorating status of African rhinos. *Oryx* **19**: 215–220.

12 · *Reassembly of the Large Predator Guild into Hluhluwe-iMfolozi Park*

MICHAEL J. SOMERS, PENNY A. BECKER, DAVE J. DRUCE, JAN A. GRAF, MICAELA SZYKMAN GUNTHER, DAVID G. MARNEWECK, MARTINA TRINKEL, MARCOS MOLEÓN, AND MATT W. HAYWARD

12.1 Introduction

As charismatic animals with high economic value for ecotourism, large African carnivores are well studied (see Kingdon and Hoffman, 2013) and are the frequent focus of re-introduction programmes. Re-introducing large predators may be important in stabilizing ecosystems and increasing ecological resilience (Wilmers and Getz, 2005; Estes *et al.*, 2011; Dalerum *et al.*, 2012). Apex predators are often critical components in the regulation of both terrestrial and marine ecosystem functioning via top-down cascades in ecosystems (Ripple and Beschta, 2012), but details of these dynamics are not yet known in African savannas where they interact with other drivers such as megaherbivores and fire (see Chapters 5 and 10). Furthermore, African savannas are more intact ecologically, species-rich, heterogeneous, and dynamic than most of the classic examples of top-down trophic cascades from temperate ecosystems. Also, in these systems each predatory species has mostly been studied as an isolated unit. Little is known about the functioning of more diverse carnivore communities, particularly how multiple large carnivore species coexist. Smaller predators are often suppressed by larger species (Prugh *et al.*, 2009), but the relationship is not always straightforward. For example, African lions

Conserving Africa's Mega-Diversity in the Anthropocene, ed. Joris P. G. M. Cromsigt, Sally Archibald and Norman Owen-Smith. Published by Cambridge University Press.

(*Panthera leo*) may suppress African wild dogs (*Lycaon pictus*) in certain areas but do not do so with cheetahs (*Acinonyx jubatus*) (Swanson *et al.*, 2014). Small reserves face special challenges in maintaining viable, self-sustaining populations of large predators (such as lion, spotted hyena *Crocuta crocuta*, brown hyena *Hyaena brunnea*, African wild dog, leopard *Panthera pardus*, and cheetah) due to the extensive space requirements of these species (Lindsey *et al.*, 2004, 2011). The Hluhluwe-iMfolozi Park (HiP) currently has a nearly complete assemblage of large carnivores (Rowe-Rowe, 1992) and herbivores (Rowe-Rowe, 1994). However, lions, wild dogs, cheetahs, and brown hyenas have all been extinct within HiP's borders at one time or another, and numbers of spotted hyenas and black-backed jackals (*Canis mesomelas*) were once severely reduced. Brown hyenas are still considered extinct, even after restoration attempts, while black-backed jackals are now seldom seen. The restoration of the large carnivore community has thus been one of the main conservation management challenges in HiP, particularly during the last 30–40 years. Here, we report on this reassembly of a large predator guild in HiP, while also updating the numbers, status, and potential carrying capacity of each species. Reporting and post-release monitoring of large carnivore re-introductions in HiP have been comprehensive for wild dogs (e.g. Gusset *et al.*, 2006a, 2006b; Somers *et al.*, 2008) and lions (e.g. Anderson, 1981; Maddock *et al.*, 1996; Stein, 1999; Trinkel *et al.*, 2008). Unfortunately, introductions of cheetahs, brown hyenas, and black-backed jackals were not well documented and even fewer data are available for other carnivore species (Rowe-Rowe, 1992). We discuss the insights derived from the successful re-introductions of lion and wild dog in HiP after their historical extinctions, and the within-guild, particularly trophic, interactions among predators, highlighting the success stories as well as failures. Given the large and continuing decline of large carnivores in Africa (e.g. Henschel *et al.*, 2010), the lessons from carnivore re-introductions in HiP will assist future conservation management.

12.2 Historical Distribution, Re-introduction History, and Population Trends of Lions

Lions were historically distributed across the whole of northern KwaZulu-Natal (Pringle, 1977; Rowe-Rowe, 1992). The arrival of European hunters with guns in the mid-1800s resulted in a large-scale decline in their numbers by the end of that century. Breeding was recorded until 1910 near Umfolozi and the nearby Mkhuze Game Reserve

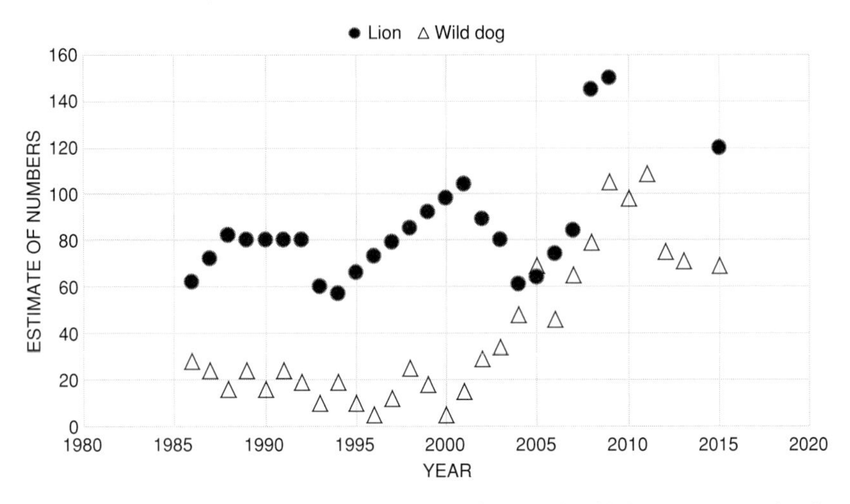

Figure 12.1 The estimated population trends of lion and wild dog in HiP (updated from Grange *et al.*, 2012).

(Vaughan-Kirby, 1916; Foster, 1955; McCracken, 2008). There was a last recorded sighting of a pair of lions in HiP in 1938 (Foster, 1955) and the last lions were shot in the Mkhuze area during the early 1940s (Player, 1997). In 1958 a single male lion entered South Africa from either Swaziland or Mozambique, killing cattle along the way, and passed through Mkhuze Game Reserve, before settling in Umfolozi Game Reserve (Steele, 1970; Player, 1997). A lioness that had been used for filming was released in HiP in 1963, but had to be shot when she tried to break into a house in an emaciated condition. In March 1965, one adult female from the Kruger National Park was unofficially introduced (Player, 1997), followed by two adult females and three cubs (two females, one male) from the Timbavati Private Nature Reserve adjoining Kruger Park (Anderson, 1981; Rowe-Rowe, 1992). The founder population was therefore three adult females and three cubs (only one of them male), plus the adult male that had settled earlier. The population increased progressively and peaked at about 140 animals in 1987 (Anderson, 1981; Maddock *et al.*, 1996), before declining to about 80 individuals in 1999 (Balfour *et al.*, unpublished data; Figure 12.1). The initial increase was despite the culling of subadult males and females in the 1970s and 1980s to restrict dispersal and associated stock killing outside the park.

By the turn of the century, however, evidence of inbreeding became apparent in the form of low genetic variation, elevated cub mortality, poor physical condition, conspicuous abscesses, and reduced immune competence (Maddock *et al.*, 1996; Stein, 1999; Trinkel *et al.*, 2008; see Chapter 13 for more on inbreeding and disease). To restore genetic variation, 16 lions sourced from Etosha National Park in Namibia, via other protected areas, were translocated into HiP between 1999 and 2001 (Trinkel *et al.*, 2008) to augment the 84 lions remaining from the founder stock. By 2004, lion numbers in HiP had shrunk to only 62 individuals; 32% were from the founding animals and their offspring (20 lions), 47% were offspring of the translocated and founder lions (29 lions), and 21% represented the translocated lions and their offspring (13 lions) (Trinkel *et al.*, 2008). By 2006, the translocated lions and their descendants had completely replaced all of the founders (Trinkel *et al.*, 2008). The 2015 population was estimated, from observations of collared or otherwise marked individuals, at about 120 lions (Figure 12.1).

12.2.1 The Re-introduction Process

The age of the translocated lions originating from Etosha ranged from 17 to 32 months, and 13 different bloodlines were represented (Trinkel *et al.*, 2008). Three translocation techniques were tested in four separate releases.

First release: three male lions (two brothers, one singleton) and three female lions (two sisters, one unrelated female) were released as a mixed-sex pride in northern HiP, where no resident lions had ranged since 1992. The intention was for these six lions to establish a pride with an attendant male coalition. However, the outcome was that two prides were formed, one consisting of two females (sisters) and two males (brothers), and the other of the unrelated female and the singleton male.

Second release: two translocated females, unfamiliar with each other, were released initially into an enclosure located within the range of a founder-population pride in western Mfolozi, joined by two females from the founder-population pride so as to bond the translocated females with the settled females. However, despite this association, the two translocated females separated from the founder females immediately following their release from the enclosure.

Third release: two translocated females, unfamiliar with each other, were released into an enclosure located just outside the range of a founder-population pride in western Mfolozi together with one female from this pride in order to bond the translocated females into the pride. However, these two females separated from the founder-population female immediately after being released from the enclosure.

Fourth release: six females (two sisters, four females unrelated to all) were released in southern Mfolozi so as to establish a single pride in a region unoccupied by a founder pride. The outcome was the establishment of two prides: one consisting of the two sisters and one unrelated female, and another pride with three unrelated females.

The translocations were meant to encourage the lionesses to form four prides, but instead they split into six prides comprising of one to three related and/or unrelated lionesses (Trinkel *et al.*, 2008). Females from the founder population had also formed larger prides in southern Mfolozi than in northern Hluhluwe (Trinkel *et al.*, 2007, 2008). The attempt to encourage new females to join founder prides was unsuccessful, but all introduced males associated with both new and founder females. The pair of males was more successful in this than the solitary male, gaining pride residence more easily and maintaining residence for longer (Trinkel *et al.*, 2008).

12.2.2 Litter Size and Cub Survival

Reproductive performance after the re-introductions provided further evidence that the founder-lion population suffered from adverse effects of inbreeding (Trinkel *et al.*, 2008). Out-crossed pairings produced significantly larger litters ($p < 0.01$) and had higher cub survival ($p < 0.02$) than inbred pairings (i.e. founder-population females with founder-population males). Inbred pairing ($n = 13$) produced 20 cubs with a mean litter size of 1.5 and cub survival of 0.31, while out-crossed pairings ($n = 13$) produced 40 cubs with mean litter size of 3.1 and cub survival of 0.67 (Trinkel *et al.*, 2008). The lone male and the pair of males sired similar numbers of offspring per male, but cub survival was significantly higher for the pair than for the singleton. Despite initial concerns about infanticide by translocated males, only two of the founder prides were taken over by the translocated males, restricting social disruption to relatively few cubs (Trinkel *et al.*, 2008).

12.2.3 Mortality Losses

Between 1999 and 2004, eight of the 16 translocated lions (seven females, one male) died. One female was killed after escaping from HiP, two other females disappeared and were presumed dead, three females died from natural causes, and a seventh female was euthanized because she was emaciated. Over this period, 15 of 84 founder-population lions were euthanized because of severe malnutrition and emaciation, and many had developed large abscesses on their elbows. At least 40 more emaciated founder animals disappeared and are assumed to have died. In contrast, all but one of the translocated lions remained in excellent body condition, suggesting that genetics rather than insufficient food resources caused the bad condition of the founder lions (see Chapter 13).

12.3 Historical Distribution, Re-introduction History, and Population Trends of Wild Dogs

African wild dogs or Cape hunting dogs (sometimes also called painted dogs) were once widespread in sub-Saharan Africa. The total number of free-ranging wild dogs in Africa is presently estimated at less than 8000 individuals surviving in only 14 of their original 39 range countries (Woodroffe *et al.*, 2004; IUCN/SSC, 2008; Davies-Mostert *et al.*, 2009; Kenya Wildlife Service, 2010). Wild dogs in KwaZulu-Natal occur primarily within HiP with a handful of smaller packs in Mkhuze Game Reserve, Tembe Elephant Park, and some private reserves. As recently as the 1980s, the species was actively persecuted because of presumed harmful effects on tourism, sport hunting, and conservation of other carnivores. At that time, the only remaining population in South Africa was in the Kruger National Park and surrounds, with occasional transients in the northern parts of South Africa bordering Mozambique, Botswana, and Namibia.

Wild dogs were originally common in northern KwaZulu-Natal with successful reproduction recorded through the 1910s and large packs sighted up to 1933 (Vaughan-Kirby, 1916; Potter, 1934; Foster, 1955; Pringle, 1977; Rautenbach *et al.*, 1980; Carruthers, 1985; McCracken, 2008). Although it was speculated that an unknown disease led to their local extinction (Foster, 1955), intensive persecution was the most likely cause (Pringle, 1977). Systematic poisoning and shooting of wild dogs was recorded from 1895 onwards (McCracken, 2008), including 89 killed by the resident Game Conservator for Zululand and his staff (Vaughan-Kirby, 1916). The last wild dogs were probably shot by farmers in the Mkhuze region around 1944. Restoration of the wild dog population in HiP began in 1980 (Table 12.1) with nine individuals, followed by 15 in 1981, released in northern Hluhluwe. A single breeding pack was formed

Table 12.1. *Wild dog re-introductions into Hluhluwe-iMfolozi Park since 1980*

Release	Pack composition	Pack size	Date of release	Comments
1	UNK	9	1980	Introduced in four stages, see Somers *et al.* (2008)
2	UNK	15	1981	Introduced in four stages, see Somers *et al.* (2008)
3	UNK	4	1986	
4	2AM, 1AF, 1YM	4	1997	
5	2AF	2	2001	Bonded with two existing males
5	1AF, 3PM	4	April 2003	Bonded with four dogs from another source. Released as single pack
6	2AM,2AF	4	April 2003	Bonded with four dogs from another source. Released as single pack
7	2AM,3YF	5	December 2005	
8	2AF	2	March 2006	
9	2AF, 1UNK	3	April 2008	
10	1AM	1	September 2008	
11	2AF	2	August 2011	
12	2AM	2	May 2012	
13	1AM[1],1AF	2	May 2013	

[1] Free-roaming.
UNK, Unknown; AM, adult male; AF, adult female; PM, pup male; YF, yearling female.

and the re-introduction was considered a success for the next decade (Maddock, 1999). Another four were re-introduced in 1986. However, the population dwindled to five adult individuals and by 1996 had formed a single pack along with another group of two females (Maddock, 1995, 1999; Somers *et al.*, 2008). There was no breeding up to 1997, because the alpha female apparently could not breed (Somers *et al.*, 2008). To stimulate population growth, additional packs were translocated to HiP in 1997 (Somers and Maddock, 1999), 2001, and 2003 (Graf *et al.*, 2006; Gusset *et al.*, 2006a) as well as elsewhere in KwaZulu-Natal in 2005 and 2006 (Davies-Mostert *et al.*, 2009). The HiP wild dog population grew substantially after these initial re-introductions and further reinforcements (Figure 12.1; Maddock *et al.*, 1996; Moehrenschlager and Somers, 2004; Graf *et al.*, 2006; Somers *et al.*, 2008; WAG-SA minutes). As of January 2015, the population of African wild dogs in HiP numbered 69 animals (24 adults, 19 yearlings, 26 pups) in seven packs (WAG-SA minutes). A comprehensive behavioural, demographic, and genetic database for HiP's wild dogs has been compiled from close monitoring since 1993 and genetic sampling from 2003 to 2008 (Somers *et al.*, 2008; Becker *et al.*, 2012). This has enabled studies of pack dynamics, inbreeding, and reproductive performance (Spiering *et al.*, 2010, 2011; Becker *et al.*, 2012).

Wild dogs have occasionally left HiP. These breakouts can be separated into two types: (1) transient long-distance movement from the park without return (i.e. dispersal), or (2) gradual movement of a pack repeatedly breaking out and returning. Transient (dispersal) groups have been found to take a distinctive route when leaving HiP where four variables (elevation, land cover, road density, and human density) best predicted the probability of presence for transient wild dogs (Whittington-Jones *et al.*, 2014). In 2014, there were two cases of wild dogs dispersing from HiP and forming new packs outside protected areas in KwaZulu-Natal. The one instance involved two dispersing groups of wild dogs leaving HiP and forming a pack 30 km west of HiP within the Opathe-Emakhosini area, utilizing an area of approximately 200 km^2 (Ezemvelo KwaZulu-Natal Wildlife, unpublished data). The other instance involved a single dispersal group of three females travelling 120 km north before meeting a dispersing male from a private game reserve and forming a pack now settled along the Mkhuze river outside formally protected areas. Both packs are being continuously monitored. The second type of breakout was exhibited by two packs that expanded their range into an area < 3 km outside the park after 2013. Contributing to these breakouts were repeated flooding of rivers, poor fence maintenance, large space requirements of wild dog packs (Creel and Creel, 2002), competition with lions (Mills and Gorman, 1997; Creel and Creel, 2002), and propensity for using fences for hunting (Van Dyk and Slotow, 2003).

Wild dogs seem to have the capacity to discriminate between kin and non-kin through 'recognition by association' (Blaustein *et al.*, 1987; Becker *et al.*, 2012). Between 1997 and 2008, inbreeding was rarely detected in HiP between parents and offspring both in the natal pack and after the death of a dominant adult (a reproductive vacancy), and between siblings after dispersal (Becker *et al.*, 2012). While only three inbreeding opportunities resulted in mating, breeding occurred in 73% of opportunities to mate with unrelated males. Within six breeding pairs that were confirmed via genetic analysis to be related at the level of third-order kin (first cousin) or higher, only a mother–son pair was familiar with one another before mating. The other five related pairings were genetic relatives, but consisted of individuals that never resided simultaneously in a common natal pack before joining together to breed. While inbreeding avoidance may maintain gene diversity, population modelling has shown that without continued augmentation, there may be severe demographic impacts on the future viability of small, isolated wild dog populations. The absence of unrelated mates may lead to population decline and extinction before the impacts of inbreeding depression are observed (Becker *et al.*, 2012). The fact that wild dogs display strong inbreeding avoidance, along with a high cost of long-distance dispersal, supports the idea that the detrimental effects of wild dog inbreeding must be severe (Kokko and Ots, 2006). Even in a limited number of generations, inbred individuals in HiP with inbreeding coefficients of greater than or equal to 0.25 may have had lifespans up to 10 months shorter than outbred contemporaries, suggesting that some deleterious effects of inbreeding may already exist (Spiering *et al.*, 2011). This trend was, however, confounded by pack-specific effects as many inbred individuals originated from the once large Mfolozi pack.

Not only do wild dogs appear to avoid mating with related kin, but recent studies have revealed that parentage within packs is more commonly shared with subordinate males and females than previously reported (Girman *et al.*, 1997; Spiering *et al.*, 2010). In packs that contained siblings and half-siblings of the alpha individuals, subordinate males sired up to 45% of pups, and subordinate females whelped litters in half of all years (Spiering *et al.*, 2010). Although facilitating the maintenance of genetic diversity in a small, re-introduced population, this shared parentage could make it challenging for offspring to distinguish parents from aunts, uncles, and non-relatives. Thus, wild dogs may primarily rely on kin recognition by association to avoid inbreeding.

To manage the growing number of small wild dog populations in South Africa, the IUCN South African Wild Dog Action Group (now the IUCN South African Wild Dog Advisory Group) was initiated in 1997. The wild dogs in all of these places, including HiP, were to be managed as a single metapopulation. The establishment of a viable, self-sustaining wild dog population requires dispersal of individuals between subpopulations (i.e. different packs) in order to maintain genetic diversity (Creel and Creel, 1998; Leigh *et al.*, 2012). To initiate new subpopulations and maintain the demographic and genetic health of the managed metapopulation, wild dogs have been periodically translocated between sites across South Africa. Consequently, the managed metapopulation in KwaZulu-Natal became more genetically diverse than the population in Kruger National Park (Edwards, 2009). In general, the heterozygosity level based on microsatellite loci of South Africa's wild dogs falls within the range of diversity reported for other canids such as grey wolves *Canis lupus*, red wolves *Canis rufus*, and golden jackals *Canis aureus*, although wild dogs are at the lower end of the observed spectrum (Edwards, 2009).

12.4 Historical Distribution, Re-introduction History, and Population Trends of Other Carnivores

12.4.1 Cheetah

Foster (1955) considered cheetahs still numerous in HiP in the first two decades of the twentieth century, but they apparently disappeared from HiP and surrounding regions during the 1920s. Sixty-four individuals were re-introduced into HiP between 1966 and 1969 (Bourquin *et al.*, 1971; Pringle, 1977; Whateley and Brooks, 1985; Rowe-Rowe, 1992). However, by 1992 only three cheetahs remained in Hluhluwe Game Reserve and 10 in Umfolozi Game Reserve (Rowe-Rowe, 1992). Further translocations to augment the population took place during 1995 from Namibia (Marker-Kraus, 1996, in Hunter, 1998). At the end of 2014, the HiP cheetah population was estimated at 30 animals (EKZNW, unpublished data). However, the results of a photographic census that was compiled in mid-2015 indicates that the cheetah population may number fewer than 15 individuals.

12.4.2 Leopard

Leopards persisted in HiP without re-introductions. Estimates of leopard numbers in the park by managers have ranged from 50 to 270 individuals. Since 2011, camera trap surveys undertaken by the NGO Panthera produced estimates of 125 individuals in 2012, 72 individuals in 2013, 67 individuals in 2014, and 46 individuals in 2015. Whether this represents an actual decrease or the result of improved survey methods is unclear.

12.4.3 Spotted Hyena

Early records on spotted hyenas from the 1920s indicate that they were rare in Umfolozi GR (Roberts, 1954), but plentiful in Hluhluwe (Potter, 1934). Numbers were subsequently reduced through large-scale poisoning inside and outside the reserve (Potter, 1941). Direct persecution of spotted hyenas was stopped in 1952, and spotted hyenas were considered common across most of HiP by the beginning of the 1960s (Deane, 1962). Spotted hyenas were sometimes shot when found along fences, in case they broke out and caused problems. In 2003–2004 the HiP population was estimated at 321 spotted hyenas based on a capture–recapture technique using call-ups (Graf *et al.*, 2009), which appeared high compared to other protected areas in southern Africa. Recent estimates have been substantially lower, possibly because hyenas have become habituated to the call-up method.

12.4.4 Brown Hyena

Roberts (1954) suggested brown hyenas were more numerous than spotted hyenas in Umfolozi GR in the 1920s. Breeding by brown hyenas was last recorded there in the 1950s (Player, 1997), with the last sighting made in 1961 (Bourquin *et al.*, 1971; Pringle, 1977; Whateley and Brooks, 1985). This decline coincided with the increase in spotted hyenas in the park and intensifying persecution outside. Re-establishment of brown hyenas was unsuccessfully attempted in the form of four tame individuals in the late 1970s (Whateley and Brooks, 1985; Rowe-Rowe, 1992) and there have been no sightings of brown hyenas since 1982 (Rowe-Rowe, 1992).

12.4.5 Black-Backed Jackal

Foster (1955) considered black-backed jackals to be numerous in the first two decades of the twentieth century, after which they declined. Owen-Smith (personal comments) stated that black-backed jackals were commonly seen and heard in western Umfolozi GR during the 1960s. They apparently disappeared from HiP during the 1970s, possibly as a result of infection by sarcoptic mange, canine distemper, or rabies (Whateley and Brooks, 1985; Rowe-Rowe, 1992; Chapter 13). Rabies infections in black-backed jackal appear to peak every 4–8 years, depending on rainfall and social behaviour (Walton and Joly, 2003). Several unsuccessful re-introduction attempts carried out during the 1990s were not well documented, except that one failure was associated with the translocation of habituated animals. Recently there have been some sightings of black-backed jackals both in Mfolozi and Hluhluwe regions of the park.

12.4.6 Other Carnivores

Other small carnivores recorded in HiP include side-striped jackal (*Canis adustus*; one photographed in HiP in 2012), African clawless otter (*Aonyx capensis*), honey badger (*Mellivora capensis*), striped polecat (*Ictonyx striatus*), large-spotted genet (*Genetta tigrina*), slender mongoose (*Galerella sanguinea*), white-tailed mongoose (*Ichneumia albicauda*), water mongoose (*Atilax paludinosus*), banded mongoose (*Mungos mungo*), and aardwolf (*Proteles cristatus*; but aardwolves have not been seen since 1982; Whateley and Brooks, 1985; Rowe-Rowe, 1992). Serval (*Felis serval*) has recently been recorded by camera traps in the central Corridor section of HiP. Caracal (*Caracal caracal*) is not listed as resident in the area by Rowe-Rowe (1992) and has not been recorded on camera traps. Striped weasel (*Poecilogale albinucha*) occurs nearby and may thus be present in the park (Rowe-Rowe, 1992).

12.5 The Diets of the Carnivores

12.5.1 Predation

We analysed the changes through time in the contributions of different prey species to the diets of lion, leopard, cheetah, spotted hyena, and wild dog. Our data set derives from the HiP carnivore kill record and spans three different periods: 1983–1990, 1991–2000, and 2001–2010. Park rangers registered all ungulate carcasses found during their patrols, noting species and cause of death when possible. Death was attributed to predation when predators or their signs were observed near the carcasses, or when carcasses presented any evidence indicating the predator species responsible for the kill (Grange *et al.*, 2012). We assume that all recorded kills were the result of the predatory activity of the ascribed carnivore, although some observations could involve scavenging from a different predator. However, biases are likely to be small, as scavenged carcasses are similar to or slightly larger than hunted prey (see below). Moreover, cheetahs and wild dogs rarely scavenge (Pereira *et al.*, 2014).

The prey species most commonly killed by lions in HiP were African buffalo (*Syncerus caffer*), blue wildebeest (*Connochaetes taurinus*), nyala (*Tragelaphus angasii*), zebra (*Equus quagga*), impala (*Aepyceros melampus*), and greater kudu (*Tragelaphus strepsiceros*) (Table 12.2). Some temporal dietary

Table 12.2. *List of all prey species recorded in the kills of large carnivores in HiP between 1983 and 2010.*

Prey	Lion	Leopard	Cheetah	Spotted hyena	Wild dog
African elephant	4	0	0	0	0
White rhinoceros	5	0	0	0	0
Black rhinoceros	6	0	1	1	0
Hippopotamus	1	0	0	0	0
Giraffe	80	1	0	1	1
Eland	2	1	0	2	0
African buffalo	933	7	0	10	3
Zebra	541	19	3	6	2
Greater kudu	260	13	11	10	7
Waterbuck	91	7	4	3	4
Blue wildebeest	692	16	2	4	1
Nyala	549	240	40	40	492
Warthog	114	41	7	1	4
Bushpig	48	11	1	4	0
Impala	287	301	144	14	309
Bushbuck	4	10	0	0	3
Reedbuck	4	2	0	3	3
Mountain reedbuck	0	5	3	0	0
Steenbok	0	0	2	0	0
Grey duiker	5	16	3	0	6
Red duiker	0	2	0	1	2
Aardvark	1	1	0	0	0
Porcupine	6	1	0	0	0
Chacma baboon	0	12	0	0	0
Vervet monkey	0	5	0	0	0
Samango monkey	0	1	0	0	0
Bushbaby	0	1	0	0	0
Scrub hare	1	1	0	0	0
Cane rat	1	0	0	0	0
Cheetah	2	1	0	0	0
Caracal	1	1	0	0	0
Wild dog	7	0	0	0	0
Black-backed jackal	0	1	0	0	0
Spotted hyena	5	0	0	0	1
Banded mongoose	1	0	0	0	0
Bald ibis	2	0	0	0	0
Crocodile	1	1	0	0	0
Python	4	0	0	2	0
Mozambique spitting cobra	1	0	0	0	0
Total	3659	718	221	102	838

changes were observed: buffalo and impala became more commonly represented in lion kills with time, while the contributions by zebra and nyala decreased (Figure 12.2). However, the proportional selection for these prey species relative to their availability changed little. Buffalo, zebra, wildebeest, and kudu were killed by lions more than expected from their abundance in all periods (except for buffalo in the first study period), impala were taken much less frequently than expected, and nyala were killed in proportion to their abundance (Figure 12.2A). These findings resemble patterns found in other areas (Hayward and Kerley, 2005).

Kills made by wild dog were almost totally composed of nyala and impala (Table 12.2). While nyala were almost solely killed during the period 1983–1990, wild dogs concentrated less on nyala thereafter, and after 2001 impala contributed a similar proportion to that of nyala (Figure 12.2E). Nevertheless, nyala remained strongly preferred relative to their abundance. In other parts of the wild dog distribution, impala are consistently the most strongly selected prey although in many of these areas nyala does not occur (Hayward *et al.*, 2006c). A detailed study on the diet of wild dogs in HiP from examination of scat contents as well as direct observations recorded a greater contribution by impala than the observations reported by park staff (Krüger *et al.*, 1999). However, it was observed that the capture success of wild dogs for nyala was much higher than for impala (Krüger *et al.*, 1999).

Leopard diet comprised mainly two species: impala and nyala (Table 12.2). The only other prey representing $> 5\%$ of the diet of leopards were warthog (*Phacochoerus africanus*) and zebra (the latter only in the first period). Impala strongly increased in leopard kills over time, following a similar increase in impala numbers in the park. Nyala decreased correspondingly over this same time, whereas warthog remained constant (Figure 12.2). Nevertheless, nyala consistently contributed proportionately more relative to their abundance. Impala is generally positively selected by leopards in other areas, but in many of these areas nyala is absent (Hayward *et al.*, 2006a).

For cheetah, impala and nyala were the first and second most important prey species, respectively, and all other prey species contributed $< 5\%$ (Table 12.2). The contribution of impala increased over time, and impala were consistently preferred more relative to their availability than nyala (Figure 12.2C). This dietary pattern is consistent with observations on cheetahs elsewhere (Hayward *et al.*, 2006b).

There were comparatively few records of kills ascribed to spotted hyena and these included similar prey species to those killed by lions, except

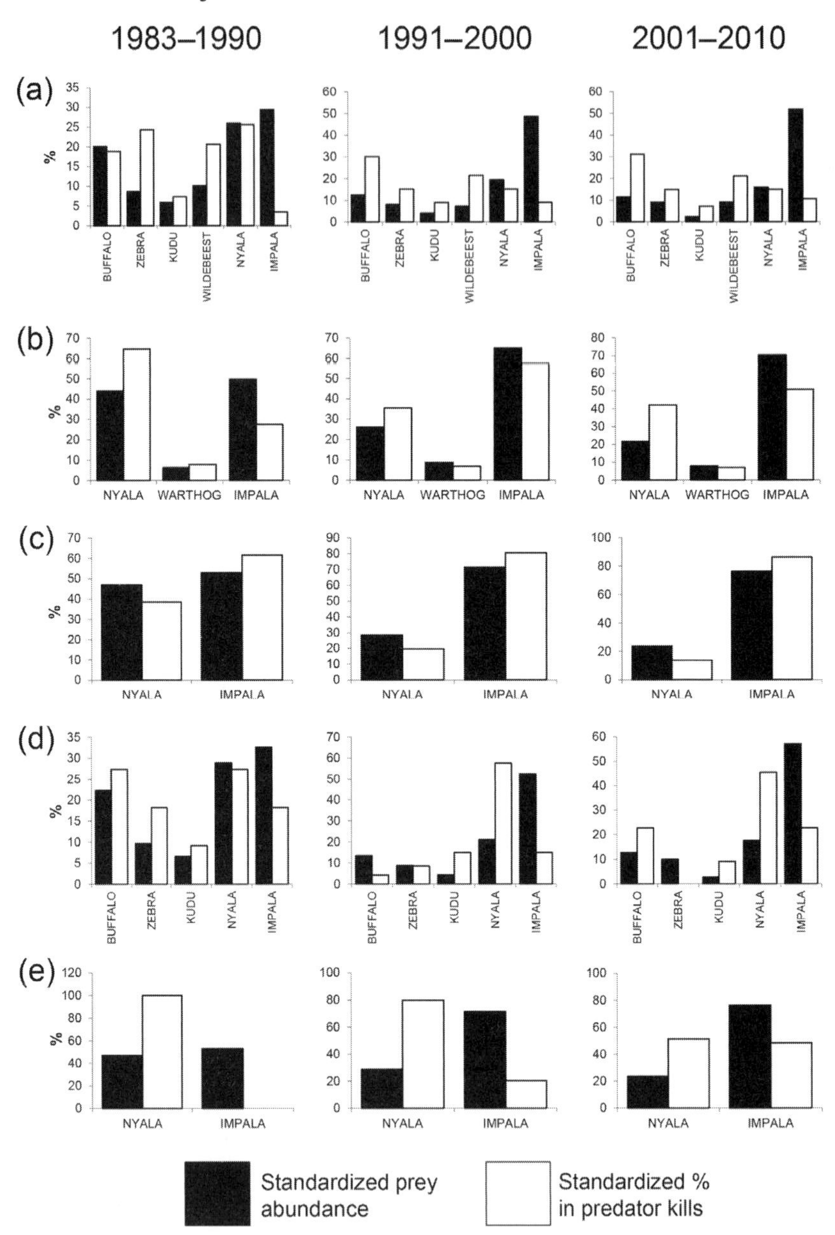

Figure 12.2 Prey selection of (A) lion, (B) leopard, (C) cheetah, (D) spotted hyena, and (E) wild dog in HiP relative to prey availability during the three time periods. Only those prey representing > 5% of the overall diet of each carnivore species are represented. For each carnivore and time period, prey abundances and representation in kills were standardized as percentages summing to 100%.

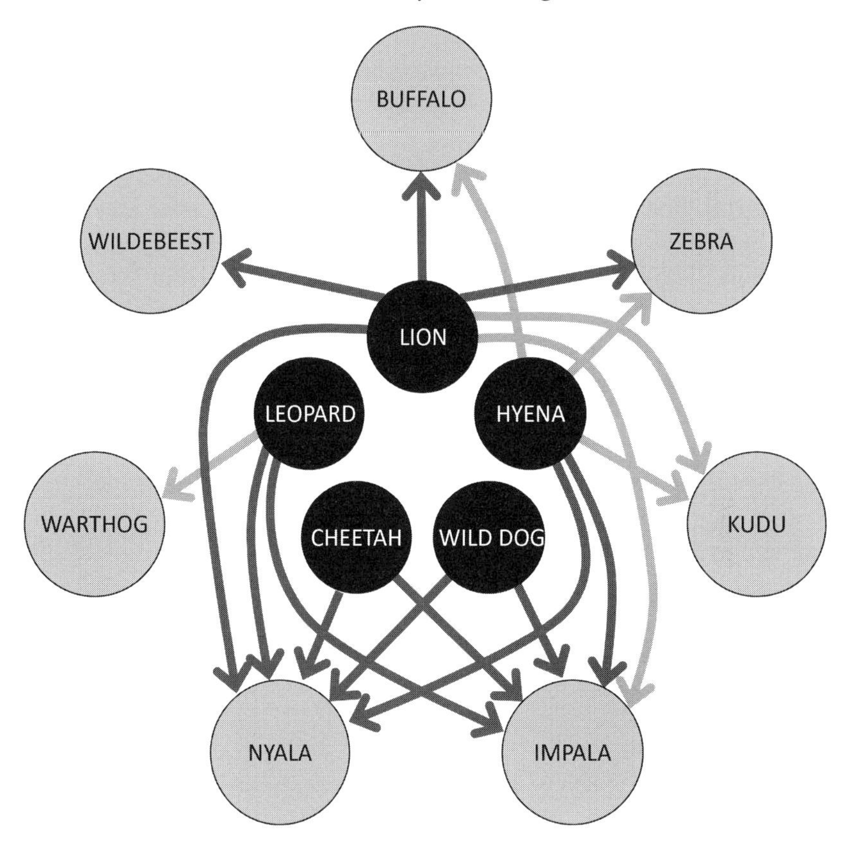

Figure 12.3 Trophic relationships between large carnivores and their main prey species in HiP. Dark grey arrows represent proportional consumption amounting to > 10% of the carnivore diet, and light grey arrows consumption amounting to < 10% but > 5% of the carnivore diet.

for a low representation of wildebeest (Table 12.2). Nyala increased in their proportional contribution with time, while the frequency of kills of zebra decreased. Buffalo were seldom represented during 1991–2000, in contrast to other two periods (Figure 12.2D). The patterns of prey consumption by spotted hyena in HiP are consistent with those described elsewhere (Hayward, 2006).

Overall, seven prey species supported most of the predation by large carnivores in HiP (Figure 12.3). Chapter 4 discusses some of the potential consequences of this predation for the different prey species. Species ranging in size from giraffe (*Giraffa camelopardalis*) down to wildebeest, plus warthog and bushpig (*Potamochoerus larvatus*), incurred predation mainly by lions, while nyala and impala were subject to predation by a diversity of carnivores (Table 12.2). There were very few records of the two rhino species (white rhino *Ceratotherium simum* and black rhino *Diceros bicornis*) and elephants (*Loxodonta africana*) being killed, most by lions, and the ages of these animals were not recorded. There were also few records of primates being killed, and these were solely by leopards. This aligns with findings in other African systems (Sinclair *et al.*, 2003; Owen-Smith and Mills, 2008). Occasional instances of intraguild predation have been documented in HiP by lions and leopards (Table 12.2). Occurrences by spotted hyenas may have been missed because the sample of observations for this species was relatively small.

Previous work has indicated that wild dogs and cheetahs may avoid close proximity to lions and spotted hyenas (Mills and Gorman, 1997; Creel and Creel, 1998). In HiP, wild dogs apparently avoid lions, especially during the denning season. However, wild dog packs in HiP were large enough to defend themselves against spotted hyenas (Darnell *et al.*, 2014). Saleni *et al.* (2007) showed that the wild dogs in HiP have very different activity patterns to lions and spotted hyenas through being mainly diurnal hunters, thus showing temporal avoidance of the dominant carnivores.

12.5.2 Scavenging

Carrion consumption has been much neglected in carnivore research, despite its common occurrence (DeVault *et al.*, 2003; Wilson and Wolkovich, 2011; Moleón *et al.*, 2014; Pereira *et al.*, 2014). Scavenging patterns within HiP were monitored using camera traps placed on a wide size range of carcasses (Moleón *et al.*, 2015). Small carcasses (< 10 kg) were represented by chicken, medium carcasses (10–100 kg) by impala and nyala, and large carcasses (> 100 kg) by wildebeest, buffalo, white rhino, and elephant. Lion, leopard, spotted hyena, large-spotted genet, white-tailed mongoose, and slender mongoose all fed at carcasses. Spotted hyenas and lions were detected at all carcass sizes, while genets and mongooses were only present at small carcasses. The spotted hyena was the main scavenger, consuming almost half of the carrion mass, depending on carcass size. Lions ate 23% of large carcasses and 5% of small carcasses. Genets and mongooses consumed 25% of the small carcasses. Wild dogs and cheetahs were not recorded scavenging at the carcasses placed. These findings confirm the primary role of spotted hyenas and lions as consumers of carrion as well as the fact that wild dogs and cheetah avoid carcasses (Pereira *et al.*, 2014).

12.6 Predicting the Carrying Capacity of Predators in HiP

Using the ungulate census data from 2006 to 2012 (Chapter 4), we predicted the population of each large predator species that could be sustained within HiP using equations derived by Hayward *et al.* (2007) using data from sites elsewhere in Africa. These predictions were based on the relationship between the density of each predator species and either the biomass of preferred prey species or the preferred prey weight range (see Hayward and Kerley, 2005 for definitions). We further partitioned the contribution that each HiP management section made to the overall predicted population for each large predator. Finally, we compared these predictions with the estimated populations of each large predator within HiP (see above).

Spotted hyenas were predicted to be the most abundant carnivore numerically based on preferred prey species abundance, whereas lions were predicted to be most abundant based on the abundance of prey species within the preferred weight range. Based on preferred prey, cheetahs and wild dogs should maintain population sizes below 50 (Figure 12.4A). Over the study period, the actual lion population was estimated to vary from 84 to 140 individuals (see above) compared with our prediction of ~130 lions. The actual spotted hyena population – estimated to be around 300 individuals – was double the prediction based on preferred prey, but close to the level predicted from preferred prey weight range. The actual leopard population is estimated to lie within the range 54–125 individuals, which was similar to our prediction of between 60 and 91. The actual cheetah population is estimated to be 30 individuals and we predict that the park can support 24–72 cheetahs. Finally, the actual wild dog population was estimated to vary between 40 and 106 individuals, and we predicted that HiP could support between 28 and 77 wild dogs. Based on the distribution of prey, the Corridor region is predicted to support the bulk of the lions and spotted hyenas, while the wilderness section of Mfolozi is predicted to support the bulk of the cheetahs and wild dogs.

12.7 Conclusion

Populations of all five of the major large mammalian carnivores have been restored in HiP, with three species (lion, cheetah, and wild dog) being re-introduced successfully to complement the two species that had persisted (leopard and spotted hyena). Brown hyenas, always rare, have

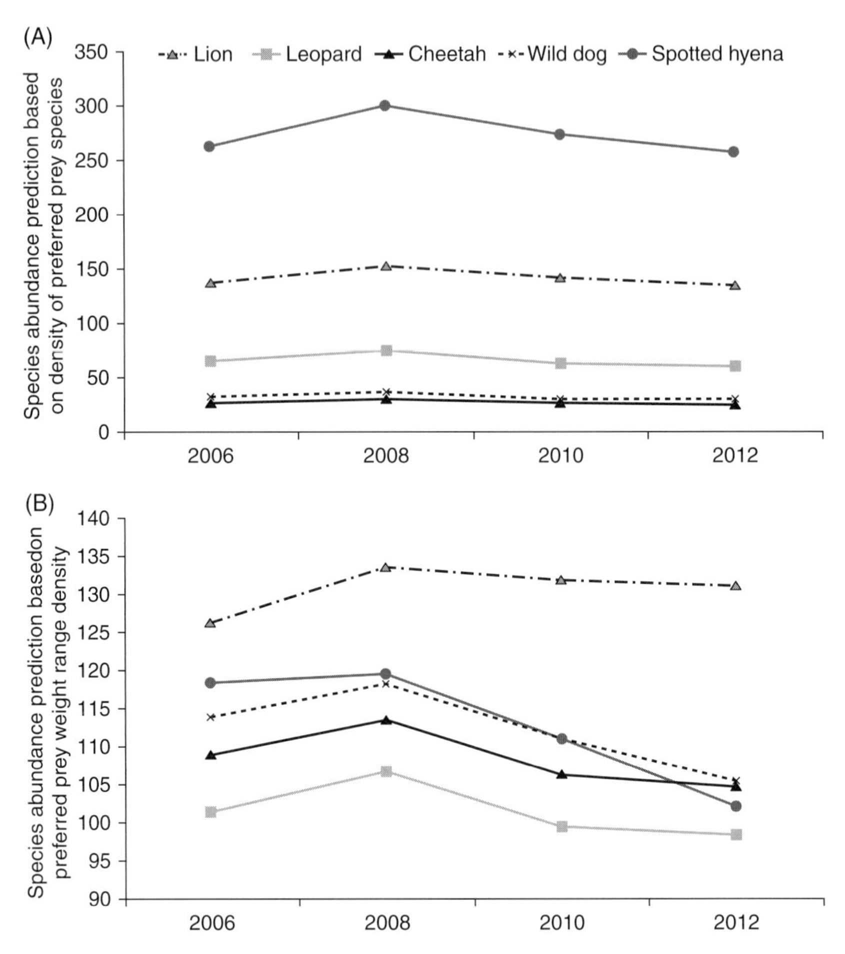

Figure 12.4 Predicted carnivore population based on (A) the abundance of preferred prey species, and (B) the abundance of prey within the preferred prey weight range (Hayward *et al.*, 2007).

become locally extinct, while black-backed jackals, formerly common, have surprisingly mostly disappeared. Continuing re-introductions were needed to sustain the wild dog and cheetah populations, and a fading lion population had to be restored through genetic augmentation. The current carnivore numbers seem to be consistent with those estimated from relationships established with the prey base in other areas.

Nevertheless, recent declines by several ungulate populations within HiP, especially impala, nyala, and wildebeest (Chapter 4), suggest that current carnivore numbers may not be sustainable. If populations of these herbivore species continue to decline, the carnivore populations will follow those of their prey. Changes in vegetation towards denser woody cover (Chapter 3) may be facilitating prey capture success particularly by lions (see Chapter 4 for more details).

Small reserves must also consider the impacts of persistent inbreeding on the viability of small carnivore populations (Mills, 1991). Two factors raise concerns about the genetic future of HiP's wild dogs. First, the relatively few wild dogs in HiP are an isolated population, with a correspondingly high potential for inbreeding, and hence for local extinction. This can, however, be counteracted by assisted movements of animals or whole packs between subpopulations as is being conducted under the South African Wild Dog Managed Metapopulation programme (Davies-Mostert *et al.*, 2009). The lion population founded by very few individuals showed clear signs of inbreeding, effectively overcome by introducing further lions from a different genetic stock. The HiP lion population still remains at probably under 140 individuals, meaning that active management will need to be ongoing to ensure its genetic viability.

Although HiP is completely fenced, carnivores still find gaps that allow them to break out – not only wild dogs but also lions, hyenas, and leopards. This has adverse consequences for neighbouring people and their livestock, compromising the community's attitude towards the protected area (Gusset *et al.*, 2008). Although there has been recent debate around the benefits and dangers of fences in conservation (see Somers and Hayward, 2012; Creel *et al.*, 2013; Packer *et al.*, 2013), the situation in HiP suggests that the outcome of fencing is context-dependent. With far-ranging and potentially dangerous species (such as wild dogs and lions) in small- to medium-sized reserves (such as HiP), fences are critical to protect the predators inside, and humans and livestock outside.

What is the future for HiP carnivores? If predation is driving substantial declines in prey numbers, some form of management action may be considered. However, it is important to recognize ancillary causes of changed predator–prey relationships, such as vegetation change, in order to effectively remedy the threat posed to conservation objectives. Modelling to project possible outcomes of management interventions (e.g. Gusset *et al.*, 2009) should continue, supported by better field estimates of carnivore populations aided by camera trapping. Ensuring a sustainable balance between predator and prey populations within protected areas as small as HiP is an ongoing challenge.

12.8 References

Anderson, J. L. (1981) The re-establishment and management of a lion *Panthera leo* population in Zululand, South Africa. *Biological Conservation* **19**: 107–117.

Becker, P. A., Miller, P. S., Gunther, M. S., *et al.* (2012) Inbreeding avoidance influences the viability of reintroduced populations of African wild dogs (*Lycaon pictus*). *PLoS ONE* **7**: e37181.

Blaustein, A. R., Bekoff, M., & Daniels, T. J. (1987) Kin recognition in vertebrates (excluding primates): empirical evidence. In: *Kin recognition in animals* (eds D. J. C. Fletcher & C. D. Michener), pp. 287–331. Wiley, Chichester.

Bourquin, O., Vincent, J., & Hitchins, P. M. (1971) The vertebrates of the Hluhluwe Game Reserve–Corridor–Umfolozi Game Reserve Complex. *Lammergeyer* **14**: 1–58.

Carruthers, E. (1985) The Pongola Game Reserve: an eco-political study. *Koedoe* **28**: 1–16.

Creel, S. & Creel, N. M. (1998) Six ecological factors that may limit African wild dogs, *Lycaon pictus*. *Animal Conservation* **1**: 1–9.

Creel S. & Creel, N. M. (2002) *The African wild dog: behavior, ecology, and conservation.* Princeton University Press, Princeton, NJ.

Creel, S., Becker, M. S., Durant, S. M., *et al.* (2013) Conserving large populations of lions – the argument for fences has holes. *Ecology Letters* **16**: 1413–e3.

Dalerum, F., Cameron, E. Z., Kunkel, K., & Somers, M. J. (2012) Interactive effects of species richness and species traits on functional diversity and redundancy. *Theoretical Ecology* **5**: 129–139.

Darnell, A. M., Graf, J. A., Somers, M. J., Slotow, R., & Gunther, M. S. (2014) Space use of African wild dogs in relation to other large carnivores. *PLoS ONE* **9**: e98846.

Davies-Mostert, H. T., Mills, M. G. L., & Macdonald, D. W. (2009) A critical assessment of South Africa's managed metapopulation recovery strategy for African wild dogs. In: *Reintroduction of top-order predators* (eds M. W. Hayward & M. J. Somers), pp. 10–42. Wiley-Blackwell, London.

Deane, N. N. (1962) The spotted hyaena *Crocuta crocuta crocuta*. *Lammergeyer* **2**: 26–44.

DeVault, T. L., Rhodes, R, Jr., & Shivik, J. A. (2003) Scavenging by vertebrates: behavioural, ecological and evolutionary perspectives on an important energy transfer pathway in terrestrial ecosystems. *Oikos* **102**: 225–234.

Edwards, J. M. (2009) Conservation genetics of African wild dogs *Lycaon pictus* (Temminck, 1820) in South Africa. MSc thesis, University of Pretoria.

Estes, J. A., Terborgh, J., Brashares, J. S., *et al.* (2011) Trophic downgrading of planet earth. *Science* **333**: 301–306.

Foster, W. E. (1955) History of the Umfolozi Game Reserve. Unpublished report, Natal Parks Board, Pietermaritzburg.

Girman, J. G., Mills, M. G. L., Geffen, E. & Wayne, R. K. (1997) A molecular genetic analysis of social structure, dispersal and interpack relationships of the African wild dog (*Lycaon pictus*). *Behavioral Ecology and Sociobiology* **40**: 187–198.

Graf, J. A., Gusset, M., Reid, C., *et al.* (2006) Evolutionary ecology meets wildlife management: artificial group augmentation in the re-introduction of endangered African wild dogs (*Lycaon pictus*). *Animal Conservation* **9**: 398–403.

Graf, J., Somers, M. J., Szykman, M., & Slotow, R. (2009) Heterogeneity in the density and distribution of spotted hyaenas in Hluhluwe-iMfolozi Park, South Africa. *Acta Theriologica* **54**: 333–343.

Grange, S., Owen-Smith, N., Gaillard, J.-M., *et al.* (2012) Changes of population trends and mortality patterns in response to the reintroduction of large predators: the case study of African ungulates. *Acta Oecologica* **42**: 16–29.

Gusset, M., Slotow, R., & Somers, M. J. (2006a) Divided we fail: the importance of social integration for the re-introduction of endangered African wild dogs (*Lycaon pictus*). *Journal of Zoology* **270**: 502–511.

Gusset, M., Graf, J., & Somers, M. J. (2006b) The re-introduction of endangered wild dogs into Hluhluwe-iMfolozi Park, South Africa: an update on the first 25 years. *Re-introduction NEWS* **25**: 31–33.

Gusset, M., Maddock, A. H., Szykman, M., *et al.* (2008) Conflicting human interests over the re-introduction of endangered wild dogs in South Africa. *Biodiversity and Conservation* **17**: 83–101.

Gusset, M., Jakoby, O., Müller, M. S., *et al.* (2009) Dogs on the catwalk: modelling re-introduction and translocation of endangered wild dogs in South Africa. *Biological Conservation* **142**: 2774–2781.

Hayward, M. W. (2006) Prey preferences of the spotted hyaena (*Crocuta crocuta*) and degree of dietary overlap with the lion (*Panthera leo*). *Journal of Zoology* **270**: 606–614.

Hayward, M. W. & Kerley, G. I. H. (2005) Prey preferences of the lion (*Panthera leo*). *Journal of Zoology* **267**: 309–322.

Hayward, M. W., Henschel, P., O'Brien, J., *et al.* (2006a) Prey preferences of the leopard (*Panthera pardus*). *Journal of Zoology* **270**: 298–313.

Hayward, M. W., Hofmeyr, M., O'Brien, J., & Kerley, G. I. H. (2006b) Prey preferences of the cheetah (*Acinonyx jubatus*) (Felidae: Carnivora): morphological limitations or the need to capture rapidly consumable prey before kleptoparasites arrive? *Journal of Zoology* **270**: 615–627.

Hayward, M. W., O'Brien, J., Hofmeyr, M. & Kerley, G. I. H. (2006c) Prey preferences of the African wild dog *Lycaon pictus* (Canidae: Carnivora): ecological requirements for conservation. *Journal of Mammalogy* **87**: 1122–1131.

Hayward, M. W., O'Brien, J., & Kerley, G. I. H. (2007) Carrying capacity of large African predators: predictions and tests. *Biological Conservation* **139**: 219–229.

Henschel, P., Azani, D., Burton, C., *et al.* (2010) Lion status updates from five range countries in West and Central Africa. *Cat News* **52**: 34–39.

Hunter, L. T. B. (1998) The behavioural ecology of reintroduced lions and cheetahs in the Phinda Resource Reserve, KwaZulu-Natal, South Africa. PhD thesis, University of Pretoria.

IUCN/SSC (2008) *Regional conservation strategy for the cheetah and wild dog in eastern Africa*. IUCN, Gland, Switzerland.

Kenya Wildlife Service (2010) *Proposal for inclusion of species on the Appendices of the Convention of the Conservation of Migratory Species of Wild Animals*. Kenya Wildlife Service, Nairobi, Kenya.

Kingdon, J. & Hoffmann, M. (2013) *Mammals of Africa volume V carnivores, pangolins, equids and rhinoceroses*. Bloomsbury Publishing, London.

Kokko, H. & Ots, I. (2006) When not to avoid inbreeding. *Evolution* **60**: 467–475.

Krüger, S. C., Lawes, M. J., & Maddock, A. H. (1999) Diet choice and capture success of wild dog (*Lycaon pictus*) in Hluhluwe-Umfolozi Park, South Africa. *Journal of Zoology* **248**: 543–551.

Leigh, K. A., Zenger, K. R., Tammen, I., & Raadsma, H. W. (2012) Loss of genetic diversity in an outbreeding species: small population effects in the African wild dog (*Lycaon pictus*). *Conservation Genetics* **13**: 767–777.

Lindsey, P. A., du Toit, J. T., & Mills, M. G. L. (2004) Area and prey requirements of African wild dogs under varying habitat conditions: implications for reintroductions. *South African Journal of Wildlife Research* **34**: 77–86.

Lindsey, P. A., Tambling, C. J., Brummer, R., *et al.* (2011) Minimum prey and area requirements of cheetahs: implications for reintroductions and management of the species as a managed metapopulation. *Oryx* **45**: 587–599.

Maddock, A. (1995) Wild dogs in Hluhluwe-Umfolozi Park. *Reintroduction News* **11**: 16–17.

Maddock, A. (1999) Wild dog demography in Hluhluwe-Umfolozi Park, South Africa. *Conservation Biology* **13**: 412–417.

Maddock, A., Anderson, A., Carlisle, F., *et al.* (1996) Changes in lion numbers in Hluhluwe-Umfolozi Park. *Lammergeyer* **44**: 6–18.

Marker-Kraus, L. (1996) Cheetah relocation. *African Wildlife – EPPINDUST* **50**: 21.

McCracken, D. P. (2008) *Saving the Zululand wilderness: an early struggle for nature conservation.* Jacana Media, Johannesburg.

Mills, M. (1991) Conservation management of large carnivores in Africa. *Koedoe* **34**: 81–90.

Mills, M. G. L. & Gorman, M. L. (1997) Factors affecting the density and distribution of wild dogs in the Kruger National Park. *Conservation Biology* **11**: 1397–1406.

Moehrenschlager, A. & Somers, M. (2004) Canid reintroductions and metapopulation management. In: *Canids: foxes, wolves, jackals, and dogs. Status survey and conservation action plan* (eds C. Sillero-Zubiri, M. Hoffmann, & D. W. Macdonald), pp. 59–67. IUCN/SSC Canid Specialist Group, Gland, Switzerland.

Moleón, M., Sánchez-Zapata, J. A., Selva, N., Donázar, J. A., & Owen-Smith, N. (2014) Inter-specific interactions linking predation and scavenging in terrestrial vertebrate assemblages. *Biological Reviews* **89**: 1042–1054.

Moleón, M., Sánchez-Zapata, J. A., Sebastián-González, E., & Owen-Smith, N. (2015) Carcass size shapes the structure and functioning of an African scavenging assemblage. *Oikos* **124**: 1391–1403.

Owen-Smith, N. & Mills, M. G. L. (2008) Predator–prey size relationships in an African large-mammal food web. *Journal of Animal Ecology* **77**: 173–183.

Packer, C., Loveridge, A., Canney, S., *et al.* (2013) Conserving large carnivores: dollars and fence. *Ecology Letters* **16**: 635–641.

Pereira, L. M., Owen-Smith, N., & Moleón, M. (2014) Facultative predation and scavenging by mammalian carnivores: seasonal, regional and intra-guild comparisons. *Mammal Review* **44**: 44–55.

Player, I. (1997) *Zululand wilderness: shadow and soul.* David Philip, Cape Town.

Potter, H. B. B. (1934) Report of Zululand Game Reserve and Parks Committee province of Natal – report of game conservator (Capt. Potter) for 1933. *Journal of the Society for the Preservation of the Fauna of the Empire* **23**: 64–70.

Potter, H. B. B. (1941) Report of Zululand Game Reserve and Parks Committee province of Natal – report of game conservator (Capt. Potter) for 1941. *Journal of the Society for the Preservation of the Fauna of the Empire* **43**: 35–41.

Pringle, J. A. (1977) The distribution of mammals in Natal. Part 2. Carnivora. *Annals of the Natal Museum* **23**: 93–115.

Prugh, L. R., Stoner, C. J., Epps, C. W., *et al.* (2009) The rise of the mesopredator. *BioScience* **59**: 779–791.

Rautenbach, I. L., Skinner, J. D., & Nel, J. A. J. (1980) The past and present status of mammals of Maputaland. In: *The ecology of Maputaland* (eds M. N. Bruton & K. H. Cooper), pp. 322–345. Rhodes University & Natal Branch of The Wildlife Society of Southern Africa, Grahamstown.

Ripple, W. J. & Beschta, R. L. (2012) Trophic cascades in Yellowstone: the first 15 years after wolf reintroduction. *Biological Conservation* **145**: 205–213.

Roberts, A. (1954) *The mammals of South Africa*, 2nd edn. Trustees of 'The mammals of South Africa' Book Fund, Johannesburg.

Rowe-Rowe, D. T. (1992) The carnivores of Natal. Unpublished report, Natal Parks Board, Pietermaritzburg.

Rowe-Rowe, D. T. (1994) The ungulates of Natal. Unpublished report, Natal Parks Board, Pietermaritzburg.

Saleni, P., Gusset, M., Graf, J. A., *et al.* (2007) Refuges in time: temporal avoidance of interference competition in endangered wild dogs. *Canid News* **10**: 1–5.

Sinclair, A. R. E., Mduma, S., & Brashares, J. S. (2003) Patterns of predation in a diverse predator–prey system. *Nature* **425**: 288–290.

Somers, M. J. & Hayward, M. (eds). (2012) *Fencing for conservation: restriction of evolutionary potential or a riposte to threatening processes?* Springer Science & Business Media, New York.

Somers, M. J. & Maddock, A. (1999) Painted dogs of Zululand. *African Wildlife* **53**: 24–26.

Somers, M. J., Graf, J. A., Szykman, M., Slotow, R., & Gusset, M. (2008) Dynamics of a small re-introduced population of wild dogs over 25 years: allee effects and the implications of sociality for endangered species' recovery. *Oecologia* **158**: 239–247.

Spiering, P. A., Somers, M. J., Maldonado, J. E., Wildt, D. E., & Gunther, M. S. (2010) Reproductive sharing and proximate factors mediating cooperative breeding in the African wild dog (*Lycaon pictus*). *Behavioral Ecology and Sociobiology* **64**: 583–592.

Spiering, P. A., Gunther, M. S., Somers, M. J., *et al.* (2011) Inbreeding, heterozygosity and fitness in a reintroduced population of endangered African wild dogs (*Lycaon pictus*). *Conservation Genetics* **12**: 401–412.

Steele, N. A. (1970) A preliminary report on the lions in the Umfolozi and Hluhluwe Game Reserves. *Lammergeyer* **11**: 68–79.

Stein, B. (1999) Genetic variation and depletion in a population of lions (*Panthera leo*) in Hluhluwe-iMfolozi Park. MAgric thesis, University of Natal.

Swanson, A., Caro, T., Davies-Mostert, H., *et al.* (2014) Cheetahs and wild dogs show contrasting patterns of suppression by lions. *Journal of Applied Ecology* **83**: 1418–1427.

Trinkel, M., van Niekerk, R. W., Fleischmann, P. H., Ferguson, N., & Slotow, R. (2007) The influence of vegetation on lion group sizes in the Hluhluwe-Umfolozi Park, South Africa. *Acta Zoologica Sinica* **53**: 15–21.

Trinkel, M., Ferguson, A., Reid, C., *et al.* (2008) Translocating new lions into an inbred lion population in the Hluhluwe-iMfolozi Park, South Africa. *Animal Conservation* **11**: 138–143.

Van Dyk, G. & Slotow, R. (2003) The effects of fences and lions on the ecology of African wild dogs reintroduced to Pilanesberg National Park, South Africa. *African Zoology* **38**: 79–94.

Vaughan-Kirby, F. (1916) Game and game preservation in Zululand. *South African Journal of Science* **13**: 375–396.

Walton, L. R. & Joly, D. O. (2003) *Canis mesomelas*. *Mammalian Species* **715**: 1–9.

Whateley, A. & Brooks, P. M. (1985) The carnivores of the Hluhluwe and Umfolozi Game Reserves: 1973–1982. *Lammergeyer* **35**: 1–28.

Whittington-Jones, B. M., Parker, D. M., Bernard, R. T. F., & Davies-Mostert, H. T. (2014) Habitat selection by transient African wild dogs (*Lycaon pictus*) in northern KwaZulu-Natal, South Africa: implications for range expansion. *South African Journal of Wildlife Research* **44**: 135–147.

Wilmers, C. C. & Getz, W. M. (2005) Gray wolves as climate change buffers in Yellowstone. *PloS Biology* **3**: e92.

Wilson, E. E. & Wolkovich, E. M. (2011) Scavenging: how carnivores and carrion structure communities. *Trends in Ecology and Evolution* **26**: 129–135.

Woodroffe, R., McNutt, J. W., & Mills, M. G. L. (2004) African wild dog. In: *Foxes, wolves, jackals and dogs: status survey and conservation action plan* (eds C. Sillero-Zubiri & D. W. Macdonald), pp. 174–183. IUCN, Gland.

13 · *Wildlife Disease Dynamics in Carnivore and Herbivore Hosts in the Hluhluwe-iMfolozi Park*

ANNA E. JOLLES, NICKI LE ROEX, GABRIELLA FLACKE, DAVID COOPER, CLAIRE GEOGHEGAN, AND MICHAEL J. SOMERS

13.1 Introduction

Parasites and pathogens are a normal part of life for wild animal populations; even at the individual level, most wild animals harbour multiple infectious parasites and pathogens most of the time (Petney and Andrews, 1998; Pedersen and Fenton, 2007; Tompkins *et al.*, 2010). Under natural circumstances, these infections usually do not warrant human intervention. Wild animals require very little doctoring, because they often tolerate their infections well and remain clinically healthy. In instances where disease does cause individual deaths or breeding failure, wildlife populations can typically cope with these losses through compensatory survival or reproduction by uninfected animals (Jolles *et al.*, 2006). More broadly, parasites and pathogens form an integral part of ecological communities. To extirpate them would disrupt ecological interactions between competing animal species, modify consumer–resource relationships, and remove one of the essential mechanisms for maintaining genetic diversity and adaptability in animal populations.

However, human activities have altered many wildlife populations and their habitat in ways that limit the ability of animals to tolerate or compensate for infectious diseases; and conflict at the wildlife–livestock–human interface is likely to intensify as long as human populations expand. Hence, a case can be made for active management of disease

Conserving Africa's Mega-Diversity in the Anthropocene, ed. Joris P. G. M. Cromsigt, Sally Archibald and Norman Owen-Smith. Published by Cambridge University Press.

risks for some wildlife populations, especially in small reserves. Wild animals confined within protected areas are restricted in their movements, which may lead to resource restrictions and nutrient imbalances that are likely to affect their ability to defend against or tolerate infections. Moreover, wildlife populations within reserves are often comparatively small, which makes resilience to disease losses a more risky gamble. Small populations may also lack the genetic adaptability to respond to new disease challenges. Furthermore, the risk of exposure to unfamiliar pathogens through contact with domestic animals and humans is greater in smaller reserves, due to their higher edge-to-area ratio. Infections that spill over into wild animals from adjacent domestic animal populations can cause drastic population declines, because parasite dynamics are decoupled from wildlife population dynamics: even if the wildlife population crashes, the parasite finds refuge in the (often much larger) domestic animal population and can continue to put pressure on the remaining wild animal hosts (DeCastro and Bolker, 2005). Wild animals in small reserves thus face a twofold risk of increased transmission and magnified effects of infectious disease. Additionally, the high edge-to-area ratio of small reserves increases the risk of disease spillover in the opposite direction – from wildlife reservoirs to people and their animals. Most small reserves lack buffer zones around them, and livestock graze right up to the reserve boundary (and sometimes beyond). The resulting conflict between conservation and agricultural or public health interests has the potential to prejudice the long-term viability of protected areas and the species they contain.

Perhaps more than any other single factor, infectious diseases have affected the management of wildlife populations in and around the Hluhluwe-iMfolozi Park (HiP) both historically and recently (see Chapters 1 and 4). Furthermore, some of HiP's most notable conservation successes have hinged upon innovative approaches to managing genetic health and infectious diseases in wildlife populations. HiP's history thus presents a kaleidoscopic view of the challenges, as well as successes, associated with managing infectious diseases at the domestic animal–wildlife interface. In this chapter, we report on insights from HiP related to wildlife disease ecology and control. Specifically, we focus on (i) managing spillover and spill-back of infections at the wildlife–domestic animal interface; (ii) the vulnerability of HiP's carnivore populations to disease; and (iii) evaluating disease control interventions. We use examples from lion (*Panthera leo*), wild dog (*Lycaon pictus*), buffalo (*Syncerus caffer*), and neighbouring livestock populations to illustrate these recurring themes.

13.2 Disease Transmission across Boundaries of Small Reserves

13.2.1 Transmission from Wildlife to Livestock and People: Wildlife as an Infection Reservoir

African wild ungulates share many pathogens with cattle and smaller ruminants, and carry several pathogens that can infect humans (zoonoses). These include protozoan haemoparasites such as *Theileria parva* (causing Corridor disease in cattle) and *Trypanosoma* spp. (causing nagana in livestock); bacterial haemoparasites like *Anaplasma marginale* (bovine anaplasmosis) and *Ehrlichia ruminantium* (heartwater); and viral infections such as foot-and-mouth disease and Rift Valley fever (a zoonosis). Wildlife may also act as a reservoir for infections originally acquired from cattle, such as bovine tuberculosis (*Mycobacterium bovis*) and brucellosis (*Brucella abortus*), which spill back from wildlife into cattle populations and even people. Here we discuss two such cases that have been particularly influential in HiP: nagana and bovine tuberculosis.

13.2.1.1 Disease Spillover by Nagana

Nagana, or trypanosomosis, is a suite of diseases affecting domestic ungulates and humans (sleeping sickness), caused by protozoan parasites of the genus *Trypanosoma* (Connor and van den Bosche, 2004). The relevant trypanosome species in Zululand are *T. congolense* and *T. vivax*, which are transmitted by tsetse flies of the genus *Glossina*. Livestock trypanosomoses are characterized by fluctuating parasitaemia and fever, weight loss, anaemia, and for some host–parasite strain combinations, high rates of mortality. The geographic distribution of nagana is determined primarily by the distribution of competent tsetse vectors, and thus the disease is controlled primarily through limiting vector abundance. As described in Chapters 1 and 4, nagana and actions to control it have shaped the history of wildlife populations in HiP, giving rise to major conflicts between agricultural and conservation objectives towards the end of the nineteenth and beginning of the twentieth centuries. This led to game elimination campaigns and eventually aerial spraying with insecticides aimed at eradicating tsetse flies. This successfully eliminated *G. pallidipes*, the main vector of nagana, from the region, but the forest-inhabiting *G. brevipalpis* and *G. austeni*, along with trypanosomes, persisted locally. Between 1955 and 1990, only small sporadic nagana outbreaks among cattle were recorded in Zululand. However, a large outbreak occurred in 1990, perhaps due to the increased availability of shady habitat preferred by the surviving *Glossina* species, in the wake of widespread planting of eucalyptus and pine plantations. During the 1990 outbreak, cattle served by 61 diptanks around HiP and Mkhuze Game Reserve were found to be infected with *T. congolense* and *T. vivax*. This prompted major investment in nagana research in South Africa, including the establishment of a research station near HiP. Four main themes have been targeted: (a) improving trypanosome diagnostics and fly traps, (b) disease surveillance, (c) assessing the role of both remaining tsetse species in nagana transmission, and (d) understanding variation in virulence among trypanosome strains. New tsetse fly traps designed specifically for monitoring populations of the two tsetse species persisting in northern KwaZulu-Natal were developed (Green and Venter, 2007). From surveys of trypanosome prevalence in cattle and tsetse flies around HiP and other game reserves in the region (van den Bossche *et al.*, 2006; Mamabolo *et al.*, 2009; Gillingwater *et al.*, 2010; Ntantiso *et al.*, 2014) it is evident that trypanosomosis appears endemic in northern KwaZulu-Natal, although infection prevalence varies widely, both spatially and temporally. *T. congolense* is the most commonly detected trypanosome species, with *T. vivax* found at lower prevalence, while *G. brevipalpis* was consistently more abundant than *G. austeni*. Nevertheless, *G. austeni* may play the more important role as a vector, because a recent study found mature trypanosomes in *G. austeni* but not *G. brevipalpis*, and *G. austeni* transmitted infection to susceptible cattle, while *G. brevipalpis* did not (Motloang *et al.*, 2012). Trypanosome strains from wild ungulates may be more virulent (i.e. damaging to the host) than strains circulating in domestic animals (van den Bossche *et al.*, 2011), but this pattern is not consistent (Chitanga *et al.*, 2013). Trypanosomes isolated from buffalo in HiP did appear to be more virulent than those found in cattle distant from the park. Cattle close to the park boundary hosted a mixture of more virulent and less virulent strains (Motloang *et al.*, 2014). Results from these studies are relevant for guiding disease-control strategies for nagana. Targeting the less-abundant tsetse species, *G. austeni*, could disproportionately reduce disease transmission. Focusing disease surveillance and control efforts near the park boundary may be most effective, because this is where the more virulent trypanosome strains circulate in cattle.

13.2.1.2 Disease Spill-Back: Bovine Tuberculosis

Bovine tuberculosis (BTB), caused by *Mycobacterium bovis*, is another infection for which wildlife may be acting as a reservoir in South Africa. The situation is subtly different than with nagana, though, in that BTB is not a native African infection, but is thought to have been introduced to wild ungulate populations through contact with cattle. The disease does not spread very rapidly, and can be eradicated from most cattle populations through surveillance and the culling of infected animals. As a result, even though BTB originated from cattle, at this point wildlife are seen as an important reservoir of the disease, and spill-back into cattle populations is a potential risk at the wildlife–livestock interface (Bengis *et al.*, 2002). BTB probably entered HiP's wild ungulate populations in the 1960s, prior to the completion of the boundary fence that now effectively separates wildlife from livestock (see Chapter 1). The disease was first detected at HiP in a black rhino in 1970 and in buffalo in 1986. BTB is seen as problematic because it limits the ability of the conservation agency to generate income from selling wildlife to other parks and reserves, and increases the potential for conflict with livestock owners around the park. BTB also affects some of HiP's wild animal populations directly, causing clinical disease (e.g. lions: Trinkel *et al.*, 2011), reducing population growth rate (e.g. buffalo: Jolles *et al.*, 2005), or reducing the ability of the host to tolerate endoparasitic infections (Caron *et al.*, 2003; Jolles *et al.*, 2008; Ezenwa and Jolles, 2015). Moreover, BTB in livestock in areas surrounding HiP puts people at risk of zoonotic infection, with potentially grave health outcomes in the context of rural KwaZulu-Natal where HIV/AIDS prevalence in adults (> 15 years of age) exceeds 20% (Tanser *et al.*, 2013). Consequently, surveillance and control of BTB in and around HiP have been a veterinary, public health, and research priority. A survey in 1996–1998 revealed that the disease was widespread throughout the park and that buffalo herds in different areas varied in prevalence from zero BTB to upward of 50% infection (Figure 13.1). This prompted the development of a test and cull programme that began in 1999 and continues at present. The BTB control programme was aimed at (a) limiting the prevalence of the disease in buffalo, the recognized maintenance host, in an attempt to protect other mammal species (considered spillover hosts) from infection, and (b) enabling continued movement of animals from HiP to other public and privately owned wildlife areas. The programme targets different areas within the park on a rotational basis. Each area is processed at least every 4–5 years, with more heavily infected areas targeted more frequently.

To implement this disease-control programme, methods and equipment suitable for mass capture in the field were developed by the Game Capture unit, in collaboration with wildlife veterinarian Dr Dave Cooper. Moveable enclosures were designed to contain the buffalo herds, which are brought in from distances up to 5 km away. A helicopter is used to

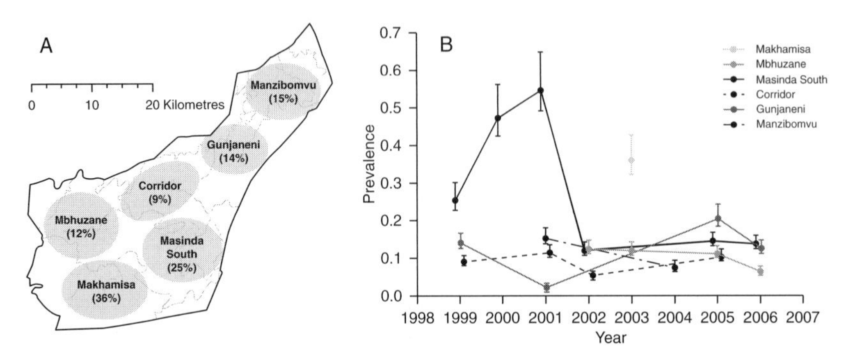

Figure 13.1 Spatial and temporal variation of BTB prevalence in HiP's buffalo population. (A) Distinctions in BTB prevalence in different regions of the park in 1998; (B) changes in prevalence, by region, between 1999 and 2006.

guide the animals into a capture funnel that is connected to the enclosure. Once in the enclosure, buffalo are partitioned into small groups and immobilized for BTB testing and data collection. Buffalo are tested for BTB using the single comparative intradermal tuberculin (SCIT) test, and upon diagnosis, BTB-positive animals are slaughtered and BTB-negative animals released back into the park.

In conjunction with the BTB control programme, a study was set up to evaluate its effectiveness in limiting BTB prevalence in the buffalo population within HiP between 1999 and 2006 (Le Roex *et al.*, 2015). During the study period, 250–950 buffalo were tested for BTB annually and BTB prevalence in the disease hot spot area at Masinda South declined dramatically, and did not expand into other areas (Figure 13.1; see map in the prelim for the location of Masinda section). BTB prevalence in moderately affected areas was maintained at 10–15%, rather than increasing as one might expect in the absence of disease control. Importantly, analysis of herd-level data showed that herds experiencing more intensive (i.e. greater fraction of the herd processed) and more frequent captures showed reduced disease transmission risk and lower increase in herd prevalence over time. In combination, these data suggest that the test and cull programme is effective at reducing BTB transmission in buffalo herds within HiP. Both the reduction in density of the buffalo population, and lowering of disease prevalence within herds, may have played a role in mediating this success. The programme, with its innovative, field-based mass capture technique for buffalo, is thus a substantial accomplishment in the management of wildlife disease in a free-ranging population.

Given this success, the challenge now lies in weighing costs and benefits of BTB control in HiP's and other buffalo populations. While frequent mass captures appear to be most effective at reducing BTB transmission, they also reduce buffalo numbers (Jolles, 2007). This imposes tight bounds on operating a BTB control programme, if buffalo are to be maintained as an important ecological component of a reserve's mammal fauna. To add to this complexity, test and cull programmes can have unintended consequences, which need to be considered when evaluating the merits of this management strategy. Captures are stressful to the animals and can disrupt their social structure – both of which may add to their infection risk after being released back into the park (Woodroffe *et al.*, 2006; McDonald *et al.*, 2008). This may be especially problematic in areas with high disease prevalence, where released susceptible animals experience a high encounter rate with infected animals. Comprehensive coverage of high-prevalence hot spots may be essential to circumvent this pitfall. In addition, removal of BTB-positive buffalo may reduce immunogenetic variation in the population (see Section 13.3.2 below). Finally, although the test and cull programme may reduce BTB transmission, eradication of this chronic, multihost species infection is not a realistic goal. Accordingly, unless wildlife populations are large enough to provide resilience to the effects of infectious diseases, their viability may depend on a long-term, costly commitment to disease control.

Complementing this disease control effort inside the park, a collaborative 'One-Health' disease prevention programme was launched among human communities living within 10 km from the HiP boundary fence in 2007. This programme brought together veterinarians, public health workers, ecologists, agriculture students, indigenous healers, and farmers to build a broad basis for designing a holistic health programme for the area. The group's efforts culminated in an integrated health campaign offering simultaneous veterinary and public health services to 26 communities. As part of this programme, 12,000 cattle were brought for voluntary health assessments, of which 400 were recruited for longer-term BTB surveillance. Complimentary vaccinations and treatments for common zoonotic infections were offered for domestic animals, and information about clinical, food safety, and zoonotic aspects of BTB were disseminated to 5500 community members. Such initiatives can contribute to shifting the balance from viewing wildlife as a threat, to recognizing them as an asset to rural communities. Local buy-in is indispensable for the long-term sustainability of wildlife conservation, and rural outreach and collaboration have potential to make a large positive impact.

However, securing long-term investment in One-Health programmes at the wildlife–livestock interface is a daunting challenge in the face of shrinking public health and conservation budgets in many regions that are of key importance to biodiversity conservation.

13.2.2 Transmission from Domestic Animals to Wildlife: Disease Risks to African Wild Dogs

With increasing human population density in rural areas throughout South Africa, more people and their domestic animals live close to the boundaries of wildlife reserves. Small populations of carnivore species living in fragmented habitats are particularly susceptible to pathogen transmission from larger populations of reservoir hosts (Prager *et al.*, 2012a). A prime example of this are disease risks to African wild dogs, which increase dramatically with proximity to human habitation, because domestic dogs can serve as either a reservoir host or transmission source for many canine pathogens (Woodroffe and Ginsberg, 1998). Disease risks to wild dogs are further heightened through their regular dispersal movements outside protected areas, leading to increased contacts with domestic dogs (Prager *et al.*, 2012b; Woodroffe *et al.*, 2012). Some spillover infections can result in disease epidemics and high mortality rates from which recovery is difficult for small or fragmented wild dog populations. Even diseases with comparatively modest impacts on mortality rates or fecundity may lead to local extinctions of such populations (Creel and Creel, 1998). Population-level effects of infectious diseases may be intensified for wild dogs due to pack structure and their highly social nature. Pathogens of particular concern for the long-term viability of wild dog populations include canine distemper virus (CDV), canine parvovirus (CPV), and rabies virus. HiP's wild dog population is closely monitored, so that unusual mortality events or breeding failures can be identified and followed up.

Rabies is a major disease threat for wild dogs across Africa (Alexander *et al.*, 1993; Gascoyne *et al.*, 1993; Mills, 1993; Kat *et al.*, 1995) and has resulted in the sudden loss of entire packs, as happened in the Serengeti ecosystem in Tanzania in 1991 (Macdonald, 1992; Gascoyne *et al.*, 1993). A rabies outbreak among domestic dogs occurred near the south-western border of HiP in 2006 during a two-month period when 20 wild dogs were lost from HiP's largest pack (Flacke *et al.*, 2013). Wild dogs frequently cross over HiP's boundaries (see Chapter 12) and, considering the highly contagious nature of the rabies virus, the ease with which the virus can be transmitted between pack members (Mills, 1993), and the rapid mortality rates observed in other wild dog packs infected with rabies (Alexander *et al.*, 1993; Hofmeyr *et al.*, 2000; Woodroffe, 2001), infections with rabies were the most likely cause of the pack's disappearance. Moreover, clinical signs consistent with rabies were observed in one pack member shortly before the pack's disappearance. Because no carcasses could be recovered for necropsy (Flacke *et al.*, 2013), other causes of acute pack mortality, such as poisoning, cannot be completely ruled out.

Similar to the rabies virus, CDV risks are particularly high for small, fragmented, and immunologically naïve populations (Mills, 1993). Numerous documented outbreaks of CDV in wild dogs across Africa indicate that the disease is of primary concern for these canids, reducing long-term viability and increasing extinction risk for already small populations (Alexander *et al.*, 1994, 1996; van de Bildt *et al.*, 2002; Goller *et al.*, 2010). Loss of wild dog packs has been associated with CDV outbreaks in the Okavango Delta, Botswana (Alexander *et al.*, 2010), Tswalu Kalahari Reserve, South Africa (G. van Dyk, personal communication 2007), and Chobe National Park, Botswana (Alexander *et al.*, 1996).

Although CPV is not considered as great a threat as rabies and CDV, it poses a risk to small populations through substantial early pup mortality, and can affect all age classes in immunologically naïve populations (Mech and Goyal, 1995; Creel *et al.*, 1997). In Selous Game Reserve (Tanzania), CPV infections among wild dogs were believed to be responsible for reductions in litter size before pup emergence (Creel *et al.*, 1997). Exposure to CPV among African wild dogs has been demonstrated across southern Africa (Prager *et al.*, 2012b), albeit with unknown population-level impacts. A CDV and CPV sero-survey of the wild dog population within KwaZulu-Natal in 2006/2007 demonstrated 0% antibody prevalence for CDV and 4% prevalence for CPV (1/24 sampled dogs), even though both infections are present in domestic dog populations throughout the region (Flacke *et al.*, 2013). Low prevalence of these pathogens in wild dogs indicates either a lack of exposure to CDV and CPV, or near 100% mortality for animals that do get exposed to these pathogens. However, such extreme mortality has not been observed during the intense monitoring programme in HiP.

In summary, spillover of viral infections from domestic dogs to HiP's wild dogs does occur, but not very frequently. There has only been one presumed rabies spillover in the HiP population to date, and the data on CDV and CPV infections are inconclusive. More regular surveillance for these pathogens could clarify the frequency and consequences of spillover of these viral pathogens from domestic dogs to wild dogs. However, surveillance involves the capture and handling of wild dogs, and the costs and risks of these interventions must be weighed against the benefits of disease detection. Given the limitations and risks associated with vaccination in wild dogs (McCormick, 1983; Durchfeld *et al.*, 1990; van Heerden *et al.*, 2002), disease control in domestic dog populations can be a more economical and practical option for protecting wild dogs from exposure to key canine pathogens (Lembo *et al.*, 2008). Alternatively, targeted vaccination of key wild dog pack members and more vulnerable packs (e.g. those located in edge habitats) can reduce the number of individuals that need to be handled to achieve effective disease protection (Vial *et al.*, 2006; Prager *et al.*, 2011). Population viability analysis suggests that vaccination of 30–40% of 'core' wild dogs in small populations every 1–2 years could effectively control rabies, thus facilitating wild dog persistence in the face of outbreaks (Vial *et al.*, 2006). This strategy might be considered for the HiP population; however, it is never easy to predict actual *in situ* outcomes, despite data from other populations and well-validated mathematical models. Disease control strategies must be implemented synergistically with management plans for other long-term threats to wild dog population viability, including continued habitat fragmentation and landscape-associated factors limiting natural dispersal events.

Besides viral pathogens, other parasite infections have been recognized quite widely (Tompkins *et al.*, 2010). Surveys for gastrointestinal (GI) parasites and haemoparasites on a sample (24/70) of HiP's wild dogs in 2006 revealed various nematodes (roundworms; *Toxocara canis*, 9/12 dogs; *Ancylostoma* spp., 6/12 dogs; *Trichuris* spp., 4/12 dogs), a cestode (tapeworm; *Dipylidium caninum*, 7/12 dogs), and two protozoan genera (*Sarcocystis*, 12/12 dogs; *Isospora*, 2/12 dogs)(Flacke *et al.*, 2010, 2013). The hemoparasitic protozoan *Ehrlichia canis* was also common (21% of dogs); but no wild dogs were found to be infected with *Babesia canis*. Most wild dogs at HiP are thus infected with multiple parasites simultaneously, which is the norm in wild animals (Petney and Andrews, 1998). The effects of these parasites on the fitness of their wild dog hosts have not been measured and are not of immediate concern.

13.3 Vulnerability of Small Populations to Disease and Inbreeding

Management interventions within HiP have been aimed at improving population genetic diversity and health as well as at countering disease infections. We focus here on one carnivore species, the African lion, and one large-bodied herbivore, the African buffalo. Both of these species are profoundly affected in their dynamics by BTB infections and BTB is, therefore, actively managed. Assessment of disease-control programmes through epidemiological research is fundamental to adaptive, evidence-based wildlife conservation management. Yet to our knowledge, HiP's BTB control programmes are the only such initiatives in sub-Saharan Africa where the effects of disease-control interventions in wildlife have been evaluated systematically.

13.3.1 Benefits and Risks of Genetic Management – Lessons from Lions

Small, isolated populations face increased risks from infectious diseases, because inbreeding can affect their ability to handle infections (Coltmann *et al.*, 1999), possibly through effects on immune function (Paterson *et al.*, 1998; Coltmann *et al.*, 2001; Reid *et al.*, 2003), and because small populations can least afford further reductions in size due to disease outbreaks. Inbred individuals may be more susceptible to infectious diseases and, therefore, inbreeding may have a significant impact on population trends. The lion population within HiP was established from seven founders during the 1960s – one male, and three females plus three cubs (Maddock *et al.*, 1996; Chapter 12). The population thus had little initial genetic variation, and further genetic losses were incurred from young animals dispersing from the park and being shot. In 1990, HiP's lion population numbered approximately 80 individuals, all descendants of the original five founders. The lions had started showing conspicuous signs of abnormalities, including abscesses and generally poor condition, attributed to inbreeding. This prompted the translocation of further lions into HiP aimed at restoring the genetic health of the HiP population (Trinkel *et al.*, 2008; Chapter 12). Between August 1999 and January 2001, 16 lions descended from genetically distant stock from Etosha National Park, Namibia, were introduced into HiP (Trinkel *et al.*, 2008; Chapter 12). The success of these introductions in improving lion health was assessed through a research project comparing infections and mortality of founding versus subsequently introduced lions and their offspring between 2000 and 2009 (Trinkel *et al.*, 2011). During this period, lions that died or were destroyed due to poor condition were subjected to a post-mortem examination for BTB. In addition, a sero-survey was conducted in 2001–2002 to assess the prevalence of exposure to viral infections within samples of lions of different origins, covering feline immunodeficiency virus (FIV), CDV, feline herpesvirus (FHV), feline parvovirus (FPV), feline coronavirus (FCoV), and feline calicivirus (FCV) (Trinkel *et al.*, 2011).

BTB was evidently a major problem among the founding lions, but not among introduced lions and their offspring. Between 2000 and 2005, more than 30% of founder lions (N = 104) either died from BTB (N = 15) or were dramatically emaciated and subsequently destroyed (N = 18), compared to no deaths from BTB and a single euthanasia due to malnutrition in later introduced lions and their offspring (N = 63). Deaths from BTB and malnutrition within the lion population subsequently declined, with only two deaths due to BTB and three deaths due to malnutrition between 2006 and 2009, by which time no pure descendants of the founding lions remained. This demonstrated dramatic benefits to lion health from introducing new genetic material into HiP (Trinkel *et al.*, 2011).

The sero-survey showed that lions in HiP were exposed to FHV, FCV, FCoV, FIV, and FPV, with seroprevalences comparable to those in other ecosystems (Spencer, 1991; Packer *et al.*, 1999). The overall antibody prevalence was highest for FHV (97%), followed by FCV (51%), FPV (41%), FIV (15%), and FCoV (10%). As with the wild dogs, CDV was not detected. The lion sero-survey detected no significant differences in seroprevalence of any of the viruses between founding and later introduced lions and their offspring (Fisher's exact test: FHV, $p = 0.79$; FCV, $p = 0.13$; FPV, $p = 0.12$; FIV, $p = 0.58$; FEC, $p = 0.37$). Seroprevalence data are hard to interpret, because they reflect both susceptibility and survival with the disease. Thus equal seroprevalence could occur if inbred lions were at higher risk of infection, but also had increased mortality rates of infected animals. To determine the population effects of these infections, the prevalence of morbid and clinically healthy lions would need to be compared for inbred and outbred populations (McCallum and Dobson, 1995). The clinical effects of these infections in lions are poorly understood, because they are usually diagnosed retrospectively with serological tests, after the infection has resolved (except for chronic infections that are not cleared). As such, it is not surprising that seropositive lions appeared clinically healthy in this study.

Another noteworthy finding of the sero-survey was that FIV may have been brought to HiP through the lion translocation. Before 2000, FIV appears to have been absent from HiP (Van Vuuren *et al.*, 2003), although it is possible, given the moderate sample size of the survey, that the pathogen was present but undetected in the population. In 2001, FIV was detected in 6 of 39 lions tested (15%) following the lion introductions, and increased to 42% (8/19 lions tested) in 2006 (Trinkel *et al.*, 2011). Two of the FIV-positive lions detected in 2001 were introduced lions that were members of the same pride as the founding lions that tested FIV-positive. In addition, some nomadic founder males that had been seen in contact with this pride tested positive.

The consequences of FIV introduction for HiP's lion population are difficult to predict. FIV in lions was until recently regarded as an innocuous endemic pathogen, causing little mortality or clinical disease (Brown *et al.*, 1994; Carpenter and O'Brien, 1995; Hofmann-Lehmann *et al.*, 1996; Packer *et al.*, 1999). However, recent research has demonstrated significant depletions of CD4+ (T-helper) immune cells in seropositive lions (Bull *et al.*, 2003; Roelke *et al.*, 2006), and biochemical, serological, and histological parameters suggest AIDS-related pathologies (Brennan *et al.*, 2006; Roelke *et al.*, 2009). As an immunosuppressive virus, the primary effects of FIV on host health may be indirect, through facilitation of co-infection by secondary parasites. In the Kruger National Park, FIV-infected lions tend to be simultaneously co-infected with a greater number of other parasite species than comparable FIV-negative individuals (Broughton *et al.*, unpublished data). On the other hand, FIV in lions does not appear to be associated with increased prevalence of BTB infection (Maas *et al.*, 2012). This is encouraging, given that one might expect strong synergism between these two infections based on the deadly codynamics of HIV and TB in humans (Kirschner, 1999; Toossi, 2003; Kwan and Ernst, 2011). The impacts of FIV on the lion population within HiP are likely to depend on the particular strains of FIV that are present, on other parasites in the system, and on the immunologic health of the lion population.

Genetic restoration, through the introduction of unrelated individuals, was thus effective at improving the health of HiP's lion population. Manifestations of inbreeding depression, including mortality and morbidity due to BTB, decreased dramatically following new lion introductions. Despite the possible inadvertent introduction of FIV with these lions, HiP's lion population has grown towards a robust total of ~120 individuals (Chapter 12), and was evaluated as 'potentially viable' by the IUCN (IUCN/SSC Cat specialist group 2006). Active genetic management is clearly essential to assure the persistence of healthy lion populations in small reserves such as HiP (Miller *et al.*, 2013), despite the risk of unwanted introduction of new pathogens. Experience from HiP thus highlights both the merits and risks of intensive management of carnivore populations in small reserves.

13.3.2 How Big Is Big Enough? The Viability of HiP's Buffalo Population

Research in conjunction with the buffalo capture programme between 2000 and 2005 revealed that BTB prevalence increased with age, as one might expect for a chronic infection: older animals have had more opportunity to become exposed to BTB than younger individuals. BTB severity has remained generally low. BTB cases that were judged to be imminently life-threatening occurred in adults of all ages, yielding an estimated 11% reduction in annual survival of infected compared to uninfected adult buffalo (Jolles, 2007). BTB also reduced pregnancy rates in the buffalo, which could impair their ability to bounce back from population reductions. The combination of elevated adult mortality and reduced fecundity due to BTB means that the disease leaves no trace, in terms of population age structure, despite strong potential impacts on the population growth rate (Jolles *et al.*, 2005). These findings raised concerns that, even at a apparently robust size of 3000–5000 animals, HiP's buffalo population may still remain somewhat vulnerable. However, census data show that the buffalo population within HiP did bounce back to previous numbers within a few years after culling pressure was reduced (Figure 13.2; see also Chapter 4). Population-level consequences of BTB in buffalo may be less severe than indicated by vital rate reductions in BTB-positive individuals, for two reasons. First, if the disease kills mainly animals that were senescent rather than prime-aged adults, such mortality could have little effect on population dynamics (Jolles *et al.*, 2006). Second, BTB-negative buffalo, adults as well as their calves, in herds with high TB prevalence seemed to survive better than BTB-negative animals in low-prevalence herds. Hence, improved survival and reproductive success by BTB-negative buffalo, likely due to improved resource availability, might compensate for losses to the population due to BTB.

Fitness reductions in BTB-positive buffalo and concomitant fitness increases in BTB-negative buffalo also suggest strong potential selection by *M. bovis* within the buffalo population. This raises the question of whether HiP's buffalo population is of adequate size to respond to selection by this and other pathogens. Ultimately, it is the genetic capacity of a population for adapting to changing biotic and abiotic conditions that determines its long-term viability. In smaller populations the effects of genetic drift are more likely to swamp adaptive changes due to natural selection. A recent study investigated selection on the locus encoding for interferon gamma (IFNγ), a signalling molecule vital in the immune response to *Mycobacterium bovis* (Lane-deGraaf *et al.*, 2015). HiP's

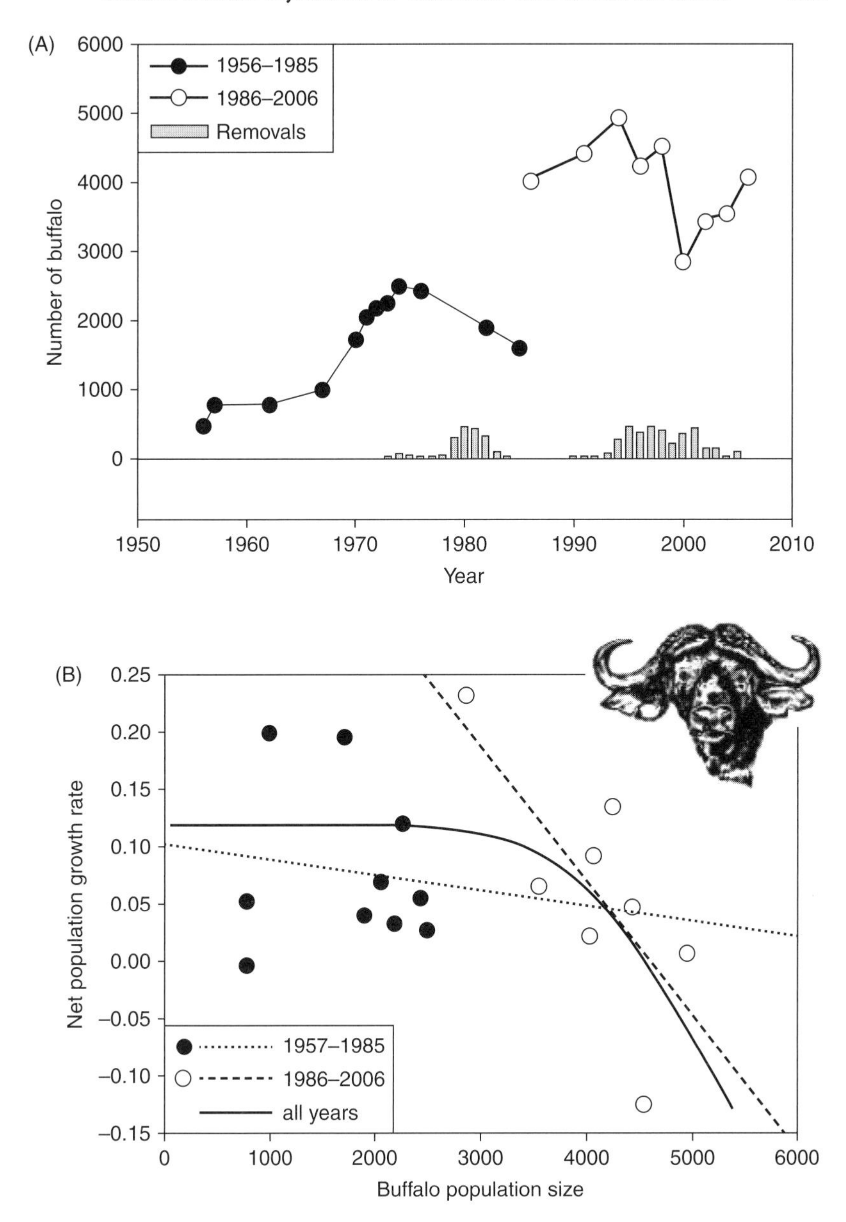

Figure 13.2 Buffalo population dynamics at HiP. (A) Buffalo population estimates and removals from 1956 to 2006. Filled circles show the period 1956–1985, when censuses were conducted as aerial counts. Open circles show the period 1986–2006, when population censuses were conducted using distance sampling. Bars show annual removals (lethal culling plus live removals). (B) Density dependence in the HiP buffalo population. The unrealistically high growth rate between 1956 and 1957 has been omitted.

buffalo population was compared to the much larger buffalo population in Kruger National Park (KNP), which is currently estimated at around 37,000 animals (Beechler *et al.*, 2013). Buffalo in KNP have also experienced BTB infection since the 1960s, but in contrast to HiP, BTB management has been limited to surveillance, with no systematic culling of BTB-infected animals. The comparison of IFNγ genetics in these two populations of contrasting size and management history is telling. Despite similar disease prevalence in both parks, Lane-deGraaf detected clear evidence for directional selection on the IFNγ gene in KNP, but not in HiP (Lane-deGraaf *et al.*, 2015). Within HiP, there was evidence of a recent bottleneck, suggesting genome-wide reductions in diversity that might have eroded the signature of natural selection in the IFNγ gene. As such, HiP's buffalo population may have limited capacity for adaptive evolutionary change in response to novel pathogen challenges, such as BTB. In addition, buffalo that were culled due to a positive TB test at HiP had significantly more rare alleles at the IFNγ locus than non-culled buffalo, suggesting that disease management at HiP may be contributing to a further erosion of immunogenetic diversity.

Overall, HiP's buffalo population appears ecologically resilient to the pathogen challenge presented by *M. bovis* invasion, as shown by compensatory survival and reproduction, and continued positive population growth rate. In contrast, the population's evolutionary resilience may be compromised due to its small size and the bottleneck imposed by a history of hunting and culling; and there are indications that management for disease control may have unintended consequences, possibly compromising the population's adaptive potential. Although genetic population management could be considered, the merits of introducing new genetic material must be balanced against the risks of attenuating local adaptation, and the potential for accidental introduction of new pathogens via animal translocation. The question 'how big is big enough for population health?' is thus complex and needs to be evaluated from ecological and evolutionary perspectives. Notably, most ungulate populations within HiP are smaller than those of buffalo (Chapter 4), and hence probably even more vulnerable to losses of genetic variability.

13.4 Understanding Disease Processes through Collaborative Endeavours

Disease processes in natural populations are not easy to study – in part because of the expense, logistic, and ethical challenges associated with catching and sampling live wild animals. Ideally, research can dovetail with disease-management activities to help alleviate some of these limitations. HIP's BTB control programme provides an example of a productive collaboration between researchers and wildlife managers. This work has allowed for the development of techniques for studying infectious diseases in buffalo, and assisted progress in understanding disease dynamics in this iconic species. For instance, the collection of blood samples from buffalo captured for BTB testing allowed testing and improvement of diagnostic assays for Corridor disease (Pienaar *et al.*, 2011) and BTB itself (Michel *et al.*, 2011), and the development of techniques for age estimation (Jolles, 2007), body condition assessment (Ezenwa *et al.*, 2009), interpretation of haematologic data (Beechler *et al.*, 2009), and the assessment of innate immunity (Beechler *et al.*, 2012) in buffalo. Culled BTB-positive buffalo at HiP were used to characterize BTB lesions in this reservoir host (Laisse *et al.*, 2011), and to compare BTB strains isolated from buffalo with strains from other wild species and cattle (Michel *et al.*, 2006).

Interactions between parasite species concurrently infecting the same hosts, and effects of these co-infections on host health, have been a strong focus of research associated with the BTB testing programme within HiP. Data on GI helminths collected in conjunction with BTB infection data revealed a striking pattern of negative association, where buffalo herds with high BTB prevalence had low GI parasite burdens, and individual buffalo infected with GI helminths were much less likely to be co-infected with BTB than their worm-free counterparts (Jolles *et al.*, 2008). These patterns suggested that GI helminths might play a key role in driving BTB dynamics in buffalo (Ezenwa *et al.*, 2010). Immune-mediated interactions between the two types of parasites may underlie these patterns, or buffalo with both infections may have reduced survival. The net outcome of these two putative mechanisms by which GI parasites mediate BTB dynamics may depend on the environmental context (Ezenwa and Jolles, 2011; Beechler *et al.*, 2012), because resource limitations can affect survival rates and immune function in buffalo (Figure 13.3). Concordant with the idea that seasonal resource availability may mediate interactions between co-infecting parasites, a study investigating co-infections by GI helminths and coccidia (a protozoan parasite infecting the GI tract) in HiP buffalo found that coccidia infection prevalence varied dramatically with season; and this variation in coccidia infection status was the strongest predictor of GI helminth infection (Gorsich *et al.*, 2014). Co-infected buffalo also suffered reduced body condition at the end of the dry season, which may affect their survival rate. In addition, it appears that

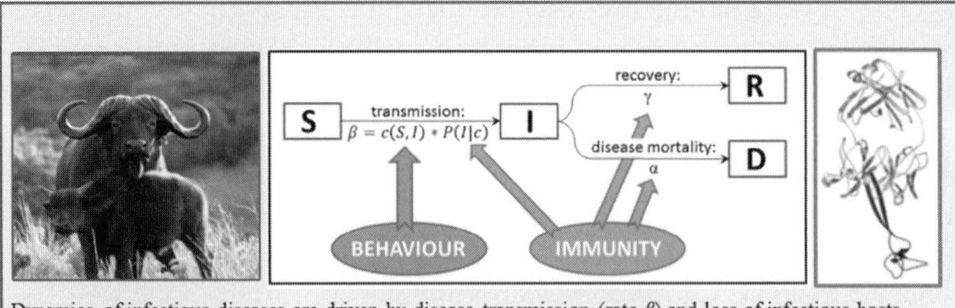

Dynamics of infectious diseases are driven by disease transmission (rate β) and loss of infectious hosts through death (rate α) or recovery (rate γ). Host behaviour determines the rate at which infectious (I) and susceptible (S) hosts come into contact ($c(S,I)$). Host immunity mediates the likelihood of transmission given a contact ($P(I|c)$), the fraction and rate at which hosts recover (R, γ) or die (D, α) from infection. Diagram is based on directly transmitted infection, but principles apply regardless of transmission pathway.

Figure 13.3 Schematic outline of how disease dynamics results from the interplay of host behaviour and immunity.

variation in parasite richness might play a role in mate selection by both sexes of buffalo. An analysis of buffalo horn size data showed that both male and female buffalo with smaller horns (for their age) harboured a greater number of parasite species. This suggests that, aside from their obvious function as weapons, buffalo horns might also serve as an honest indicator of health in both sexes, providing a possible target for parasite-mediated sexual selection (Ezenwa and Jolles, 2008).

Data collected in conjunction with HiP's BTB control programme have thus provided a wealth of new information about the biology and dynamics of BTB in buffalo, along with a comprehensive toolkit of diagnostic, immunological, and ecological protocols that position buffalo as a model species for studying disease processes in natural populations (Jolles and Ezenwa, 2015). This investment in developing novel methods has already provided fundamental insights into mechanisms and health outcomes of multiparasitic co-infection (Ezenwa and Jolles, 2015), and opens opportunities for investigating ecological and eco-immunological questions of disease biology that can only be addressed in free-ranging wildlife populations.

13.5 Conclusion

HiP's origins as a protected area and the management of its wildlife populations have been intertwined with managing disease issues affecting both wildlife and domestic animal populations around the park. As a small reserve, HiP has a high boundary-to-area ratio, which increases the risk of disease transmission between wildlife and domestic animals. Spillover of rabies from domestic dogs to wild dogs, and spillover of nagana and BTB from cattle to wildlife and back to cattle, have posed significant management challenges. HiP's populations of most of the larger ungulates are sufficiently small to limit both population resilience to disease, and the ability to control the spread of infections through host removal programmes. In addition, small populations, particularly of the larger carnivores, can require animal translocations to maintain their genetic health, which introduce yet another source of disease transmission risk. The founding lion nucleus showed obvious signs of inbreeding depression, including high susceptibility to BTB. The introduction of unrelated lions solved the BTB issue, but brought FIV into the park. Compared to more extensive conservation areas, small parks thus face the double jeopardy of increased disease risks and limited capacity of their wildlife populations to withstand the setbacks they incur from disease outbreaks.

On the other hand, small reserves also present special opportunities for developing and testing innovative disease-control programmes, because accessing a large proportion of any target species is more logistically feasible for smaller populations. HiP's management and wildlife veterinary staff pioneered a new approach for BTB control in buffalo, involving the capture of whole buffalo herds in their home ranges. Further lion introductions were used to ameliorate high morbidity associated with BTB infection. Partnerships between park management, veterinary and medical staff, and rural communities surrounding HiP have paved the way for cooperative management of disease spillover risk from wildlife to livestock. This marks a huge change from the indiscriminate culling of wildlife and park deproclamation that featured during the first decades of the park's existence (Chapters 1 and 4). What is perhaps most remarkable about these disease-management initiatives at HiP is that each of them operated in conjunction with intensive research, starting with the tsetse flies and trypanosomosis (Chapter 1). Wildlife disease management within HiP represents a rare case of evidence-based population medicine. Innovative approaches to disease control go hand in hand with scientific studies evaluating their efficacy, which in turn suggest new directions for improving the efficiency of disease control. Adaptive management is thus harnessed for wildlife disease control: a powerful approach that other small reserves could benefit from. In addition, these collaborations between wildlife managers and scientists have provided unparalleled opportunities for research into the dynamics and impacts of infectious diseases in free-ranging mammal populations. In these ways, small reserves like HiP can serve as a platform for innovation in adaptive wildlife disease management.

13.6 References

Alexander, K. A. & Appel, M. J. G. (1994) African wild dogs (*Lycaon pictus*) endangered by a canine distemper epizootic among domestic dogs near the Maasai-Mara National Reserve, Kenya. *Journal of Wildlife Diseases* **30**: 481–485.

Alexander, K. A., Smith, J. S., Macharia, M. J., & King, A. A. (1993) Rabies in the Maasai-Mara, Kenya – preliminary report. *Onderstepoort Journal of Veterinary Research* **60**: 411–414.

Alexander, K. A., Kat, P. W., Munson, L. A., Kalake, A., & Appel, M. J. G. (1996) Canine distemper related mortality among wild dogs (*Lycaon pictus*) in Chobe National Park, Botswana. *Journal of Zoo and Wildlife Medicine* **27**: 426–427.

Alexander, K. A., McNutt, J. W., Briggs, M. B., *et al.* (2010) Multi-host pathogens and carnivore management in southern Africa. *Comparative Immunology, Microbiology and Infectious Diseases* **33**: 249–265.

Beechler, B., Ezenwa, V. O., & Jolles, A. E. (2009) Evaluation of hematologic values in free-ranging African buffalo (*Syncerus caffer*). *Journal of Wildlife Disease* **45**: 57–66.

Beechler, B. R., Broughton, H., Bell, A., Ezenwa, V. O., & Jolles, A. E. (2012) Innate immunity in free-ranging African buffalo (*Syncerus caffer*): associations with parasite infection and white blood cell counts. *Physiological & Biochemical Zoology* **85**: 255–264.

Beechler, B. R., Paweska, J. T., Swanepoel, R., *et al.* (2013) Rift Valley fever in Kruger National Park: do buffalo play a role in the interepidemic circulation of virus? *Transboundary Emerging Diseases* **62**: 24–32.

Bengis, R. G., Kock, R. A., & Fischer, J. (2002) Infectious animal diseases: the wildlife/livestock interface. *Revue Scientifique et Technique – Office International des Epizooties* **21**: 53–65.

Brennan, G., Podell, M. D., Wack, R., *et al.* (2006) Neurologic disease in captive lions (*Panthera leo*) with low-titer lion lentivirus infection. *Journal of Clinical Microbiology* **44**: 4345–4352.

Brown, E. W., Yuhki, N., Packer, C., & O'Brien, S. J. (1994) A lion lentivirus related to feline immunodeficiency virus: epidemiologic and phylogenetic aspects. *Journal of Virology* **68**: 5953–5968.

Bull, M. E., Kennedy-Stoskopf, S., Levine, J., *et al.* (2003) Evaluation of T lymphocytes in captive African lions (*Panthera leo*) infected with feline immunodeficiency virus. *American Journal of Veterinary Research* **64**: 1293–1300.

Caron, A., Cross, P. C., & du Toit, J. T. (2003) Ecological implications of bovine tuberculosis in African buffalo herds. *Ecological Applications* **13**: 1338–1345.

Carpenter, M. A. & O'Brien, S. J. (1995) Coadaptation and immunodeficiency virus –lessons from the Felidae. *Current Opinion in Genetics & Development* **5**: 739–745.

Chitanga, S., Namangala, B., De Deken, R., & Marcotty, T. (2013) Shifting from wild to domestic hosts: the effect on the transmission of *Trypanosoma congolense* in tsetse flies. *Acta Tropica* **125**: 32–36.

Coltmann, D. W., Pilkington, J. G., Smith, J. A., & Pemberton, J. M. (1999) Parasite-mediated selection against inbred Soay sheep in a free-living, island population. *Evolution* **53**: 1259–1267.

Coltmann, D. W., Wilson, K., Pilkington, J. G., Stear, M. J., & Pemberton, J. M. (2001) A microsatellite polymorphsm in the gamma interferon gene is associated with resistance to gastrointestinal nematodes in a naturally-parasitized population of Soay sheep. *Parasitology* **122**: 571–582.

Connor, R. J. & van den Bossche, P. (2004) African animal trypanosomoses. In: *Infectious diseases of livestock* (eds J. A. W. Coetzer & R. C. Tustin), pp. 251–296. Oxford University Press Southern Africa, Cape Town.

Creel, S. & Creel, N. M. (1998) Six ecological factors that limit African wild dogs, *Lycaon pictus*. *Animal Conservation* **1**: 1–9.

Creel, S., Creel, N. M., Munson, L., Sanderlin, D., & Appel, M. J. G. (1997) Serosurvey for selected viral diseases and demography of African wild dogs in Tanzania. *Journal of Wildlife Diseases* **33**: 823–832.

De Castro, F. & Bolker, B. (2005) Mechanisms of disease-induced extinction. *Ecology Letters* **8**: 117–126.

Durchfeld, B., Baumgartner, W., Herbst, W., & Brahm, R. (1990) Vaccine-associated canine distemper infection in a litter of African hunting dogs (*Lycaon pictus*). *Journal of Veterinary Medicine Series B – Infectious Diseases and Veterinary Public Health* **37**: 203–212.

Ezenwa, V. O. & Jolles, A. E. (2008) Horns honestly advertise parasite infection in both male and female African buffalo (*Syncerus caffer*). *Animal Behaviour* **75**: 2013–2021.

Ezenwa, V. O & Jolles, A. E. (2011) From host immunity to pathogen invasion: how do within-host mechanisms scale up to disease dynamics? *Integrative and Comparative Biology* **51**: 540–551.

Ezenwa, V. O. & Jolles, A. E. (2015) Opposite effects of anthelmintic treatment on microbial infection at individual vs. population scales. *Science* **347**: 175–177.

Ezenwa, V. O., Jolles, A. E., & O'Brien, M. (2009) A reliable body condition scoring technique for estimating condition in African buffalo. *African Journal of Ecology* **47**: 476–481.

Ezenwa, V. O., Etienne, R. S., Luikhart, G., Beja-Perreira, A., & Jolles, A. E. (2010) Hidden consequences of living in a wormy world: nematode-induced immune-suppression facilitates TB invasion in African buffalo. *American Naturalist* **176**: 613–624.

Flacke, G., Spiering, P., Cooper, D., *et al.* (2010) A survey of internal parasites in free-ranging African wild dogs (*Lycaon pictus*) from KwaZulu-Natal, South Africa. *South African Journal of Wildlife Research* **40**: 176–180.

Flacke, G., Becker, P., Cooper, D., *et al.* (2013) An infectious disease and mortality survey in a population of free-ranging African wild dogs and sympatric domestic dogs. *International Journal of Biodiversity* e497623.

Gascoyne, S. C., Laurenson, M. K., Lelo, S., & Borner, M. (1993) Rabies in African wild dogs (*Lycaon pictus*) in the Serengeti region, Tanzania. *Journal of Wildlife Diseases* **29**: 396–402.

Gillingwater, K., Mamabolo, M. V., & Majiwa, P. A. O. (2010) Prevalence of mixed *Trypanosoma congolense* infections in livestock and tsetse in KwaZulu-Natal, South Africa. *Journal of the South African Veterinary Association* **81**: 219–223.

Goller, K. V., Fyumagwa, R. D., Nikolin, V., *et al.* (2010) Fatal canine distemper infection in a pack of African wild dogs in the Serengeti ecosystem, Tanzania. *Veterinary Microbiology* **146**: 245–252.

Gorsich, E. E., Ezenwa, V. O., & Jolles, A. E. (2014) Costs of co-infection in a seasonal environment: gastrointestinal parasites in African buffalo. *International Journal for Parasitology: Parasites & Wildlife* **3**: 124–134.

Green, K. K. & Venter, G. J. (2007) Evaluation and improvement of sticky traps as monitoring tools for *Glossina austeni* and *G. brevipalpis* (Diptera: Glossinidae) in north-eastern KwaZulu-Natal, South Africa. *Bulletin of Entomological Research* **97**: 545–553.

Hofmann-Lehmann, R., Fehr, D., Grob, M., *et al.* (1996) Prevalence of antibodies to feline parvovirus calicivirus, herpesvirus, coronavirus, and immunodeficiency virus and of feline leukemia virus antigen and the interrelationship of these viral infections in free-ranging lions in East Africa. *Clinical and Diagnostic Laboratory Immunology* **3**: 554–562.

Hofmeyr, M., Bingham, J., Lane, E. P., Ide, A., & Nel, L. (2000) Rabies in African wild dogs (*Lycaon pictus*) in the Madikwe Game Reserve, South Africa. *Veterinary Record* **146**: 50–52.

Jolles, A. E. (2007) Population biology of African buffalo (*Syncerus caffer*) at Hluhluwe-iMfolozi Park, South Africa. *African Journal of Ecology* **45**: 398–406.

Jolles, A. E. & Ezenwa, V. O. (2015) Ungulates as model systems for the study of disease processes in natural populations. *Journal of Mammalogy* **96**: 4–15.

Jolles, A. E., Cooper, D., & Levin, S. A. (2005) Hidden effects of chronic tuberculosis in African buffalo. *Ecology* **86**: 2358–2364.

Jolles, A. E., Etienne, R. S., & Olff, H. (2006) Independent and competing disease risks: implications for host populations in variable environments. *American Naturalist* **167**: 745–757.

Jolles, A. E., Ezenwa, V. O., Etienne, R. S., Turner, W. C., & Olff, H. (2008) Interactions between macroparasites and microparasites drive patterns of infection in free-ranging African buffalo. *Ecology* **89**: 2239–2250.

Kat, P. W., Alexander, K. A., Smith, J. S., & Munson, L. (1995) Rabies and African wild dogs in Kenya. *Proceedings of the Royal Society B* **262**: 229–233.

Kirschner, D. (1999) Dynamics of co-infection with *M. tuberculosis* and HIV-1. *Theoretical Population Biology* **55**: 94–109.

Kwan, C. K. & Ernst, J. D. (2011) HIV and tuberculosis: a deadly human syndemic. *Clinical Microbiology Reviews* **24**: 351–376.

Laisse, C. J. M., Gavier-Widen, D., Ramis, G., *et al.* (2011) Characterization of tuberculosis lesions in naturally infected African buffalo (*Syncerus caffer*). *Journal of Veterinary Diagnostic Investigation* **23**: 1022–1027.

Lane-deGraaf, K. E., Amish, S. J., Gardipee, F., *et al.* (2015) Testing for signatures of natural and artificial disease-induced selection at immune loci in a free-ranging wildlife population. *Conservation Genetics* **16**: 289–300.

Lembo, T., Hampson, K., Haydon, D. T., *et al.* (2008) Exploring reservoir dynamics: a case study of rabies in the Serengeti ecosystem. *Journal of Applied Ecology* **45**: 1246–1257.

Le Roex, N., Cooper, D., van Helden, P. D., Hoal, E. G., & Jolles, A. E. (2015) Disease control in wildlife: evaluating a test and cull programme for bovine tuberculosis in African buffalo. *Transboundary Emerging Diseases* DOI: 10.1111/tbed.12329.

Maas, M., Keet, D. F., Rutten, V. P. M. G., Heesterbeek, J. A. P., & Nielen, M. (2012) Assessing the impact of feline immunodeficiency virus and bovine tuberculosis co-infection in African lions. *Proceedings of the Royal Society B* **279**: 4206–4214.

Macdonald, D. W. (1992) Cause of wild dog deaths. *Nature* **360**: 633–634.

Maddock, A., Anderson, A., Carlisle, F., *et al.* (1996) Changes in lion numbers in Hluhluwe-Umfolozi Park. *Lammergeyer* **44**: 6–18.

Mamabolo, M. V., Ntantiso, L., Latif, A., & Majiwa, P. A. O. (2009) Natural infection of cattle and tsetse flies in South Africa with two genotypic groups of *Trypanosoma congolense*. *Parasitology* **136**: 425–431.

McCallum, H. & Dobson, A. (1995) Detecting disease and parasite threats to endangered species and ecosystems. *Trends in Ecology and Evolution* **10**: 190–194.

McCormick, A. E. (1983) Canine distemper in African Cape hunting dogs (*Lycaon pictus*) – possibly vaccine induced. *Journal of Zoo and Wildlife Medicine* **14**: 66–71.

McDonald, R. A., Delahay, R. J., Carter, S. P., Smith, G. C., & Cheeseman, C. L. (2008) Perturbing implications of wildlife ecology for disease control. *Trends in Ecology & Evolution* **23**: 53–56.

Mech, L. D. & Goyal, S. M. (1995) Effects of canine parvovirus on gray wolves in Minnesota. *Journal of Wildlife Management* **59**: 565–570.

Michel, A. L., Bengis, R. G., Keet, D. F., *et al.* (2006) Wildlife tuberculosis in South African conservation areas: implications and challenges. *Veterinary Microbiology* **112**: 91–100.

Michel, A. L., Cooper, D., Jooste, J., de Klerk, L.-M., & Jolles, A. E. (2011) Approaches towards optimising the gamma interferon assay for diagnosing *Mycobacterium bovis* infection in African buffalo (*Syncerus caffer*). *Preventive Veterinary Medicine* **98**: 142–151.

Miller, S. M., Bissett, C., Burger, A., *et al.* (2013) Management of reintroduced lions in small, fenced reserves in South Africa: an assessment and guidelines. *South African Journal of Wildlife Research* **43**: 138–154.

Mills, M. G. L. (1993) Social systems and behavior of the African wild dog *Lycaon pictus* and the spotted hyena *Crocuta crocuta* with special reference to rabies. *Onderstepoort Journal of Veterinary Research* **60**: 405–409.

Motloang, M., Masumu, J., Mans, B., van den Bossche, P., & Latif, A. (2012) Vector competence of *Glossina austeni* and *Glossina brevipalpis* for *Trypanosoma congolense* in KwaZulu-Natal, South Africa. *Onderstepoort Journal of Veterinary Research* **79**: 1–6.

Motloang, M. Y., Masumu, J., Mans, B. J., & Latif, A. A. (2014) Virulence of *Trypanosoma congolense* strains isolated from cattle and African buffaloes (*Syncerus caffer*) in KwaZulu-Natal, South Africa. *Onderstepoort Journal of Veterinary Research* **81**: 7 pp.

Ntantiso, L., De Beer, C., Marcotty, T., & Latif, A. A. (2014) Bovine trypanosomosis prevalence at the edge of Hluhluwe-iMfolozi Park, KwaZulu-Natal, South Africa. *Onderstepoort Journal of Veterinary Research* **81**: 8 pp.

Packer, C., Altizer, S., Appel, M., *et al.* (1999) Viruses of the Serengeti: patterns of infection and mortality in African lions. *Journal of Animal Ecology* **68**: 1161–1178.

Paterson, S., Wilson, K., & Pemberton, J. M. (1998) Major histocompatibility complex variation associated with juvenile survival and parasite resistance in a large unmanaged ungulate population (*Ovis aries* L.). *Proceedings of the National Academy of Sciences* **95**: 3714–3719.

Pedersen, A. B. & Fenton, A. (2007) Emphasizing the ecology in parasite community ecology. *Trends in Ecology & Evolution* **22**: 133–139.

Petney, T. N. & Andrews, R. H. (1998) Multiparasite communities in animals and humans: frequency, structure and pathogenic significance. *International Journal for Parasitology* **28**: 377–393.

Pienaar, R., Potgieter, F. T., Latif, A. A., Thekisoe, O. M. M., & Mans, B. J. (2011) Mixed *Theieria* infections in free-ranging buffalo herds: implications for diagnosing *Theileria parva* infections in Cape buffalo (*Syncerus caffer*). *Parasitology* **138**: 884–895.

Prager, K. C., Woodroffe, R., Cameron, A., & Haydon, D. T. (2011) Vaccination strategies to conserve the endangered African wild dog (*Lycaon pictus*). *Biological Conservation* **144**: 1940–1948.

Prager, K. C., Mazet, J. A. K., Dubovi, E. J., *et al.* (2012a) Rabies virus and canide distemper virus in wild and domestic carnivores in northern Kenya: are domestic dogs the reservoir? *Ecohealth* **9**: 483–498.

Prager, K. C., Mazet, J. A. K., Munson, L., *et al.* (2012b) The effect of protected areas on pathogen exposure in endangered African wild dog (*Lycaon pictus*) populations. *Biological Conservation* **150**: 15–22.

Reid, J. M., Arcese, P., & Keller, L. F. (2003) Inbreeding depresses immune response in song sparrows (*Melospiza melodia*): direct and intergenerational effects. *Proceedings of the Royal Society B* **270**: 2151–2157.

Roelke, M. E., Pecon-Slattery, J., Taylor, S., *et al.* (2006) T-lymphocyte profiles in FIV-infected wild lions and pumas reveal CD4 depletion. *Journal of Wildlife Diseases* **42**: 234–248.

Roelke, M. E., Brown, M. A., Troyer, J. L., *et al.* (2009) Pathological manifestations of feline immunodeficiency virus (FIV) in wild African lions. *Virology* **390**: 1–12.

Spencer, J. A. (1991) Survey of antibodies to feline viruses in free-ranging lions. *South African Journal of Wildlife Research* **21**(2): 59.

Tanser, F., Barnighausen, T., Grapsa, E., Zaidi, J., & Newell, M. L. (2013) High coverage of ART associated with decline in risk of HIV acquisition in rural KwaZulu-Natal, South Africa. *Science* **339**: 966–971.

Tompkins, D. M., Dunn, A. M., Smith, M. J., & Telfer, S. (2010) Wildlife diseases: from individuals to ecosystems. *Journal of Animal Ecology* **80**: 19–38.

Toossi, Z. (2003) Virological and immunological impact of tuberculosis on human immunodeficiency virus type 1 disease. *Journal of Infectious Diseases* **188**: 1146–1155.

Trinkel, M., Ferguson, N., Reid, A., *et al.* (2008) Translocating lions into an inbred lion population in the Hluhluwe-iMfolozi Park, South Africa. *Animal Conservation* **11**: 138–143.

Trinkel, M., Cooper, D., Packer, C., & Slotow, R. (2011) Inbreeding depression increases susceptibility to bovine tuberculosis in lions: an experimental test using an inbred–outbred contrast through translocation. *Journal of Wildlife Diseases* **47**: 494–500.

Van de Bildt, M. W. G., Kuiken, T., Visee, A. M., Lema, S., Fitzjohn, T. R., & Osterhaus, A. D. M. E. (2002) Distemper outbreak and its effect on African wild dog conservation. *Emerging Infectious Diseases* **8**: 211–213.

Van den Bossche, P., Esterhuizen, J., Nkuna, R., *et al.* (2006) An update of the bovine trypanosomosis situation at the edge of Hluhluwe-iMfolozi Park, KwaZulu-Natal Province, South Africa. *Onderstepoort Journal of Veterinary Research* **73**: 77–79.

Van den Bossche, P., Chitanga, S., Masumu, J., Marcotty, T., & Delespaux, V. (2011) Virulence in *Trypanosoma congolense* Savannah subgroup. A comparison between strains and transmission cycles. *Parasite Immunology* **33**: 456–460.

Van Heerden, J., Bingham, J., van Vuuren, M., Burroughs, R. E. J., & Stylianides, E. (2002) Clinical and serological response of wild dogs (*Lycaon pictus*) to vaccination against canine distemper, canine parvovirus infection, and rabies. *Journal of the South African Veterinary Association* **73**: 8–12.

Van Vuuren, M., Stylianides, E., Kania, S. A., Zuckerman, E. E., & Hardy, W. D. (2003) Evaluation of an indirect enzyme-linked immunosorbent assay for the detection of feline lentivirus-reactive antibodies in wild felids, employing a puma lentivirus-derived synthetic peptide antigen. *Onderstepoort Journal of Veterinary Research* **70**: 1–6.

Vial, F., Cleaveland, S., Rasmussen, G., & Haydon, D. T. (2006) Development of vaccination strategies for the management of rabies in African wild dogs. *Biological Conservation* **131**: 180–192.

Woodroffe, R. (2001) Assessing the risks of intervention: immobilization, radio-collaring and vaccination of African wild dogs. *Oryx* **35**: 234–244.

Woodroffe, R. & Ginsberg, J. R. (1998) Edge effects and the extinction of populations inside protected areas. *Science* **280**: 2126–2128.

Woodroffe, R., Donnelly, C. A., Jenkins, H., *et al.* (2006) Culling and cattle controls influence tuberculosis risk for badgers. *Proceedings of the National Academy of Sciences* **103**: 14713–14717.

Woodroffe, R., Prager, K. C., Munson, L., *et al.* (2012) Contact with domestic dogs increases pathogen exposure in endangered African wild dogs (*Lycaon pictus*). *PLoS ONE* **7**: e30099.

14 · *Elephant Management in the Hluhluwe-iMfolozi Park*

DAVE J. DRUCE, HELEEN DRUCE, MARISKA TE BEEST, JORIS P. G. M. CROMSIGT, AND SUSAN JANSE VAN RENSBURG

14.1 Introduction

Elephants (*Loxodonta africana*) were formerly widespread throughout South Africa (Carruthers *et al.* 2008). However, with the arrival of Europeans and Arabs came huge exploitation of elephant populations, particularly for their ivory. Exports of ivory through Delagoa Bay (present-day Maputo) and later Port Natal (present-day Durban) reached a peak around 1860. Between 1844 and 1895, 869 tons of ivory was exported, much of it coming from the Zululand hinterland (McCracken, 2008). The last elephant remaining in Zululand was reportedly shot in 1890 by John Dunn on the banks of the Black Umfolozi river (Vincent, 1970), although some elephants did still remain in the far north-east near the Mozambique border. Hence by the end of the nineteenth century, elephants had been almost completely eliminated from the current KwaZulu-Natal province.

This chapter outlines lesson learnt from the re-establishment of elephants in the Hluhluwe-iMfolozi Park (HiP), as well as the consequences and lessons learnt from the ecological restoration and subsequent growth of the elephant population. Elephants were re-established in the park for the following main reasons (Wills, 1986):

1. to reinstate a species that had previously occurred in the area;
2. to re-introduce the ecological role of elephants within the area, and in particular redress the problem of shrub and tree encroachment into more open savanna, grassland and wetland habitats; and

Conserving Africa's Mega-Diversity in the Anthropocene, ed. Joris P. G. M. Cromsigt, Sally Archibald and Norman Owen-Smith. Published by Cambridge University Press.

3. to improve the conservation status of elephants as a threatened species and to add to the biological and conservation value of the two game reserves.

By 2001, various small state and private wildlife reserves (< 100 km^2) in South Africa had also introduced elephants, mostly obtained from Kruger National Park (Slotow *et al.*, 2005). Ethical concerns and the growing perception for the need to manage small and large populations led to the development of National Norms and Standards for the Management of Elephants in South Africa in 2008. This document specified the need for conservation agencies and land owners to develop an elephant management plan, with that for HiP being approved in 2012. The second half of the chapter includes an overview of the park plan and proposed future interventions aimed at ensuring that there are few negative effects on other biodiversity elements that have a higher priority ranking in the park than elephants.

14.2 Re-introduction and Subsequent Growth of the Elephant Population

In 1981 the first three groups of elephants totalling 26 animals were brought into the Hluhluwe Game Reserve from the Kruger National Park (Dominy *et al.*, 1998). A further 30 elephants were released in two groups in Umfolozi Game Reserve in 1985 (Wills, 1986). All of these elephants were orphans from multiple family groups left over from culling operations, ranging in age from 2–5 years (Dominy *et al.*, 1998). In total, 174 elephants were successfully moved into HiP between 1981 and 1996 (Dominy *et al.*, 1998). Because only young animals could be transported at that time, introductions were staggered in time in an attempt to produce a diverse spread of ages in the introduced population. In 2000, 10 mature male elephants were translocated from Kruger Park into HiP to address the problem of young elephant males killing rhinos (see Section 14.4.1).

The first calf was born in 1990 (Dominy *et al.*, 1998). Between 1990 and 1996, when many animals were still reproductively immature, the elephant population increased at an annual rate of 4.7% (Mackey *et al.*, 2006). Between 2004 and 2014, the annual increase rate averaged 6.7%, at the upper limit of the growth rate that can be sustained by an elephant population (van Aarde and Jackson, 2007). With abundant food and water, ages at first reproduction have been as young as 8–12 years and inter-calving intervals sometimes as brief as 3 years (Slotow *et al.*, 2005).

Even though the introduced individuals were all juveniles (apart from the large bulls introduced in 2000) and hence the population relatively youthful (none older than 40 years), natural mortalities have occurred. Between 1992 and the end of 2014, a total of 39 animals had been recorded as dying of natural causes, with known reasons for death including disease or malnutrition (four), drowned or stuck in mud (four), old age (three of the introduced mature males), predation by unknown predator (three), fighting (two), injured (two), and one calf stillborn. For the remaining 20 individuals, the cause of death was unknown. By 2014 the HiP elephant population was estimated at 698 animals, from ground-based observations, resulting in an effective density of 0.73 elephants per km^2 (see Figure 4.1C in Chapter 4). The population was structured in 21 herd units (range 8–58 individuals) which showed loose associations. There were approximately 144 adult males that moved independently of the herds, but only one of the 10 mature males introduced in 2000 was potentially still alive. This individual was last photographed in August 2011 and would then have been between 50 and 60 years of age. Because of the staggered introductions, the population structure now resembles that of other stable populations, with the oldest matriarch being approximately 38 years old.

14.3 Population Management

In 1985, A. J. Wills (unpublished report) proposed a management plan that outlined the above-mentioned goals of the re-introduction programme and envisaged the potential problems, which were, first, that elephants might break out of the park and either themselves cause damage or allow other animals to escape and cause damage and, second, that the elephants may damage the vegetation within the park to an unacceptable level. In the light of these foreseen impacts, an upper limit of 320 elephants was proposed which was based on the density being implemented in Kruger National Park at the time. Provision was given to change this upper limit based on the effects elephants had on the vegetation. Vegetation monitoring was deemed necessary to determine a realistic upper limit based on sound data. However, in practice, the elephant population within HiP has expanded with little human intervention thus far, due in part to lack of evidence of undesirable impacts on vegetation or to other priority biodiversity elements. The moratorium on culling and, from 2008, prescriptive nature of the 'National Norms and Standards' also played a role (see Section 14.6).

The only routine management intervention implemented up to 2014 has been the control of so-called 'problem animals'. A problem animal is defined as an individual that posed a risk to or endangered human lives or livelihoods. Problem animals shot in HiP have included animals that have repeatedly broken out of the park (at least three exits), animals that have repeatedly shown signs of aggression towards vehicles or have lost their fear of vehicles and made contact with vehicles on numerous occasions, and animals that have repeatedly damaged infrastructure or chased cars and people. As of 2014, 21 adult elephants (all but one males) had been shot as problem animals, while two males were shot in self-defence on wilderness trails. In addition, some males were shot that had been identified as having killed rhino (see Section 14.4.1). The only other intervention was the translocation of 24 elephants from HiP to the iSimangaliso Wetland Park during 2001. Three of the young males broke out shortly after introduction and were returned to HiP. However, in 2014 an immunocontraception programme was initiated as a form of more proactive population management (see Section 14.6.3).

14.4 Ecological Consequences

14.4.1 Elephants Killing Rhinos

A problem that was not envisaged when orphan elephants were introduced into various protected areas in South Africa was the killing of rhinos, mainly by male elephants approaching maturity. A total of 58 white rhinos and five black rhinos were killed in HiP by elephants from 1991 to 2001 (Slotow *et al.*, 2001). In Pilanesberg National Park where the same phenomenon occurred, males introduced as orphans came into musth at a younger age than normal and stayed in musth for longer periods than usual (Slotow *et al.*, 2000). It was suggested that this resulted from the absence of older, more dominant males. Accordingly, six fully mature male elephants were relocated from Kruger Park to Pilanesberg in late 1998. This resulted in reduced musth duration in younger males and the cessation of rhino killings (Slotow *et al.*, 2000). However, the individuals that had been observed attacking rhinos were all destroyed. Following the Pilanesberg experience, 10 mature male elephants (up to 45 years of age) were introduced into HiP from Kruger National Park in May 2000 (Slotow *et al.*, 2001). The number of rhinos killed by elephants in HiP declined from an average 10 annually from June 1996 to May 2001 to around four per year from June 2001 to May 2009 and only about one per year thereafter. In HiP, the introduction of mature male elephants did not immediately stop rhino killing by elephants and it took almost 10 years for numbers of rhinos killed to drop substantially.

14.4.2 Elephant Movement Patterns

With the re-introduction of elephants into HiP, there were concerns that elephants would leave the park and cause damage outside, or allow other animals, like lions, to follow (Wills, 1986). Consequently, individuals in each of the first three groups released in Hluhluwe GR were fitted with radio collars. These elephants all initially settled within a very small home range (25–50 ha), which after 3 years had expanded to approximately 60 km^2 following joining by the elephants released in Umfolozi GR (Wills, 1986). Subsequently, these elephants split into smaller family units led by matriarchs, as described above.

In October 2006, adult female elephants in six different groups were fitted with GPS collars recording five locations per day and their movements tracked until September 2008 (Bodasing, 2011). Additionally, two females representing a group of eight individuals in Mfolozi and a group of 33 individuals in northern Hluhluwe carried GPS collars recording one to two locations per hour beyond 2008. Some of the findings included: (1) groups collared in the Mfolozi region remained in small home ranges in the south throughout the year, while groups collared in Hluhluwe used both the northern and southern part of the park in the wet season, but stayed predominantly in the north in the dry season; (2) herds favoured relatively green vegetation in both seasons, with scarp forest and thicket being positively selected in the dry season, while scarp forest and riparian vegetation were favoured most in the wet season; (3) locations were concentrated closer to rivers in the wet season than during the dry season, although there was no effect of time of day on the distance elephants were from water; (4) wider use of the landscape in the wet season became locally concentrated in the dry season (Bodasing, 2011). Groups within Mfolozi retained home ranges within a few kilometres of the two Mfolozi rivers year-round. Groups within Hluhluwe appeared to concentrate their movements during the winter dry season in valleys, riverine forests and scarp forests. The group initially collared in Hluhluwe bearing the long-lasting collar restricted its movements to the north of Hluhluwe (including extensive use of the scarp forest) and south-east of Mfolozi during the dry months, but during the wet season ranged more widely (Figure 14.1A,B). The Mfolozi group restricted its movement to

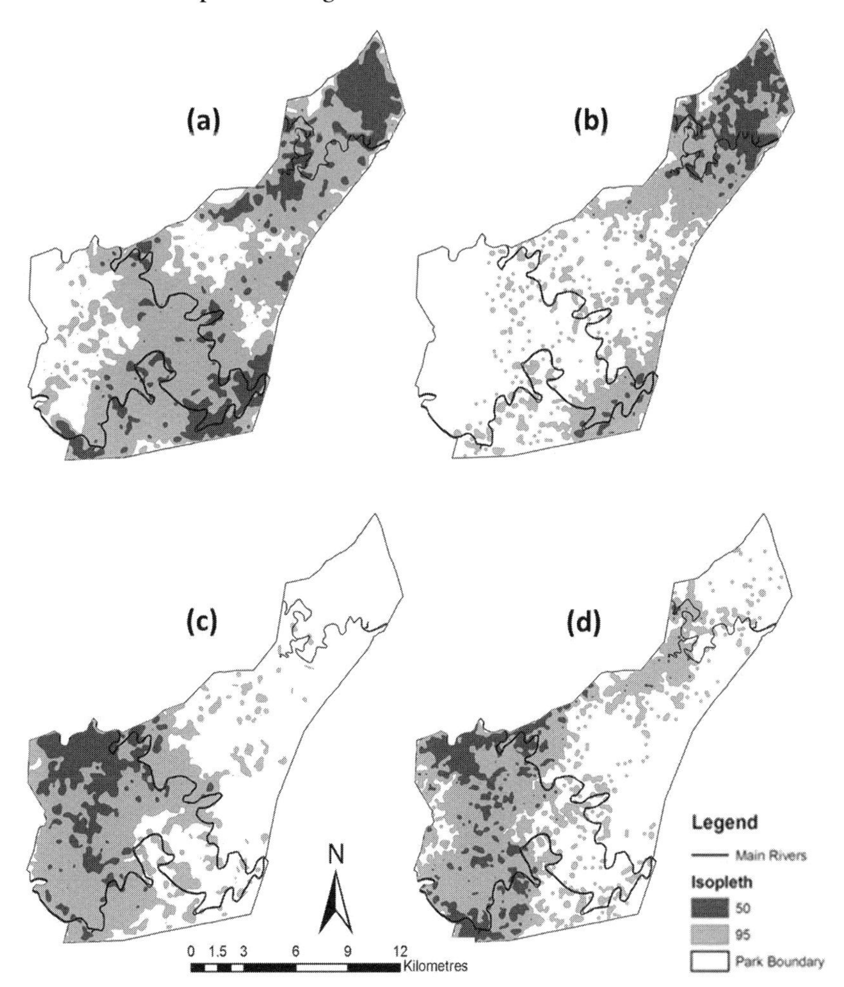

Figure 14.1 Home range use of two female elephants in different herds in 50% and 95% kernels generated from four GPS locations a day from November 2006 to March 2014. (A) BH071 wet season range, (B) BH071 dry season range, (C) BH178 wet season range, and (D) BH178 dry season range.

Mfolozi in the wet season, but wandered into Hluhluwe during the dry season, although not as far as the scarp forests (Figure 14.1C,D). Based on the movements of just these two herds, it is evident that elephant groups utilize the whole extent of the park.

Elsewhere, surface water sources may restrict space use by elephants, which usually need to drink every two days. Within HiP, permanent water is readily accessible from the various rivers, a few springs, and

Figure 14.2 An elephant in the Hluhluwe section of HiP with a debarked, dead marula tree in the background (photo: Norman Owen-Smith).

two large natural pans near the White Mfolozi river, while in the wetter summer months additional surface water is available in numerous small natural pans. As a consequence, no point in HiP is further than 8 km from a river (Boundja and Midgley, 2009), so that water is thus always within daily travelling range of elephants. Hence managing water sources cannot be used to shift elephant distribution in HiP (because there is only one artificial, pumped waterpoint) in contrast to other larger parks (Chamaillé-Jammes *et al.*, 2007). One resulting open question is whether HiP has enough spatial refuges for plant species that do not fare well with repeated elephant impact.

14.4.3 Elephant Effects on Vegetation

Prior to the re-introduction of elephants into HiP, there was concern about the effect they would have on the vegetation (Figure 14.2). While it was hoped that elephants would assist in reversing the trend of bush encroachment, there was a worry that they would select endangered or endemic plant species with a higher conservation value than elephants, such as *Albizia suluensis* and *Warburgia salutaris*, among others (Wills, 1986). Consequently, the feeding behaviour of the first group of elephant released into Hluhluwe Game Reserve was studied both before

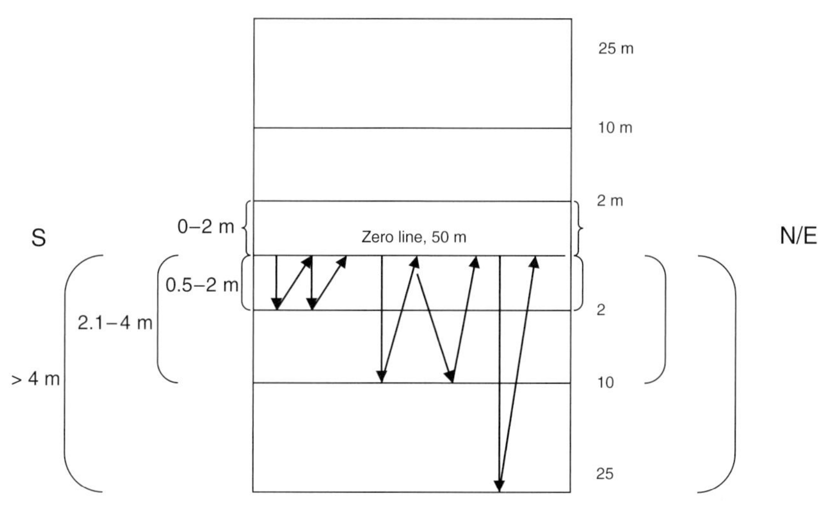

Figure 14.3 Elephant utilization sample plot design. Numerals to the left of the diagram indicate the height classes that were sampled in the particular block area, while numerals to the right of the diagram indicate the distance (in metres) off the central transect line (zero line). Each height class was sampled on both sides of the central transect line.

and after they were released from the holding pens (Wills, 1986). Initial results indicated that their diet consisted of 50% grass and 50% browse, which was similar to that shown by elephants in Kruger National Park (Wills, 1986) and elsewhere (see Kerley *et al.*, 2008). Moreover, the young elephants showed no preference for any endangered or endemic plant species within the first few years after their release (Wills, 1986).

However, concerns continued as the elephant population grew. This prompted further monitoring of vegetation changes induced by elephants. In 1999 a survey, including 369 plots distributed across the park, was set up and a subset of these plots were reassessed in 2003 (186 plots) and 2007 (175 plots). The first two surveys were conducted with the assistance of Earthwatch Institute volunteers, while staff, students, and other volunteers undertook the 2007 survey. Because of a lack of support and funds, the programme was discontinued. Sampling was conducted in 0.25-ha plots placed randomly in all habitat types in HiP distinguished by Whateley and Porter (1983) using a stratified design. Sampling entailed different size areas for different tree height classes (Figure 14.3). In the plots the species, height, diameter, and number of stems of each tree individual were recorded. For each tree, damage by elephants was recorded as proportions of the tree showing branch breakage, bark stripping, or

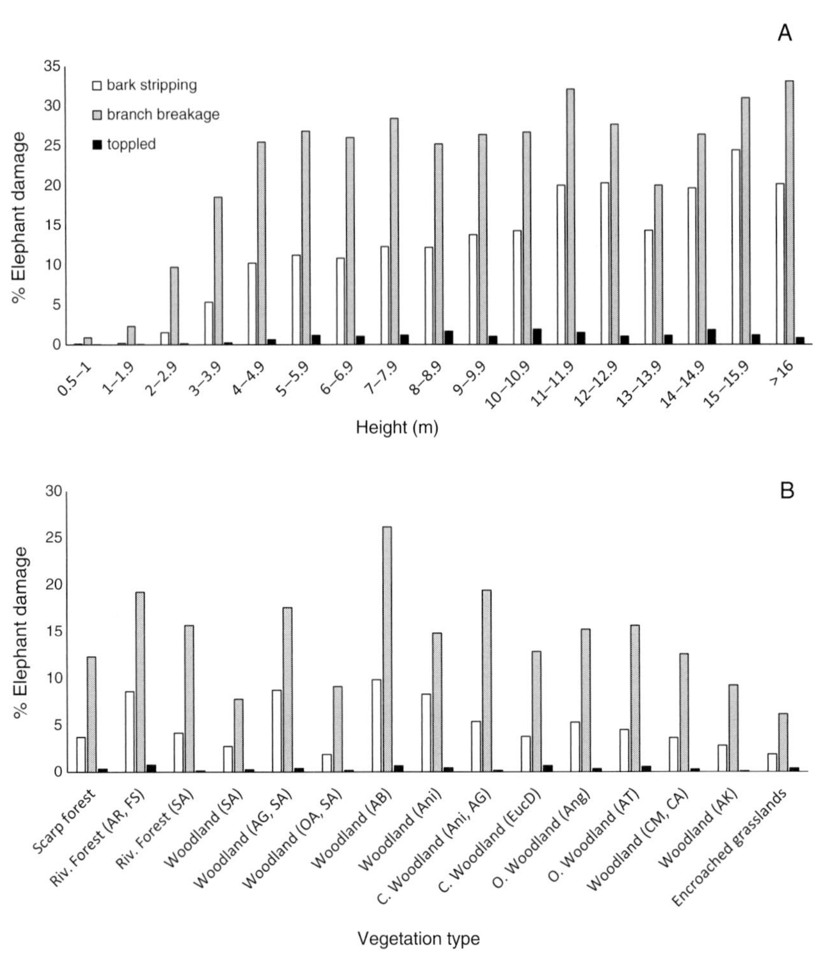

Figure 14.4 Damage inflicted by elephants on (A) woody plants in different height classes, and (B) woody plants growing in different vegetation types. The response measure was the proportion of plants showing either bark-stripping (black bars), branch breakage (white bars), or toppling (grey bars). Abbreviations in parentheses refer to dominant tree species in the vegetation type (see also Table 2.2). AR: *Acacia robusta*; FS: *Ficus sycamorus*; SA: *Spirostachys africana*;AG: *Acacia gerrardii*; OA: *Olea africana*; AB: *Acacia burkei*; Ani: *Acacia nilotica*; EUCD: *Euclea divinorum*; Ang: *Acacia nigrescens*; AT: *Acacia tortilis*; CM: *Combretum molle*; CA: *Combretum apiculatum*; AK: *Acacia karroo*.

having been toppled (Figure 14.4). Signs of browsing by other herbivores were recorded as well. Most damage was due to branch breakage. For trees over 3 m, 20–25% of trees in each height class showed branch breakage. Trees taller than 11 m were more frequently bark-stripped than smaller trees; on average, 20% of trees in these largest height classes showed signs of bark stripping. Toppled amounted to only 1–2% of trees taller than 5 m. *Acacia burkei* woodlands were utilized most, followed by *Acacia robusta/Ficus sycomorus* riverine forests, *Acacia grandicornuta/Spirostachys africana* woodlands, and *Acacia nilotica* woodlands (Figure 14.4). Elephants used different woody species for feeding on leaves and branches rather than for feeding on bark or toppling (Table 14.1).

Overall, woody density remained little changed in HiP during the study period, but some shifts in representation of different size classes occurred (Figure 14.5). Trees smaller than 4 m made up 61% of the total sample in 1999 but had increased to 71% by 2007, mostly via trees in the height class 2–3 m. The representation of trees 4–8 m in height had declined from 31% of the total sample in 1999 to 21% in 2007. Trees taller than 8 m remained unchanged over this period at 8% of the total sample. The species most strongly selected by elephants in the 3 survey years included several common species, such as *Ziziphus mucronata*, *Acacia burkeii*, *Acacia nilotica*, and *Acacia robusta*, as well as less-common species, such as *Sclerocarya birrea*, *Schotia brachypetala*, and *Sideroxylon inerme* (see Table 14.2). Relatively rare species that were strongly selected by elephant included *Cussonia* spp., *Albizia versicolor*, *Ficus* spp., and *Garcinia livingstonei* (Table 14.2). Among the common tree species, a decline in the number of trees in larger height classes (4–10 m) was shown by *Acacia burkei*, *A. nigrescens*, *Sclerocarya birrea*, and *A. robusta*, while *A. nigrescens* also showed a declining trend among smaller size classes (Figure 14.6). For *A. nilotica*, all size classes taller than 2 m declined over time. However, *Ziziphus mucronata* maintained a stable size-class representation. The progressive disappearance of tall acacia trees of various species is of some concern because they are favoured for nest sites by both lappet-faced and white-headed vultures in HiP (D. J. Druce, personal observation).

Among the rare, endangered, and priority plant species identified in HiP, only *Albizia suluensis* was recorded in the sample plots. Twenty-two individuals of the latter were sampled in total, six of which showed damage by elephant. Among vegetation types, scarp forest has conservation importance because it contains endemic species and distinct species assemblages. Sample plots within scarp forest showed levels of elephant damage (4% bark stripping, 12% branch breakage, and 0.3% toppled) similar to those of other vegetation types.

An important objective of introducing elephants back into HiP was that they would contribute to halting ongoing woody encroachment in the park. The species most responsible for woody encroachment in HiP are *Dichrostachys cinerea*, *Acacia karroo*, and *Gymnosporia senegalensis*. However, *D. cinerea* and *G. senegalensis* were both strongly avoided by

Table 14.1. *Levels of damage by elephants of different forms incurred by particular tree and shrub species*

Species	Number of trees measured	Proportion of trees broken (%)
Cussonia zuluensis	5	60.0
Albizia versicolor	52	11.5
Antidesma venosum	66	9.1
Strychnos madagascariensis	145	8.3
Rhus chirindensis	197	7.6
Celtis africana	388	7.5
Acacia nilotica	4445	7.4
Englerophytum natalense	165	7.3
Ziziphus mucronata	1394	6.6
Pancovia golungensis	144	6.3
Acacia robusta	1839	6.1
Schotia brachypetala	815	5.2
Canthium inerme	428	5.1
Scolopia zeyheri	119	5.0
	Number of trees measured	Proportion of trees bark stripped (%)
Schotia brachypetala	815	7.6
Acacia nilotica	4445	5.2
Acacia grandicornuta	2962	5.1
Acacia robusta	1839	5.1
Celtis africana	388	4.9
Olea woodiana	125	2.4
Schotia capitata	227	2.2
Berchemia zeyheri	1492	2.1
Cordia caffra	262	1.5
Acacia gerrardii	1932	1.3
Combretum apiculatum	551	1.3
Acacia nigrescens	2357	1.1
Acacia burkeii	1790	1.0
Sideroxylon inerme	898	1.0
	Number of trees measured	Proportion of population toppled (%)
Maytenus undata	67	4.5
Olea capensis	152	3.3
Berchemia zeyheri	1492	2.7
Ziziphus mucronata	1394	2.2
Rhus chirindensis	197	2.0
Sideroxylon inerme	898	1.9
Canthium spinosum	338	1.8
Rhus pentheri	1827	1.1
Sclerocarya birrea	523	1.1
Acacia robusta	1839	1.1
Acacia burkeii	1790	1.1
Chaectacme aristata	363	1.1
Acacia nilotica	4445	1.1
Celtis africana	388	1.0

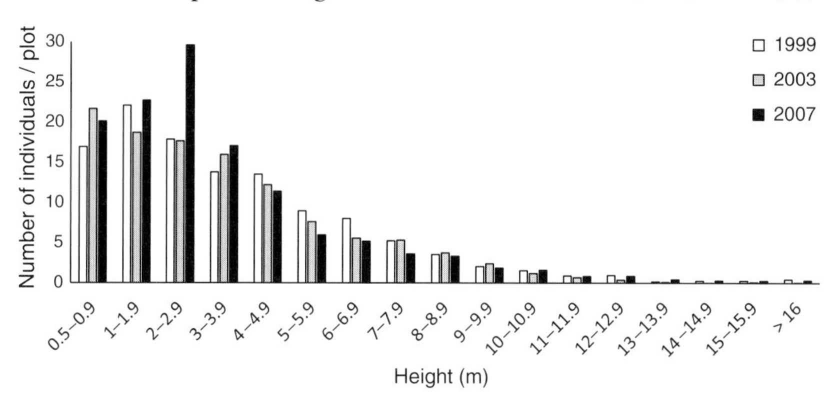

Figure 14.5 Size structure distribution of all woody plants recorded in the sample plots during successive surveys, 1999 (white bars), 2003 (grey bars), 2007 (black bars).

elephants relative to their availability, while *A. karroo* was used according to its availability (Table 14.2). Also, the size-class distributions do not show a decreasing trend for these species (Figure 14.7). On the contrary, trees in the height class 2–3 m have increased over the 3 sample years. Hence, the objective of re-introducing elephant to halt woody encroachment is not currently being realized.

These first analyses of the vegetation monitoring and elephant impact data have provided some idea of the effects that elephants may be having on the park's vegetation and plant diversity. However, it also indicated areas where data are lacking or not able to provide answers to questions that have been identified as important for managing the elephant populations. Although the current data provide much information on the browsing behaviour of elephants (e.g. which species, height classes, and vegetation types are selected for feeding or bark stripping) and which tree species and height classes are damaged (e.g. toppled or severely broken), we do not know how these effects translate to changes in habitat structure or composition. One problem is that the current data do not inform us what elephant impact means for plant survival. High utilization may lead to mortality for one species but be tolerated by another (O'Connor *et al*., 2007; Cromsigt and Kuijper, 2011). Secondly, other factors besides elephants (fire, shrub encroachment, and interactions between fire and elephants) have to be taken into account in interpreting trends in woody plant populations (Druce *et al*., 2008; Shannon *et al*., 2011). Moreover, small browsers may also have strong impacts, particularly on the smallest height classes (O'Kane *et al*., 2012). The nature of the elephant impact data

Table 14.2. *Tree species most selected for by elephant ($D > 0.30$) and selection indices for the three main encroaching species. Selection is calculated as the Jacobs selection index $D = (U/N - A/N) / (U/N + A/N - 2 \times U/N \times A/N)$. These data are based on counts summed over the 3 sampling years ($N =$ 92,562 individuals) and where more than 15 individuals of a particular species were sampled*

Species	Total no. trees utilized by elephant (U)	Total no. trees recorded (A)	Jacobs selection index (D)
Cussonia spp.	8	15	0.60
Albizia versicolor	23	52	0.53
Ficus spp.	23	52	0.53
Garcinia livingstonei	6	15	0.50
Sapium interrimum	16	44	0.46
Thespesia acutiloba	8	22	0.46
Sclerocarya birrea	176	523	0.43
Strychnos madagascariensis	49	145	0.43
Ziziphus mucronata	443	1394	0.41
Acacia burkeii	509	1790	0.37
Commiphora africana	15	51	0.37
Acacia nilotica	1203	4445	0.36
Acacia robusta	507	1839	0.35
Rhus pentheri	498	1827	0.35
Schotia brachypetala	225	815	0.35
Albizia suluensis	6	22	0.34
Sideroxylon inerme	238	898	0.33
Rhus chirindensis	51	197	0.32
Trichilia emetica	5	19	0.32
Dovyalis caffra	51	201	0.31
Maytenus acuminata	9	35	0.31
Main encroaching species			
Dichrostachys cinerea	435	10,181	–0.55
Acacia karroo	513	4079	–0.04
Maytenus senegalensis	152	2666	–0.41

described above unfortunately does not allow us to differentiate among all those factors. In order to address these shortfalls, certain changes to the monitoring will be necessary. Future vegetation monitoring could use detailed GPS-collar data on elephant movement to enable comparisons to be made between areas that are visited often and places that are rarely visited. A sample of permanently marked large trees should be assessed annually for condition and elephant utilization over prolonged time periods.

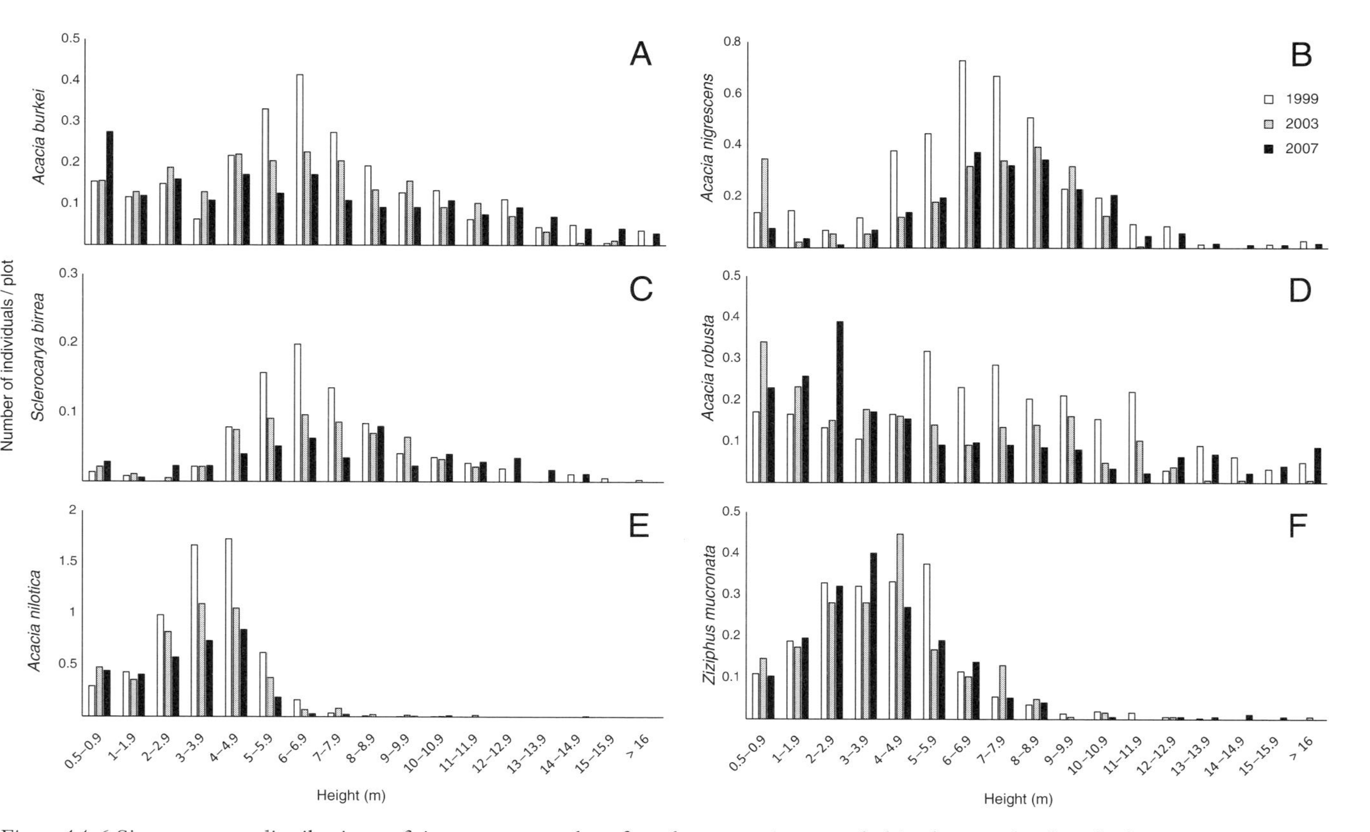

Figure 14.6 Size structure distributions of six common and preferred tree species recorded in the sample plots during successive surveys, 1999 (white bars), 2003 (grey bars), 2007 (black bars). (A) *Acacia burkei*; (B) *Acacia nigrescens*; (C) *Sclerocarya birrea*; (D) *Acacia robusta*; (E)*Acacia nilotica*; (F) *Ziziphus mucronata*.

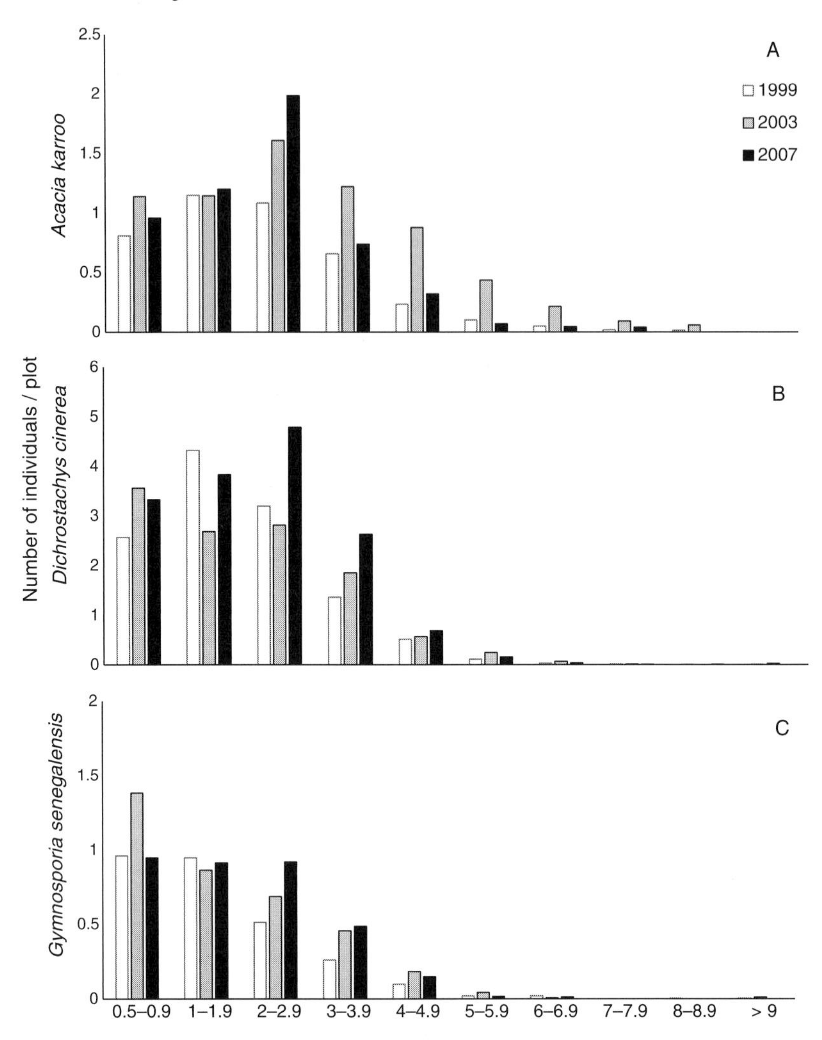

Figure 14.7 Size structure distributions of three encroaching species recorded in the sample plots during successive surveys, 1999 (white bars), 2003 (grey bars), 2007 (black bars). (A) *Acacia karroo*; (B) *Dichrostachys cinerea*; (C) *Gymnosporia senegalensis*. Data are shown as number of individuals per sampling plot for 10 different height classes (height in metres).

14.5 Implementation of the National Norms and Standards

In 2008, elephant management in South Africa moved into a new phase with the publication of the National Norms and Standards for the

Management of Elephant in South Africa as part of the National Environmental Management Biodiversity Act of 2004. This act requires that an elephant management plan is developed and signed off by the relevant authorities for any area containing elephants. In accordance with this requirement, Ezemvelo KZN Wildlife produced an elephant management plan (EMP) for HiP for the five-year period from 2011 to 2015, which was signed off in December 2012. The key objectives listed in priority order were to:

(i) maintain the elephant population in a state that does not jeopardize the conservation of biodiversity elements, priority biological assets (see below), or the maintenance of ecological processes within the park;
(ii) maintain a genetically viable population of elephants with a demographic structure (age structure and sex ratio) reflective of natural, mature elephant populations; and
(iii) ensure that risks to human infrastructure and livelihoods, both inside and outside the park, are minimized.

The most immediate perceived threat from the growing elephant population to the integrity of biodiversity in HiP is to the scarp forest, which contains numerous endemic species. Other perceived threats are to black and white rhinos, nesting sites for vultures and other rare raptors, and rare and endangered plant species (*Diospyros glandulifera*, *Albizia suluensis*, *Stangeria eriopus*, *Warburgia salutaris*, *Protea roupelliae*, *Protea caffra*, *Adenia natalensis*, *Adenia gummifera*, and *Encephalartos natalensis*). The EMP lists several surveillance and monitoring programmes that need to be in place as well as research that is required to better understand the effects of elephants on these various priorities. In addition, the EMP contains action tables which include columns for the particular element of concern, what the current management and scientific team would consider are poor (undesired), fair, good, and very good (desired) ratings for the state that the element is in, and the target, current, and management trigger ratings (see Table 14.3 for an example). Monitoring programmes need to be at a level that will indicate when the trigger rating is reached and at which point the various management actions or options (as listed in the table) should be implemented. The objective is for the management action to move the level back to an acceptable state which may be below the target setting. However, for many of the plant-related elements such as the scarp forest (priority habitat), the various rare, endangered, or priority plant species, and tall trees, the current ratings still need to be determined

Table 14.3. *Extract from the action tables within the HiP Elephant Management Plan (EMP) relevant to black and white rhino. When the trigger rating is reached, it triggers the first listed management action which in the case below is to identify the culprit, assess, and monitor the situation. If the rating then reaches the next level (poor for white rhino, fair for black rhino), the next management action should be taken (in the case below, the culprit is to be destroyed). All priority species and various other elements of concern (e.g. infrastructure damage, impact scarp forest, human livelihoods, etc.) have action tables in the EMP*

Element of concern	Poor rating	Fair rating	Good rating	Very good rating	Target	Current rating	Trigger	Management actions/options
Mortality of white rhino caused by elephant	Any elephant-induced mortality	Non-fatal injury to rhino	Negative interactions between elephant and rhino	No negative interactions	Very good	Very good	Fair	Identify culprit Assess and monitor situation If rating reaches poor, destroy culprit
Elephant attack on black rhino	Any elephant-induced mortality	Non-fatal injury to rhino	Negative interactions between elephant and rhino	No negative interactions	Very good	Very good	Good	Identify culprit Assess and monitor situation If rating reaches fair, destroy culprit

and research priorities have been listed that will assist in determining effective monitoring programmes.

Although the HiP management philosophy is not based around fixed upper limits, an ecological carrying capacity of elephants for HiP was estimated in accordance with the requirements of the National Norms and Standards. Based on the size of the park and its mean annual rainfall, this was projected to lie between 960 elephants (mesotrophic soil nutrient status) and 1140 elephants (high soil nutrient status, from Fritz and Duncan, 1994). Such numbers are seen only as a guideline, with the monitoring of the priority biodiversity assets and the relevant elements of concern in the action tables of the EMP dictating most elephant-related management actions.

14.6 Proposed Future Management Interventions

If either an element of concern or the ecological carrying capacity of elephants is exceeded, the EMP provides detailed timelines for management intervention. Moreover, the EMP gives the following priority order for the implementation of the main management interventions:

(1) explore options for park expansion;
(2) develop and implement an immunocontraception plan and research project;
(3) translocating or culling a select number of younger males;
(4) implement various management options to maintain or improve priority species and habitats;
(5) translocate whole family groups a year before the set ecological carrying capacity limits are reached;
(6) develop a culling plan a year before the set triggers are reached, if there are no areas to which elephants can be translocated; and then
(7) implement a culling plan once the set triggers are reached. Within the constraints of this culling plan, the EMP includes the possibility of sustained harvesting of elephants in order to satisfy certain resource needs of adjacent communities besides contributing to slowing or halting population growth.

It is important to note that the decisions that lead to the above management interventions have to be based on monitoring and research on the elements of concern. At the time of writing of this chapter, there has been some form of action within the first two interventions listed above, which we will explain in more detail below.

14.6.1 Park Expansion

Although HiP is surrounded by densely settled human communities, some potential has been identified for expansion of the park, mainly to the south and west. Fences between Hluhluwe and the Mpempeni Community Reserve (364 ha) to the east of Hlabisa town were dropped in late 2014. In addition, three communal areas bordering Mfolozi (Obuka, 2490 ha; Somopho, 2200 ha; and Mandlakazi, 1208 ha) have been declared nature reserves and are in the process of formation. There is some negotiation around an additional two areas which may also become nature reserves. The idea is that these areas will be fenced and can thereafter be incorporated into HiP, thereby increasing the park size. Efforts to form corridors, especially along routes that elephants would follow, such as rivers, would be supported by Ezemvelo KZN Wildlife, but need to be based on close consultation with other stakeholders and landowners. In addition, efforts are currently underway to designate buffer areas around certain parts of the park, which will limit certain types of development along the borders of the park.

14.6.2 Implementation of Immunocontraception

As proposed in the EMP, the aim is to slow the growth rate of the elephant population through immunocontraception so as to ensure that the limits set for the various elements of concern are not transgressed. Immunocontraception has a 2–3-year lead time and can only be applied on a long-term basis. Nevertheless, it would buy time for the preferred option, which is increasing land available to the elephants. In order to create and maintain a genetically viable population of elephants with a natural demographic structure, immunocontraception is applied to certain individuals or family groups on a rotational or proportional basis in order to reduce the reproductive rate. In 2014, the first elephants in HiP were given the porcine zona-pelucida (PZP) immunocontraception vaccine via darting by helicopter. Seventy adult females were successfully darted in early October, while booster darting was undertaken in mid-November. A relatively low number of female elephants was targeted because of the difficulty in locating the same individuals at both darting events. As a result, animals that were usually associated with one of the seven GPS-collared females were targeted. However, not all of the individuals that were darted during the initial application were found during the booster darting operation. This was because the booster darting took place after the first rains and the groups had already begun to split.

Furthermore, the immunocontraception plan was updated to eventually encompass 80% of the reproductively mature female elephants (approximately 160 individuals in 2014) in HiP. This level of immunocontraception application would need to continue at least until the population reached the estimated ecological carrying capacity of 960–1140 animals, which is envisaged as being achieved in around 2023–2024. At that point, monitoring and research on priority plant species, the scarp forest, and other conservation priorities, such as rhinos and vultures, should indicate what further action is needed to achieve the set biodiversity objectives.

14.6.3 Population Structure

A ground-based monitoring programme, focused on identifying individuals and their associations, has yielded the population demographic structure that the park aims at maintaining into the future. It is assumed that because the elephants that were introduced into the park came from numerous family groups, there is currently sufficient genetic diversity in the population. This will be monitored over time through the collection and analysis of genetic material from elephants that are found dead, darted for collaring, or shot as problem animals. Currently this material is being stored and will be analysed when the need arises.

14.7 Summary Overview: Past, Present, and Future

Elephants and their ecological role as transformers of vegetation have been restored in HiP, but numerous lessons have been learnt, some of which have been used to inform statements made in the National Norms and Standards for the Management of Elephants in South Africa. Although young orphan elephants were brought to establish the population, the population now contains distinct family groups and individuals that are almost always found together. Matriarchs have established themselves and the population has grown at a high rate. The oldest animals are now approximately 35–38 years old and there is an even spread of younger individuals. The introduction of older males almost immediately reduced the killing of rhinos by younger males, although it took 10 years before killing of rhinos by elephant had declined to a minimum. Although elephants move over the whole park, there have been very few breakouts through park fences. Elephants that repeatedly broke through intact fences, mostly males, were shot.

Vegetation monitoring has shown ongoing changes in the size structure of certain tree populations, although these cannot be clearly ascribed to the effects of elephants due to the nature of the monitoring design. Additional monitoring is needed to determine the levels of threat to particular plant species, and to the scarp forests. Because data are lacking to justify large management interventions, the elephant population in HiP has been subjected to little interference. Immunocontraception aimed at approximately 80% of the adult female segment should slow the growth rate of the population, and allow more time for monitoring programmes to identify elephant impact on the overall objectives of the protected area for biodiversity conservation. Protected areas of the size of HiP and smaller face the most acute challenges for managing elephants if neither surface water restrictions nor inaccessible regions limit the spatial extent of the impacts of this largest extant megaherbivore. As a result, these small parks (< 1000 km^2) need different, more proactive, approaches to elephant management than do larger ones (Owen-Smith *et al.*, 2006).

14.8 References

Bodasing, T. (2011) Determinants of elephant spatial use, habitat selection and daily movement patterns in Hluhluwe-iMfolozi Park. MSc thesis, University of KwaZulu-Natal.

Boundja, R. P. & Midgely, J. J. (2009) Patterns of elephant impact on woody plants in the Hluhluwe-iMfolozi Park, KwaZulu-Natal, South Africa. *African Journal of Ecology* **48**: 206–214.

Carruthers, J., Boshoff, A., Slotow, R., *et al.* (2008) The elephant in South Africa: history and distribution. In: *Elephant management: a scientific assessment for South Africa* (eds R. J. Scholes and K. G. Mennell), pp. 84–145. Wits University Press, Johannesburg.

Chamaillé-Jammes, S., Valeix, M., & Fritz, H. (2007) Managing heterogeneity in elephant distribution: interactions between elephant population density and surface-water availability. *Journal of Applied Ecology* **44**: 625–633.

Cromsigt, J. P. G. M. & Kuijper, D. P. J. (2011) Revisiting the browsing lawn concept: evolutionary interactions or pruning herbivores? *Perspectives in Plant Ecology, Evolution and Systematics* **13**: 207–215.

Dominy, N. J., Ferguson, N. S., & Maddock, A. (1998) Modeling elephant (*Loxodonta africana africana*) population growth in Hluhluwe-Umfolozi Park to predict and manage limits. *South African Journal of Wildlife Research* **28**: 61–67.

Druce, D., Shannon, G., Page, B. R., *et al.* (2008) Ecological thresholds in the savanna landscape: developing a protocol for monitoring the change in composition and utilization of large trees. *PLoS ONE* **3**: e3979.

Fritz, H. & Duncan, P. (1994) On the carrying capacity for large ungulates of African savanna ecosystems. *Proceedings of the Royal Society of London Series B* **256**: 77–82.

Kerley, G. I. H., Landman, M., Kruger, L., *et al.* (2008) Effects of elephants on ecosystems and biodiversity. In: *Elephant management: a scientific assessment for South Africa* (eds R. J. Scholes & K. G. Mennell), pp. 84–145. Wits University Press, Johannesburg.

Mackey, R. L., Page, B. R., Duffy, K. J., *et al.* (2006) Modelling elephant population growth in small, fenced, South African reserves. *South African Journal of Wildlife Research* **36**: 33–43.

McCracken, D. P. (2008) *Saving the Zululand wilderness. An early struggle for nature conservation.* Jacana Media, Johannesburg.

O'Connor, T. G., Goodman, P. S., & Clegg, B. (2007) A functional hypothesis of the threat of local extirpation of woody plant species by elephant in Africa. *Biological Conservation* **136**: 329–345.

O'Kane, C. A. J., Duffy, K. J., Page, B. R., *et al.* (2012) Heavy impact on seedlings by the impala suggests a central role in woodland dynamics. *Journal of Tropical Ecology* **28**: 291–297.

Owen-Smith, N., Kerley, G. I. H., Page, B., *et al.* (2006). A scientific perspective on the management of elephants in the Kruger National Park and elsewhere. *South African Journal of Science* **102**: 389–394.

Shannon, G., Thaker, M., Vanak, A. T., *et al.* (2011) Relative impacts of elephant and fire on large trees in a savanna ecosystem. *Ecosystems* **14**: 1372–1381.

Slotow, R., van Dyk, G., Poole, J., *et al.* (2000) Older bull elephants control young males. *Nature* **408**: 425–426.

Slotow, R., Balfour, D., & Howison, O. (2001) Killing of black and white rhinoceroses by African elephants in Hluhluwe-Umfolozi Park, South Africa. *Pachyderm* **31**: 14–20.

Slotow, R., Garaï, M. E., Reilly, B., *et al.* (2005) Population dynamics of elephants re-introduced to small fenced reserves in South Africa. *South African Journal of Wildlife Research* **35**: 23–32.

Van Aarde, R., & Jackson, T. (2007) Megaparks for megapopulations: addressing the causes of locally high elephant numbers in southern Africa. *Biological Conservation* **134**: 289–297.

Vincent, J. (1970) The history of Umfolozi Game Reserve, Zululand, as it relates to management. *Lammergeyer* **11**: 7–49.

Whateley, A. & Porter, R. N. (1983) The woody vegetation communities of the Hluhluwe–Corridor–Umfolozi Game Reserve Complex. *Bothalia* **14**: 745–758.

Wills, A. J. (1986) Re-establishment of elephant in the Hluhluwe and Umfolozi Game Reserves, Natal, South Africa. *Pachyderm* **7**: 12–13.

15 · *Successful Control of the Invasive Shrub* Chromolaena odorata *in Hluhluwe-iMfolozi Park*

MARISKA TE BEEST, OWEN HOWISON, RUTH A. HOWISON, L. ALEXANDER DEW, MANDISA MGOBOZI POSWA, LIHLE DUMALISILE, SUSAN JANSE VAN RENSBURG, AND COLETTE TERBLANCHE

15.1 Introduction

Invasions of alien species have increased dramatically in the past century and now constitute one of the greatest threats to biodiversity (Sala *et al.*, 2000). Invasions are characterized by the proliferation, spread, and persistence of species in new areas that are often very distant from their native ranges (Mack *et al.*, 2000). Invasive species can have profound impacts on the systems they invade, ecologically (Ehrenfeld, 2010) as well as economically (Pejchar and Mooney, 2009) and directly affect people's livelihoods by invading communal grazing lands or using up water supplies (McWilliam, 2000). In South Africa it is estimated that invasive alien plants alone waste 7% of the national water resources (Van Wilgen *et al.*, 1998). The science of invasion ecology is highly relevant for conservation biology, specifically if a species invades relatively small conservation areas that aim to protect species with high conservation value. A prime example of this is an invasion that unfolded over the last decades in the savannas of KwaZulu-Natal, including Hluhluwe-iMfolozi Park (HiP), by the alien shrub *Chromolaena odorata* (L.) King and Robinson (Asteraceae), commonly called triffid weed but henceforth referred to simply as

Conserving Africa's Mega-Diversity in the Anthropocene, ed. Joris P. G. M. Cromsigt, Sally Archibald and Norman Owen-Smith. Published by Cambridge University Press.

Chromolaena (Figure 15.1). This species was first recorded in Durban in the 1940s and reached HiP in 1961 (Macdonald, 1983). The invasion progressed rapidly in HiP, and by 2001, 20% of the northern sections of HiP were covered in dense monocultures (O. E. Howison, 2009). Today the species is largely under control thanks to an extensive control programme, 'Impi ka Sandanezwe' (isiZulu for 'war against Chromolaena'), which was launched towards the end of 2003 and modelled after the national Working for Water (WfW) campaign aimed at controlling invasive alien plants affecting ground water levels (Van Wilgen *et al.*, 1998). The control programme in HiP is arguably one of the most successful examples of this WfW approach. The invasion and the control programme have been well documented and supported with scientific studies. This case study of Chromolaena in HiP therefore serves as a leading example of an alien plant invasion and its effective management in a protected area.

In this chapter we discuss the patterns and processes of the invasion of Chromolaena and the consequences for plant diversity and herbivores in HiP as well as the implementation and lessons learnt from the control programme. We will conclude by looking forward. What have we learned from the control of Chromolaena? Will it prove successful in the long term? We believe that Chromolaena could be considered a model species in the study of biological invasions, and its detailed history in HiP will be helpful for understanding the long-term effects of species invasions in general (Strayer *et al.*, 2006) and the control of such invasions in (small) conservation areas in particular.

15.2 Features of *Chromolaena odorata*

Chromolaena odorata is listed among the world's worst alien invasive species (Lowe *et al.*, 2000) and is globally a high-impact invasive species (Gaertner *et al.*, 2014). Because it generally occurs in remote tropical areas, its invasion has been studied much less than invading plants in the temperate zone (Pyšek *et al.*, 2008). Chromolaena is a perennial, semi-lignified, scrambling shrub averaging 1.5–2 m in height but reaching up to 6 m as a climber on other plants. The species is part of a large genus (> 165 species) that is native to the Americas (King and Robinson, 1970). Its distribution is confined to tropical and subtropical biomes and ranges from southern USA to northern Argentina (Raimundo *et al.*, 2007; Zachariades *et al.*, 2009). Chromolaena has invaded the humid tropics and subtropics of the Old World over the past century, including a wide variety of ecosystems from tropical rainforests to savannas (Zachariades *et al.*, 2009).

Figure 15.1 (A) *Chromolaena odorata* infestation in a *Euclea racemosa* woodland in Maphumulo area of northern Hluhluwe showing the dense tangle of stems. The path through the Chromolaena monoculture was created by elephants (June 2003). (B) Chromolaena infestation in the same area during the wet season (March 2005). (C) Chromolaena infestation of the south bank of the Hluhluwe river opposite Maphumulo picnic site, preventing access for buffalo (August 2004). (D) Detail of Chromolaena plant (March 2005). (E) Cut-stump clearing treatment (August 2003). (F) Chromolaena stem residues remaining after clearing operation (August 2003). Photos: Mariska te Beest. (For the colour version, please refer to the plate section. In some formats this figure will only appear in black and white.)

It invades not only human-altered environments, like road verges and abandoned agricultural fields, but also protected areas (Macdonald, 1983; Foxcroft *et al.*, 2009). Here, it forms dense, monospecific stands in savanna woodlands and along river courses and forest margins (Goodall and Erasmus, 1996). In South Africa, this species is highly invasive in mesic savannas, even though climate models show that these savannas are at the margins of what is its climatically suitable habitat (Kriticos *et al.*, 2005; Raimundo *et al.*, 2007; Robertson *et al.*, 2008; te Beest *et al.*, 2013). The origin of this southern African biotype has been determined as the Greater Antilles in the Caribbean (Zachariades *et al.*, 2004; Robertson *et al.*, 2008). Chromolaena has also invaded most of West Africa and is currently expanding into East Africa, where the first specimen has been collected along the Mara river in the Serengeti system in Tanzania (Zachariades *et al.*, 2013; Beale *et al.*, 2013).

In its native range, Chromolaena is typically a plant of secondary succession, growing in forest gaps and along rivers and the edges of savannas (Cruttwell McFadyen, 1988). This common and widely distributed species succeeds pioneering ephemeral herbs and is subsequently displaced by small trees and bushes. It disappears completely after the forest canopy begins to close (Kassi and Decocq, 2008). Where agriculture and human activity prevent forest regeneration, Chromolaena persists along forest edges and paths, in abandoned fields, pastures and building sites, and along roads, railways and streams (Cruttwell McFadyen, 1988).

15.3 Mechanisms of Invasion

The invasive success of Chromolaena is thought to depend on its high reproductive capacity, high relative growth rate, and high net assimilation rate. These three factors combined are a measure of the photosynthetic efficiency of plants (Ramakrishnan and Vitousek, 1989), as well as their ability to effectively shade out native vegetation (Kushwaha *et al.*, 1981; Honu and Dang, 2000; te Beest *et al.*, 2015b) and to survive severe disturbances (te Beest *et al.*, 2012). Chromolaena can reproduce asexually through seeds (apomixis) (Gautier, 1992) and has a prolific production of light wind-dispersed seeds, which are also easily dispersed by mammals or vehicles (Blackmore, 1998). A single shrub can produce as many as 800,000 seeds (Witkowski and Wilson, 2001). In the seedling stage, Chromolaena invests mainly in stems and height growth (te Beest *et al.*, 2009, 2013), often using trees as support, which allows it to quickly grow taller than native vegetation and to shade out competitors. A study on plant–soil interactions showed that investment in stems and height growth was larger on soil with invasive-range biota than on soil with native range biota (te Beest *et al.*, 2009). Canopy light interception measured in a greenhouse study showed that the horizontally placed leaves of seedlings intercepted 90% of the available light (te Beest *et al.*, 2013). A field study in HiP measured similar levels of light interception by adult shrubs. Over 80% of the available light was intercepted by the canopy, leading to a 45% reduction in grass biomass compared to an adjacent open patch (te Beest *et al.*, 2015b). The same study showed that the specific leaf area (SLA) of Chromolaena, which is a proxy for relative growth rate, was about twice as high as for co-occurring native species (te Beest *et al.*, 2015b). These studies show how the plant architecture, biomass allocation pattern, and high growth rates of Chromolaena play roles in its invasiveness in South African savannas.

Chromolaena is a vigorous sprouter and adult shrubs are able to survive severe disturbances, like fire or cutting, or abiotic stress by quickly regrowing from the basal stems (te Beest *et al.*, 2012). The survival of large Chromolaena shrubs was surprisingly not affected by intense firestorms that resulted in nearly 100% top-kill of surrounding savanna trees (te Beest *et al.*, 2012). Chomolaena seedlings are more sensitive and have been shown to be weak competitors against grasses (te Beest *et al.*, 2013). Low-intensity grassland fires reduce the seedling survival of Chromolaena (te Beest *et al.*, 2015a). Hence, establishment in grasslands is hampered but not prevented by fire. A seedling transplant experiment in HiP showed that for Chromolaena seedlings to establish in grass-dominated habitats, they need either disturbance of the grassland, e.g. by herbivores, or protection by 'nurse trees' (te Beest *et al.*, 2015a). Fire greatly reduced seedling survival, from 26% in undisturbed grasslands to 5% post-fire. Seedling survival in disturbed grasslands (mimicked by grass clipping and soil disturbance) increased to 45% and remained high post-fire (22%). Nurse trees increased seedling survival in the absence of fire (44%), but greatly reduced post-fire survival (6%). Thus, small-scale disturbances facilitate Chromolaena seedling establishment in savanna grassland both directly, by creating micro-sites for establishment, and indirectly, by reducing the negative effect of fire.

15.4 Invasion History and Habitat Association

Multiple pathways have been suggested for the introduction of Chromolaena to South Africa. Some reports state that seeds were accidentally brought in during World War II in packaging material offloaded at

Durban harbour, 350 km south of HiP (Liggitt, 1983; Goodall and Erasmus, 1996; Witkowski and Wilson, 2001). Others note that Chromolaena had been introduced to South Africa as early as the mid-nineteenth century, when it was found growing in the Cape Town Botanic Garden, Kirstenbosch (Zachariades *et al.*, 2004). The first identification of the species in HiP was in 1961 (Macdonald, 1983). The development of the invasion in HiP has been particularly well documented, with its distribution mapped periodically from the 1960s to current times. Few invasions in tropical habitats have been mapped with the same level of spatial and temporal resolution from an early stage of invasion. The presence of Chromolaena was first systematically mapped in Hluhluwe Game Reserve in 1978 in grid cells of 0.25 km^2. Mapping was done by personnel surveying on foot and from vehicles. Mapping of the spatial distribution using this method was undertaken in 1978, 1980, 1981, 1982, 1983, 1987, 1998, and 2001 (Figure 15.2). No sustained control operations were undertaken during this time. While in 1978 only 5% of grid cells in Hluhluwe GR were invaded, this had increased to 73% by 2001, an average increase of 5% per year (O. E. Howison, 2009).

Table 15.1 shows the initial rate of spread between 1978 and 1983 for each vegetation type as classified and mapped by Whateley and Porter (1983; see Chapter 2 for description of vegetation types). This was used to assess habitat suitability based on the assumption that the higher the initial rate of spread, the more suitable the habitat is (O. E. Howison, 2009). The habitat types with the highest initial rates of spread were the scarp forests, the *Sporobolus africana/Cyperus textiles* sedge/grasslands and the *Acacia caffra* thickets, as well as *Combretum* savanna. The sedge/grasslands and the *Acacia caffra* thickets are habitat types that are closely associated with the scarp forests and occur along its margins (Whateley and Porter, 1983). These habitats (forests margins and gaps) correspond closely to the habitats where Chromolaena can be found in its native range and where it invades in West Africa and south-east Asia. In these forest habitats, gap dynamics play an important role and Chromolaena is displaced by small trees and disappears after the forest canopy begins to close (De Rouw, 1991; Honu and Dang, 2000; Kassi and Decocq, 2008). Similarly, in HiP Chromolaena does not occur under the closed canopy of scarp forests. However, in contrast to its native range, Chromolaena reaches very high densities and even monocultures in savanna woodlands and certain riverine habitats. Savanna woodlands have a more open tree canopy than forests, which allows for enough light for Chromolaena to invade these woodlands in their natural state without the necessity for canopy disturbance. Therefore, while eventual gap closure inhibits large-scale

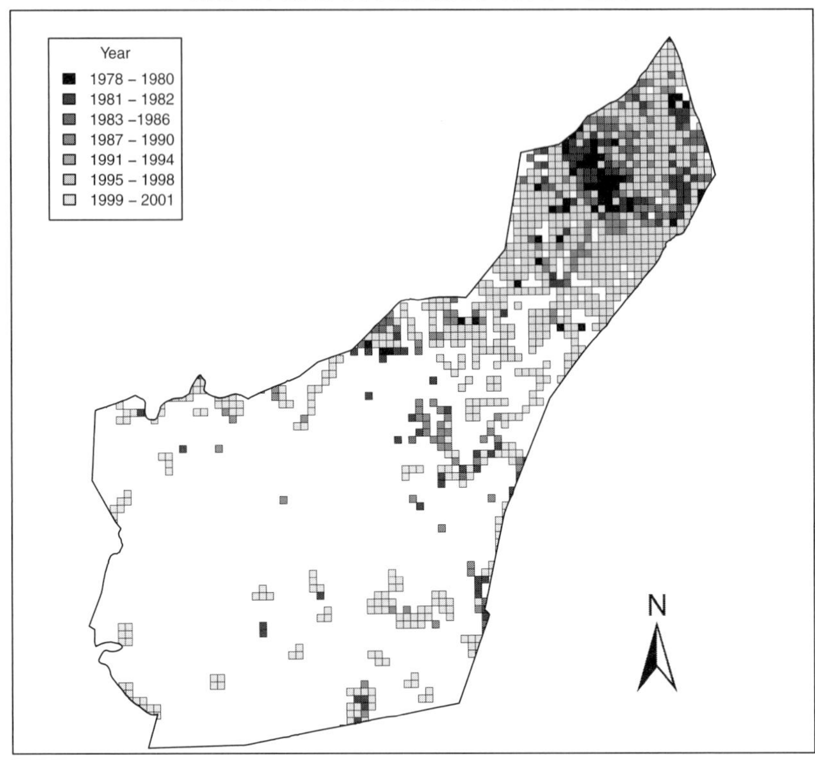

Figure 15.2 Expanding spread of *Chromolaena odorata* in grid cells (0.25 km^2) within HiP from 1978 to 2001 (for Hluhluwe from O. E. Howison, 2009; for Mfolozi from EKZNW unpublished data).

Chromolaena invasion in forests, this control mechanism does not play a role in savanna woodlands (te Beest, 2010). Levels of invasion are lowest in savanna grasslands, where 'nurse trees' or disturbance (fire or herbivores) of the grassland is required for Chromolaena to establish (te Beest *et al.*, 2015a; see Section 15.3).

Soil characteristics also determine where Chromolaena is likely to invade. Between 2003 and 2006, Chromolaena densities and environmental variables were measured on 58 transects distributed across HiP (McLennan, 2006). Findings showed that Chromolaena could colonize and establish on a range of soil types, but was most commonly found in nutrient-rich and clayey soils and on duplex soils, where higher levels of water in the upper horizon may favour the shallow-rooting

Table 15.1 *Average rate of spread of* Chromolaena odorata *for each vegetation community within Hluhluwe as identified by Whateley and Porter (1983) (see Chapter 2, Table 2.2, for more detailed descriptions for each vegetation community) during the beginning of the invasion process (1978–1983). Data are for Hluhluwe section only*

Vegetation community	Average rate of invasion 1978–1983 (%)	Area (ha)
Celtis africana–Euclea racemosa forest	7.7	1063
Celtis africana–Harpephyllum caffrum forest	7.4	1603
Ficus sycamorus–Schotia brachypetala riverine forest	4.4	1877
Spirostachys africana–Euclea racemosa riverine forest	0.9	1158
Euclea divinorum woodland	2.6	3201
Combretum molle woodland	7.5	537
Spirostachys africana woodland	2.2	784
Acacia caffra thicket	7.2	985
Acacia karroo–Dichrostachys cinerea induced thicket	2.1	6703
Acacia burkei woodland	1.8	2714
Acacia karroo woodland	3.8	890
Acacia nilotica woodland	2.4	2
Sporobulus africanus–Cyperus textilis sedge grassland	7.7	107
Themedia triandra grassland	3.8	944

Chromolaena. Well-drained sandy soils tended to show low abundance of Chromolaena (McLennan, 2006). An extension of this study, comprising 112 transects, found higher percentages of total carbon and nitrogen on transects with higher densities of Chromolaena (te Beest *et al.*, 2015b). These studies show that soil fertility and organic carbon content are important predictors of Chromolaena abundance, together with rainfall and the availability of water in the landscape, e.g. subsoil run-off, seepage, drainage lines, and gullies (Liggit, 1983; te Beest *et al.*, 2013).

15.5 Impacts of the Invasion by *Chromolaena odorata*

The impact of invasive species is often highly variable and therefore difficult to quantify (Vila *et al.*, 2011). The impact is generally greatest when invasive species alter major ecosystem processes, such as N dynamics or fire regimes (Brooks *et al.*, 2004; Ehrenfeld, 2010; Gaertner *et al.*, 2014). A whole range of negative impacts have been associated with Chromolaena in HiP and elsewhere.

15.5.1 Socioeconomic Impact

Numerous papers deal with the socioeconomic impact of Chromolaena and show that the invasion of this species creates major socioeconomic problems, for example through impairing opportunities for agriculture, livestock husbandry, and forestry (Lucas, 1989). Because in most of the tropics people still depend on subsistence farming, the invasion of Chromolaena directly impairs people's livelihoods by invading pastures and decreasing carrying capacity for livestock (McWilliam, 2000). In extreme cases in the Philippines, this has led to villages being abandoned (McWilliam, 2000). Ironically, Chromolaena is also valued as a fallow species in slash-and-burn agriculture in West and Central Africa, where the species forms a nutrient sink and suppresses weeds during years when the field is not planted with crops (Slaats, 1995; Norgrove *et al.*, 2000; Koutika and Rainy, 2010). The shallow-rooting Chromolaena suppresses deep-rooting invasive grasses, such as *Imperata cylindrica*, which are difficult to remove once the field is prepared for planting (Akobundu and Ekeleme, 1996). Effective control of Chromolaena at this stage requires good tillage to remove the rootstocks before planting food or crops.

15.5.2 Impacts on Biodiversity

Several studies document the impact of Chromolaena on biodiversity in HiP and surroundings. Te Beest *et al.* (2015b) describe how Chromolaena reduces grass biomass and increases litter fall and show that these impacts are related to specific traits of the species. These include high specific leaf area, resulting in fast growth rates, and high leaf area index, resulting in efficient outshading of native vegetation and drying out of the soil due to high transpiration rates. This results in reduced understorey diversity in dense stands of Chromolaena (M. te Beest, unpublished data). In India and West Africa, Chromolaena reduces local diversity of native vegetation (Murali and Setty, 2001; Mangla *et al.*, 2008), suppresses native vegetation through shading and physical smothering (De Rouw, 1991; Slaats, 1995), and halts natural succession by creating dense thickets (Honu and Dang, 2000). Other studies have shown that Chromolaena may act as a host for pest species (Moder, 1984; Boppre and Fischer, 1994), is allellopathic (Ambika, 2002; Sangakkara *et al.*, 2008), and accumulates seed- and soil-borne fungi, *Fusarium* spp. (Esuruoso, 1971; Mangla *et al.*, 2008).

Several studies suggest a negative impact of Chromolaena on animal communities. Small-mammal species richness and diversity was lower in areas in HiP that were invaded by Chromolaena than in uninvaded areas with similar habitat (Dumalisile, 2008). Certain large mammals were excluded from dense Chromolaena monocultures in Hluhluwe, most notably grazers (Rozen-Rechels, 2015). Another study used spiders as ecological indicators in HiP and showed that dense Chromolaena infestations altered spider assemblage patterns and reduced spider abundance, species diversity, and species richness (Mgobozi *et al.*, 2008). At Lake St Lucia, located 50 km east of HiP, shading of Nile crocodile (*Crocodylus niloticus*) nesting sites by Chromolaena produced a female-biased sex ratio, which could pose a serious threat to the survival of this species in the area (Leslie and Spotila, 2001). Chromolaena may also reduce food availability for large mammals as has been suggested for black rhino (*Diceros bicornis*) in HiP (R. A. Howison, 2009) and gorilla (*Gorilla gorilla*) in equatorial Africa (Van der Hoeven, 2007).

15.5.3 Impacts through Affecting the Fire Regime

Chromolaena may have severe ecosystem impacts through its ability to alter fire regimes. Interestingly, it may reduce or increase the role of fire. Once established, Chromolaena shades out grasses (te Beest *et al.*, 2015b), which may lead to a reduction in fuel loads, and thus reduce fire frequencies. In contrast, during dry, fire-prone conditions, Chromolaena can ignite easily and spread fire into otherwise fire-excluding habitats, such as forest or closed woodland, by increasing the fire intensity and by growing as a scrambler into tree canopies. In this way, Chromolaena may create infrequent but high-intensity canopy fires (te Beest *et al.*, 2012), which can have large ecological impacts (see also the firestorms section in Chapter 10). For rare vegetation types, such as scarp forests, the fire risk imposed by Chromolaena is of particular concern. Chromolaena regularly occurs on forest margins, linking the grass canopy with the forest canopy, and representing a risk to fire-sensitive forest trees (Macdonald, 1983; te Beest *et al.*, 2012). During the last decade, several intense fire events have damaged parts of HiP's scarp forest, most likely in part fuelled by Chromolaena. Chromolaena forms a similar threat to riverine forest habitat. A high-intensity fire in 2003 in the Maphumulo area of

Hluhluwe destroyed large tracts of this habitat. This fire was triggered by accumulated dry Chromolaena stems left after clearing operations (Figure 15.1E,F) at the inception of a large-scale management experiment that was designed to test the interaction between clearing and fire (see Section 15.7.3; te Beest *et al.*, 2012). As another example, in 2008 an intense fire burned through most of the northern section of Hluhluwe. The relative role of Chromolaena versus extreme weather conditions in the 2008 firestorm is debated (see Chapter 10 for more details). However, data from the Chromolaena Clearing Project show that in 2008 the whole area that was burned had been under clearing control before the burn, including dense infestations that were cleared for the first time, leaving behind large amounts of dry Chromolaena stems. We suggest that these piles of dried Chromolaena stems most likely played a role in intensifying the fire and carrying it into the tree canopies.

15.6 Biocontrol

In its native range, numerous insects and pathogens are found on Chromolaena (Cruttwell McFadyen, 1988; Barreto and Evans, 1994) and are believed to control its densities. In contrast, outside its native range, only a few phytophagous insects have been recorded to feed on Chromolaena (Kluge and Caldwell, 1992). This suggested good prospects for biocontrol of Chromolaena and led to much research (Zachariades *et al.*, 1999; Muniappan *et al.*, 2005). Unfortunately, the leaf-mining moth *Pareuchaetes insulata*, which had proven a successful agent in Ghana and Indonesia (Zachariades *et al.*, 1999), has caused only localized damage to Chromolaena in South Africa (Zachariades *et al.*, 2009). Between 2001 and 2009, 1.9 million *Pareuchaetes* larvae were released at 30 sites in KwaZulu-Natal province, including at Maphumulo picnic site in HiP, but only one population successfully established, and this was at Umkomaas, 50 km south of Durban (Zachariades *et al.*, 2011). This moth has subsequently spread along the wetter coastal belt and is causing localized damage to Chromolaena. The reason for the low success achieved in the seasonal savannas of KZN is believed to be due to their relatively cold winter temperatures in combination with fires and/or dry seasons (Zachariades *et al.*, 1999; Robertson *et al.*, 2008). However, efforts to identify regions of the Neotropics that are climatically similar to South African savannas have allowed for better climatic matching of potential biological control agents (Robertson *et al.*, 2008) and research on a wide species range of potential insects and pathogens is ongoing (Zachariades *et al.*, 2011).

15.7 Successful Control of Chromolaena in HiP

The Chromolaena Clearing Project ('Impi ka Sandanezwe') was initiated in 2003 by Ezemvelo KZN Wildlife (EKZNW), the provincial nature conservation authority, and the KwaZulu-Natal Department of Economic Development, Tourism and Environmental Affairs (EDTEA). It was renamed to 'Invasive Alien Species Programme' (IASP) in 2008 when similar projects were initiated in all EKZNW areas. Control of Chromolaena actually started in 2000 with WfW clearing teams but with more limited funding. At a national level, between its inception in 1995 and 2008, the WfW programme spent 171.8 million ZAR (~US$24.5 million) controlling Chromolaena alone (van Wilgen *et al.*, 2012). However, despite this effort, Chromolaena increased in terms of invaded area, ranking 14th in 2000 and 4th in 2010 (van Wilgen *et al.*, 2012). In HiP, however, during approximately the same period, the Chromolaena invasion strongly decreased to manageable densities, albeit at a local cost of 103 million ZAR (US$15 million).

15.7.1 Mechanical Clearing, Effectiveness, and Costs

Clearing was prioritized based on previous mapping exercises (see Section 15.4) and was started from the least invaded areas (Mfolozi) working towards the area with high densities (Hluhluwe) to limit risk of further spreading. Riverine areas were also prioritized. A further intention had been to clear along the boundary in the adjoining communal lands to create a 1-km buffer zone around HiP. However, due to funding constraints, this was not achieved everywhere and only priority areas were cleared, such as upstream from water courses flowing through HiP as well as along roads entering HiP. At the height of infestation in 2003, almost 40,000 ha of HiP were infested with Chromolaena, which is 43% of the whole park and > 75% of the Hluhluwe section (Figure 15.3). Mechanical clearing consisted of hand-pulling and spraying seedlings and slashing established plants, followed by herbicide application to the remaining stumps (Figure 15.1E,F; Van Gils *et al.*, 2004; Marais and Wannenburgh, 2008). After the initial clearing, a series of follow-up controls was undertaken to limit reinvasion, requiring more than 10 repetitions in the most densely invaded areas. The time between follow-up clearings was 7–9 months in the areas with the densest infestations, but up to 2 years in areas that were less densely invaded. Each contract area was assessed by the Project Manager for the density of the reinvasion on a regular basis

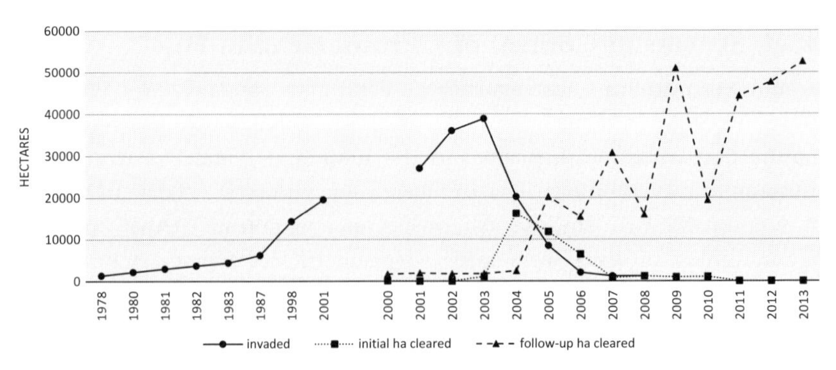

Figure 15.3 Area invaded (ha) by *Chromolaena odorata* (black line with circles). Period 1978–2001 area invaded for Hluhluwe only (data from O. E. Howison, 2009). Period 2001–2008 area invaded for the whole of HiP (data from Invasive Alien Species Programme). Grey lines represent the area cleared from Chromolaena between 2000 and 2013. Light grey line with squares: initial hectares cleared; dark grey line with triangles: follow-up hectares cleared.

(6 months to a year depending on initial densities) and the follow-up clearings were based on these assessments. More than 50 person days per ha were necessary to control the densest areas, which were virtually monocultures of Chromolaena (Dew, 2015). The achievement was astounding. Between 2003 and 2013 the clearing programme cleared almost 40,000 initial hectares and 307,000 follow-up hectares, totalling 755,000 person days (~2000 man years!). The clearing efficacy expressed in proportional reduction of Chromolaena cover was high in most areas. Monitoring undertaken along a total of 200 km of transects across HiP (biannual herbivore census transects, see Chapter 4) in 2004, 2010, and 2014 (te Beest, unpublished data) showed complete eradication of the highest density infestations (monocultures) over this period (Figure 15.4). Moreover, the cover of Chromolaena on the sampled transects declined from 22.3% in 2004 to 7.8% in 2014 (Dew, 2015).

15.7.2 Organizational Characteristics

Being a poverty-relief programme, the Chromolaeana Clearing Project had very specific criteria for employment. People contracted to clear Chromolaena were hired from local communities surrounding the park. There are 10 tribal authorities around HiP and each tribal authority was given equal employment opportunities. Transport logistics generally required that teams conducted clearing in areas that were nearest to

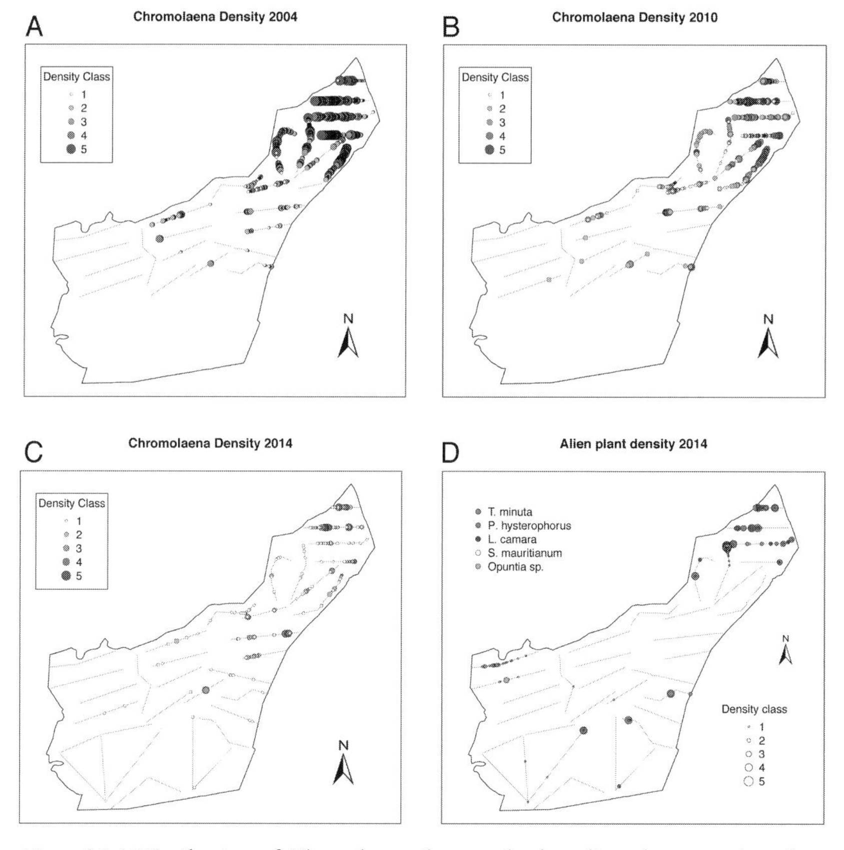

Figure 15.4 Distribution of *Chromolaena odorata* and other alien plants monitored along biannual game census transects (grey lines) in 2004 (A), 2010 (B), and 2014 (C). The different sizes and shades of the circles refer to different cover estimates (density class) according to a modified Braun-Blanquet scale: 1, 5% cover; 2, 25% cover; 3, 50% cover; 4, 75% cover; and 5, 100% cover.

their respective communities. Each clearing team consisted of a contractor and 14 workers. They were not directly employed by the Clearing Project, but contracted to clear a certain area and given a certain number of days to clear it, based on a prior estimate of the density of infestation. This so-called *contractor development model* was developed by WfW and aims to promote small businesses and entrepreneurship (Coetzer and Louw, 2012). Contract areas with higher initial Chromolaena densities were generally smaller and teams were given more time to clear than for follow-up clearings. The team was only paid following completion of a contract and after a thorough inspection of the quality of the work by the Project Manager. If the quality was inadequate, i.e. Chromolaena shrubs were still present, teams had to go back in and clear remaining plants before they were paid for the contract. At the height of operations, nearly 1000 labourers were clearing in the park at a time.

15.7.3 Fire, Rainfall, and the Control of Chromolaena

The successful control of Chromolaena in HiP was first and foremost due to the enormous clearing effort and supporting financial commitment by the provincial government, but other factors also played a role. Ecological factors contributed to the clearing efficacy, notably low rainfall and fire (Dew, 2015). Chromolaena is very sensitive to dry conditions (te Beest *et al.*, 2013) and clearing efficacy increased during years with low rainfall (Dew, 2015). The combined use of clearing and fire strongly increased the efficacy of Chromolaena control and reduced the costs. This was shown through a management experiment that was set up as a close interaction between park managers, the clearing project team, and researchers. This collaborative experiment was set up in 2003 in the Maphumulo area of Hluhluwe to investigate the efficacy of the cut-stump clearing practice and the role of fire in increasing this efficacy (see te Beest *et al.*, 2012 for details). The experiment included high-density Chromolaena stands that received four treatments; clearing only, burning only, clearing prior to burning, and a control (i.e. no clearing or fire). The clearing treatment was undertaken by the clearing teams, while the burning treatment was carried out by the park management. An important finding was that fire alone *did not* control Chromolaena, and often worsened the invasion. Resprouts appeared within weeks after even the most intense fires, and without competition from the grasses Chromolaena could quickly regain or even increase dominance. The timing of burning during seed set in August/September, the main burning season, may contribute to the use of fire without clearing worsening the invasion, because seeds often remain intact on the shrubs above the flame zone (M. te Beest, personal observations). The clearing-only treatment also did not control Chromolaena densities. The clearing treatment was generally most effective for the largest plants. Due to its multi-stemmed growth form, small stems of medium-sized plants were easily missed during herbicide application and quickly resprouted. The clearing prior to burning treatment, however, proved very effective, as most plants of all size classes (> 90%) as well as seeds in the soil were killed.

In retrospect, the timing of the experiment a few months before the inception of the Chromolaena clearing programme turned out to be ideal, as results from the experiment could directly feed into the clearing protocols that were being developed. For example, a few months after the initiation of the treatments it was clear that large Chromolaena bushes were tolerant to fire and resprouted again. This knowledge changed the prioritization within the clearing programme and whenever an area was burned, clearing teams were sent in. Over the years this became standard practice and the planning of the clearing was linked to fire management plans. This practice of follow-up controls after fires also made it possible to use fire as an initial control, which reduced the costs of clearing. This integration between management, clearing, and science led to an effective science-based control of Chromolaena. Similar protocols have proved effective in iSimangaliso Wetland Park, 40 km east of HiP (personal comments of Carl Myhill to Mariska te Beest, 26 March 2015).

15.7.4 Summarizing the Dos and Don'ts for the Use of Fire in Chromolaena Control

Research in HiP has shown that fire is a powerful but difficult tool to control Chromolaena densities. Burning following clearing is a very effective way to control Chromolaena. However, the wrong timing or intensity of fires, particularly without clearing, may worsen its invasions. Low-intensity fires during early spring (September) are arguably the worst because plants that grow above the flame zone will release their undamaged seeds while the competitive grass layer has been removed. Moreover, the management experiment clearly showed that even high fire intensity without clearing does not prevent adult shrubs from resprouting (te Beest *et al.*, 2012). If Chromolaena stands are accidently burned without prior clearing, following up the burns with clearing teams is essential for controlling the invasion (Dew, 2015).

15.7.5 Key Features Determining the Success of the Control Programme

To summarize, several features contributed to the successful control of Chromolaena in HiP. (1) A **Steering Committee** to oversee the clearing included park management, research management, a community representative and a public representative (tour operator), as well as the project coordinator. All role players were thus kept informed regarding employment, field strategy, budgets, etc., which created a feeling of joint ownership. (2) A **Rapid Response Team** consisting of four people drove around HiP throughout the year and cleared emerging weeds in HiP. They were also responsible for clearing all invasive plants around gates, lodges and staff accommodation. (3) **Low infestation** areas were treated by issuing 500 ml herbicide sprayers to Ezemvelo KZN Wildlife staff (management and field rangers) to clear areas where it would be too expensive to send a team in. (4) **Very flexible management** was a key factor, with unplanned fires being followed up with clearing. (5) **A good database** kept in HiP was used to manage all invasive alien plant data. (6) The project was given sufficient **space** to house 10 staff members, an office, and enough storage space for herbicide, protective clothing and equipment. (7) Last, but not least, a **large amount of funding** was allocated for a sufficient amount of time to allow for a large-scale approach and long-term planning.

15.7.6 Maintenance of Chromolaena Clearing in the Longer Term

At some point in each successful control programme there should be a transition into a long-term 'maintenance programme'. When densities become too low, it is no longer worthwhile to send in expensive clearing teams to find the last few remaining plants. Nevertheless, it remains essential to monitor and control the area to prevent reinvasion. With the transitioning of the Chromolaena Clearing Project to the Invasive Alien Species Programme, the first step was made towards a longer-term maintenance programme that focuses on all invasive alien species in HiP. However, different methodologies should be adopted too to reduce costs, which is the next step to be made. One possibility is to equip field rangers with a Personal Digital Assistant that allows them to report any encountered invasive plant, automatically recording time, date, and location. Such devices have been successfully implemented in Kruger National Park to record alien species (Foxcroft *et al.*, 2009). A permanent *rapid response team* could then deal with the larger infestations.

15.8 Other Alien Plants

The first alien plant survey within HiP was conducted in 1978 (Macdonald, 1983). Twenty alien tree, shrub, and creeper species were recorded and five of these had already formed extensive infestations: *Chromolaena odorata*, *Melia azedarach*, *Opuntia* spp., *Psidium guajava*, and *Solanum mauritianum* (Macdonald, 1983). All have since been controlled.

Another survey of alien plant species conducted in HiP in 2004 (Henderson, unpublished data) recorded a total of 54 species, most of which belong to the family Asteraceae. The vast majority of species originated from the Americas (70%), followed by other parts of Africa (18.5%) and Asia (5.5%). Woody shrubs (24%) plus annual (20%) and perennial herbs (22%) were the most prominent life forms. Remarkably, no single alien grass species has invaded HiP, which is generally true for all African savanna grasslands (Foxcroft *et al.*, 2010). The species were categorized with different priorities of control. The species with highest priorities are listed in Table 15.2.

In 2014 the distribution of five alien invasive species that are considered to pose the greatest threat of becoming problematic were mapped along the biannual game census transects (Dew, 2015). Apart from Chromolaena, these were *Parthenium hysterophorus* L., *Lantana camara* L., *Opuntia* spp. (*O. ficus-indica* and *O. monacantha*), *Solanum mauritianum* Scop., and *Tagetes minuta* L. None of these species had reached the levels of the Chromolaena infestation in the early 2000s. The current challenge is to find measures that are effective in controlling them. *Parthenium* is of particular concern because it seems to be spreading rapidly in the region and forms a risk for human, and perhaps animal, health.

15.9 Synthesis

The successful control of Chromolaena in HiP has only been possible thanks to extensive government funding and enormous amounts of labour in a relatively confined area. However, this support will not be available to other landowners, such as communal or private farmers, making successful control at a provincial or even national scale much more challenging. The poverty-relief structure of the programme, which was initially developed by WfW (Van Wilgen *et al.*, 1998; Coutzer and Louw, 2012), is a great model for providing the necessary labour to combat invasive species as well as improve people's livelihoods. This aspect is implemented widely in South Africa with many spin-offs (wetland and ecosystem rehabilitation, control of wildland fires) and the model could be a great example for other medium- to low-income countries.

Mapping of Chromolaena in 2004, 2010, and 2014 showed that, even though the control programme has strongly reduced densities,

Table 15.2 *Invasive and potentially invasive alien plants in Hluhluwe-iMfolozi Park based on the 2004 alien plant survey (Henderson, unpublished data). Species with the highest priority ratings are shown. Priority ratings: (1) concerted control campaign needed; very serious invasive alien species threatening biodiversity; (2) eradicate or start control programme; potentially very serious invasive alien species; (3) eradicate, few individuals present, but potentially invasive alien species; (4) requires ongoing control efforts, will continue to invade from outside HiP. Species marked with an asterix were not prioritized in 2004, but were identified as priority species in 2014 (te Beest, unpublished data). C, central; S, south*

Scientific name	Family	Origin	Growth form	Legal status	Priority for control
Chromolaena odorata	Asteraceae	C, S America, Caribbean	Woody shrub	Cat. 1	1
Caesalpinia decapetala	Fabaceae	Asia	Woody shrub	Cat. 1	2
Ipomoea carnea	Convolvulaceae	C, S America, Caribbean	Climber		2
Jacaranda mimosifolia	Bigoniaceae	S America	Tree	Cat. 3	2
Lantana camara	Verbenaceae	C, S America	Woody shrub	Cat. 1	2
Melia azedarach	Meliaceae	India	Tree	Cat. 3	2
Montanoa hibiscifolia	Asteraceae	C America	Tree	Cat. 1	2
Parthenium hysterophorus	Asteraceae	S America, Caribbean	Annual herb	Cat. 1	2
Psidium guajava	Myrtaceae	C, S America	Tree	Cat. 2	2
Senna bicapsularis	Fabaceae	S America, Caribbean	Woody shrub	Cat. 3	2
Senna didymobotrya	Fabaceae	Tropical Africa	Woody shrub	Cat. 3	2
Solanum mauritianum	Solanaceae	S America	Woody shrub	Cat. 1	2
Solanum seaforthianum	Solanaceae	C, S America	Climber	Cat. 1	2
Tithonia diversifolia	Asteraceae	C America	Woody shrub	Cat. 1	2
*Tagetes minuta**	Asteraceae	S America	Annual herb		4
*Opuntia ficus-indica**	Cactaceae	Mexico	Succulent	Cat. 1	4
*Opuntia monacantha**	Cactaceae	S America	Succulent	Cat. 1	4

Chromolaena is still distributed across HiP (Figure 15.4), and hence the risk of reinvasion remains high. Control programmes are thus likely to be a long-term commitment, making it important to find a cost-effective way of managing the invasion. Costs of clearing may be reduced and efficacy increased if clearing programmes take advantage of ecological factors, such as fire and low rainfall periods (Dew, 2015). Fire is a particularly effective tool if used during the right season and with high enough intensity, although it should be used with particular care of vulnerable habitats, such as riverine and scarp forests (te Beest *et al.*, 2012).

In conclusion, we stress that it is imperative that prevention, control, and, whenever possible, eradication of alien species be prioritized by managers, particularly when biodiversity is at risk in small protected areas, in order to reduce or prevent loss of indigenous species and natural habitats. The 'Chromolaena story' we presented could provide valuable lessons especially for protected areas in eastern Africa where the issue of alien invasive species is just emerging and where the spread of Chromolaena and other alien plants (e.g. *Lantana camara* and *Parthenium hysterophorus*) is increasing (Beale *et al.*, 2013). Ensuring that these invading plants are recognized and dealt with promptly reduces future costs and impacts.

15.10 Acknowledgements

We would like to thank the Invasive Alien Species Programme, in particular Zanele Jele and Mbuyiselwa Mthembu, for making their data available.

15.11 References

Akobundu, O. & Ekeleme, F. E. (1996) Potentials for Chromolaena odorata (L.) R. M. King and H. Robinson in fallow management in West and Central Africa. In: *Proceedings of the 3rd International Workshop on Biological Control and Management of Chromolaena odorata, 15–19 Nov. 1993, Cote d'Ivoir* (eds U. K. Prasad, R. Muniappan, P. Ferrar, J. P. Aeschliman, & H. De Foresta). Agricultural Experiment Station, University of Guam, Publication No. 202, Mangilao, Guam.

Ambika, S. R. (2002) Allelopathic plants. 5. *Chromolaena odorata* (L.) King and Robinson. *Allelopathy Journal* **9**: 35–41.

Barreto, R. W. & Evans, H. C. (1994) The mycobiota of the weed *Chromolaena odorata* in southern Brazil with particular reference to fungal pathogens for biological control. *Mycological Research* **98**: 1107–1116.

Beale, C. M., van Rensberg, S. J., Bond, W. J., *et al.* (2013) Ten lessons for the conservation of African savannah ecosystems. *Biological Conservation* **167**: 224–232.

Blackmore, A. C. (1998) Seed dispersal of *Chromolaena odorata* reconsidered. In: *Proceedings of the 4th International Workshop on Biological Control and Management of*

Chromolaena odorata (eds P. Ferrar, R. Muniappan & K. P. Jayanth), pp. 16–21. Agricultural Experiment Station, University of Guam, Publication No. 216, Mangilao, Guam.

Boppre, M. & Fischer, O. W. (1994) Zonocerus and Chromolaena in West Africa. In: *New trends in locust control* (eds S. Krall & H. Wilps), pp. 107–126. GTZ, Eschborn.

Brooks, M. L., D'Antonio, C. M., Richardson, D. M., *et al.* (2004) Effects of invasive alien plants on fire regimes. *Bioscience* **54**: 677–688.

Coetzer, A. & Louw, J. (2012) An evaluation of the Contractor Development Model of Working for Water. *Water SA* **38**: 793–801.

Cruttwell McFadyen, R. E. (1988) Ecology of *Chromolaena odorata* in the Neotropics (1). In: *Proceedings of the 1st International Workshop on Biological Control and Management of Chromolaena odorata* (ed. R. Muniappan), pp. 13–20. Agricultural Experiment Station, University of Guam, Mangilao, Guam.

De Rouw, A. (1991) The invasion of *Chromolaena odorata* (L.) King and Robinson (ex *Eupatorium odoratum*), and competition with the native flora, in a rain forest zone, south-west Cote d'Ivoire. *Journal of Biogeography* **18**: 13–23.

Dew, L. A. (2015) Monitoring and managing *Chromolaena odorata* in a South African savanna reserve. Evaluating the efficacy of current control programs in response to ecological factors and management protocols. MSc thesis, Umeå University.

Dumalisile, L. (2008) Effects of *Chromoleana odorata* on mammalian biodiversity in Hluhluwe-iMfolozi Park, South Africa. MSc thesis, University of Pretoria.

Ehrenfeld, J. G. (2010) Ecosystem consequences of biological invasions. *Annual Reviews of Ecology, Evolution and Systematics* **41**: 59–80.

Esuruoso, O. F. (1971) Seed-borne fungi of the Siam weed, *Eupatorium odoratum* in Nigeria. *Pans* **17**: 458–460.

Foxcroft, L. C., Richardson, D. M., Rouget, M., & MacFadyen, S. (2009) Patterns of alien plant distribution at multiple spatial scales in a large national park: implications for ecology, management and monitoring. *Diversity and Distributions* **15**: 367–378.

Foxcroft, L. C., Richardson, D., Rejmánek, M. & Pyšek, P. (2010) Alien plant invasions in tropical and sub-tropical savannas: patterns, processes and prospects. *Biological Invasions* **12**: 3913–3933.

Gaertner, M., Biggs, R., te Beest, M., *et al.* (2014) Invasive plants as drivers of regime shifts: identifying high-priority invaders that alter feedback relationships. *Diversity and Distributions* **20**: 733–744.

Gautier, L. (1992) Taxonomy and distribution of a tropical weed: *Chromolaena odorata* (L.) R. King and H. Robinson. *Candollea* **47**: 645–662.

Goodall, J. M. & Erasmus, D. J. (1996) Review of the status and integrated control of the invasive alien weed, *Chromolaena odorata*, in South Africa. *Agriculture, Ecosystems and Environment* **56**: 151–164.

Honu, Y. A. K. & Dang, Q. L. (2000) Responses of tree seedlings to the removal of *Chromolaena odorata* Linn. in a degraded forest in Ghana. *Forest Ecology and Management* **137**: 75–82.

Howison, O. E. (2009) The historical spread and potential distribution of the invasive alien plant *Chromolaena odorata* in Hluhluwe-iMfolozi Park. MSc thesis, University of Kwazulu-Natal.

Howison, R. A. (2009) Food preferences and feeding interactions among browsers, and the effect of an exotic invasive weed (*Chromolaena odorata*) on the endangered black rhino (*Diceros bicornis*), in an African savanna. MSc thesis, University of KwaZulu-Natal.

Kassi, N. J. K. & Decocq, G. (2008) Successional patterns of plant species and community diversity in a semi-deciduous tropical forest under shifting cultivation. *Journal of Vegetation Science* **19**: 809–812.

Kluge, R. L. & Caldwell, P. M. (1992) Phytophagous insects and mites on *Chromolaena odorata* (Compositae, Eupatoreae) in Southern Africa. *Journal of the Entomological Society of Southern Africa* **55**: 159–161.

Koutika, L. S. & Rainey, H. (2010) *Chromolaena odorata* in different ecosystems: weed or fallow plant? *Applied Ecology and Environmental Research* **8**: 131–142.

Kriticos, D. J., Yonow, T., & McFadyen, R. C. (2005) The potential distribution of *Chromolaena odorata* (Siam weed) in relation to climate. *Weed Research* **45**: 246–254.

Kushwaha, S. P. S., Ramakrishnan, P. S., & Tripathi, R. S. (1981) Population dynamics of *Eupatorium odoratum* in successional environments following slash and burn agriculture. *Journal of Applied Ecology* **18**: 529–535.

Leslie, A. J. & Spotila, J. R. (2001) Alien plant threatens Nile crocodile (*Crocodylus niloticus*) breeding in Lake St Lucia, South Africa. *Biological Conservation* **98**: 347–355.

Liggitt, B. (1983) The invasive alien plant *Chromolaena odorata*, with regards to its status and control in Natal. *Monograph 2*, pp. 41. Institute of Natural Resources, University of KwaZulu-Natal, Pietermaritzburg, South Africa.

Lowe, S. J., Browne, M. & Boudjelas, S. (2000) *100 of the world's worst invasive alien species*. IUCN/SSC Invasive Species Specialist Group, Auckland.

Lucas, E. O. (1989) Siam weed (*Chromolaena odorata*) and crop production in Nigeria. *Outlook on Agriculture* **18**: 133–138.

King, R. M. & Robinson, H. (1970) Studies in the Eupatorieae (Compositae). XXIX. The genus *Chromolaena.Phytologia* **20**: 196–209.

Macdonald, I. A. W. (1983) Alien trees, shrubs and creepers invading indigenous vegetation in the Hluhluwe-Umfolozi Game Reserve Complex in Natal. *Bothalia* **14**: 949–959.

Mack, R. N., Simberloff, D., Lonsdale, W. M., *et al.* (2000) Biotic invasions: causes, epidemiology, global consequences, and control. *Ecological Applications* **10**: 689–710.

Mangla, S., Inderjit & Callaway, R. M. (2008) Exotic invasive plant accumulates native soil pathogens which inhibit native plants. *Journal of Ecology* **96**: 58–67.

Marais, C. & Wannenburgh, A. (2008) Restoration of water resources (natural capital) through the clearing of invasive alien plants from riparian areas in South Africa – costs and water benefits. *South African Journal of Botany* **74**: 526–537.

McLennan, S. (2006) Determining the effect of environmental factors (rainfall and topo-edaphic) on the abundance of *Chromolaena odorata* in Hluhluwe-iMfolozi Park, KwaZulu-Natal. BSc thesis, University of KwaZulu-Natal.

McWilliam, A. (2000) A plague on your house? Some impacts of *Chromolaena odorata* on Timorese livelihoods. *Human Ecology* **28**: 451–469.

Mgobozi, M. P., Somers, M. J. & Dippenaar-Schoeman, A. S. (2008) Spider responses to alien plant invasion: the effect of short-and long-term *Chromolaena odorata* invasion and management. *Journal of Applied Ecology* **45**: 1189–1197.

Moder, W. W. D. (1984) The attraction of *Zonocerus variegatus* (L.) (Orthoptera, Pyrgomorphidae) to the weed *Chromolaena odorata* and associated feeding behavior. *Bulletin of Entomological Research* **74**: 239–247.

Muniappan, R., Reddy, G. V. P. & Po-Yung Lai (2005) Distribution and biological control of *Chromolaena odorata*. In: *Invasive plants: ecological and agricultural aspects* (ed. Inderjit), pp. 223–233. Birkhauser Verlag, Basel.

Murali, K. S. & Setty, R. S. (2001) Effect of weeds *Lantana camara* and *Chromolaena odorata* growth on the species diversity, regeneration and stem density of tree and shrub layer in BRT sanctuary. *Current Science* **80**: 675–678.

Norgrove, L., Hauser, S., & Weise, S. F. (2000) Response of *Chromolaena odorata* to timber tree densities in an agrisilvicultural system in Cameroon: aboveground biomass, residue decomposition and nutrient release. *Agriculture Ecosystems and Environment* **81**: 191–207.

Pejchar, L. & Mooney, H. A. (2009) Invasive species, ecosystem services and human well-being. *Trends in Ecology and Evolution* **24**: 497–504.

Pyšek, P., Richardson, D. M., Pergl, J., *et al.* (2008) Geographical and taxonomic biases in invasion ecology. *Trends in Ecology and Evolution* **23**: 237–244.

Raimundo, R. L. G., Fonseca, R. L., Schachetti-Pereira, R., Townsend Peterson, A., & Lewinsohn, T. M. (2007) Native and exotic distributions of siamweed (*Chromolaena odorata*) modeled using the genetic algorithm for rule-set production. *Weed Science* **55**: 41–48.

Ramakrishnan, P. S. & Vitousek, P. M. (1989) Ecosystem-level processes and the consequences of biological invasions. In: *Biological invasions – a global perspective* (eds J. A. Drake, H. A. Mooney, F. di Castri, R. H. Groves, F. J. Kruger, M. Rejmánek, & M. Williamson), pp. 281–300. John Wiley and Sons, New York.

Robertson, M. P., Kriticos, D. J. & Zachariades, C. (2008) Climate matching techniques to narrow the search for biological control agents. *Biological Control* **46**: 442–452.

Rozen-Rechels, D. (2015) The repulsive shrub. Impact of an invasive shrub on habitat selection by African large herbivores. MSc thesis, Swedish University of Agricultural Sciences.

Sala, O. E., Chapin, F. S., Armesto, J. J., *et al.* (2000) Biodiversity – global biodiversity scenarios for the year 2100. *Science* **287**: 1770–1774.

Sangakkara, U. R., Attanayake, K. B., Dissanayake, U. & Bandaranayake, P. R. S. D. (2008) Allelopathic impact of *Chromolaena odorata* (L.) King and Robinson on germination and growth of selected tropical crops. *Journal of Plant Diseases and Protection*: 323–326.

Slaats, J. J. P. (1995) *Chromolaena odorata* fallow in food cropping systems: an agronomic assessment in south-west Ivory Coast. PhD thesis, Wageningen Agricultural University.

Strayer, D. L., Eviner, V. T., Jeschke, J. M., & Pace, M. L. (2006) Understanding the long term effects of species invasions. *Trends in Ecology and Evolution* **30**: 1–7.

te Beest, M. (2010) The ideal weed? Understanding the invasion of *Chromolaena odorata* in a South African savanna. PhD thesis, University of Groningen.

te Beest, M., Stevens, N., Olff, H., & Van der Putten, W. H. (2009) Plant–soil feedback induces shifts in biomass allocation in the invasive plant *Chromolaena odorata*. *Journal of Ecology* **97**: 1281–1290.

te Beest, M., Cromsigt, J. P. G. M., Ngobese, J., & Olff, H. (2012) Managing invasions at the cost of native habitat? An experimental test of the impact of fire on the invasion of *Chromolaena odorata* in a South African savanna. *Biological Invasions* **14**: 607–618.

te Beest, M., Elschot, K., Olff, H., & Etienne, R. S. (2013) Invasion success in a marginal habitat: an experimental test of competitive ability and drought tolerance in *Chromolaena odorata*. *PLoS ONE* **8**: e68274.

te Beest, M., Mpandza, N. J., & Olff, H. (2015a) Fire and simulated herbivory have antagonistic effects on resistance of savanna grasslands to alien shrub invasion. *Journal of Vegetation Science* **26**: 114–122.

te Beest, M., Esler, K., & Richardson, D. (2015b) Linking functional traits to impacts of invasive plant species: a case study. *Plant Ecology* **216**: 293–305.

Van der Hoeven, C. A. (2007) The missing link: bridging the gap between science and conservation. PhD thesis, Wageningen University.

Van Gils, H., Delfino, J., Rugege, D., & Janssen, L. (2004) Efficacy of *Chromolaena odorata* control in a South African conservation forest. *South African Journal of Science* **100**: 251–253.

Van Wilgen, B. W., Le Maitre, D. C., & Cowling, R. M. (1998) Ecosystem services, efficiency, sustainability and equity: South Africa's Working for Water programme. *Trends in Ecology and Evolution* **13**: 378.

Van Wilgen, B. W., Forsyth, G. G., Le Maitre, D. C., *et al.* (2012) An assessment of the effectiveness of a large, national-scale invasive alien plant control strategy in South Africa. *Biological Conservation* **148**: 28–38.

Vilá, M., Espinar, J., Hejda, M., *et al.* (2011) Ecological impacts of invasive alien plants: a meta-analysis of their effects on species, communities and ecosystems. *Ecology Letters* **14**: 702–708.

Whateley, A. & Porter, R. N. (1983) The woody vegetation communities of the Hluhluwe–Corridor–Umfolozi Game Reserve complex. *Bothalia* **14**: 745–758.

Witkowski, E. T. F. & Wilson, M. (2001) Changes in density, biomass, seed production and soil seed banks of the non-native invasive plant, *Chromolaena odorata*, along a 15 year chronosequence. *Plant Ecology* **152**: 13–27.

Zachariades, C., Strathie-Korrubel, L. W., & Kluge, R. L. (1999) The South African programme on the biological control of *Chromolaena odorata* (L.) King and Robinson (Asteraceae) using insects. *African Entomology Memoir* No. 189–102.

Zachariades, C., Von Senger, I., & Barker, N. P. (2004) Evidence for a northern Caribbean origin for the southern African biotype of *Chromolaena odorata*. In: *Proceedings of the 6th International Workshop on Biological Control and Management of Chromolaena odorata* (eds M. D. Day & R. C. McFadyen), pp. 25–27. ACIAR Technical Reports No. 55, Cairns, Australia.

Zachariades, C., Day, M., Muniappan, R., & Reddy, G. V. P. (2009) *Chromolaena odorata* (L.) King and Robinson (Asteraceae). In: *Biological control of tropical weeds using arthropods* (eds R. Muniappan, G. V. P. Reddy, & A. Raman), pp. 130–160. Cambridge University Press, Cambridge.

Zachariades, C., Strathie, L. W., Retief, E., & Dube, N. (2011) Progress towards the biological control of *Chromolaena odorata* (L.) R. M. King and H. Rob. (Asteraceae) in South Africa. *African Entomology* **19**: 282–302.

Zachariades, C., van Rensburg, S. J., & Witt, A. B. R. (2013) Recent spread and new records of *Chromolaena odorata* in Africa. In: *Proceedings of the Eighth International Workshop on Biological Control and Management of Chromolaena odorata and other Eupatorieae* (eds C. Zachariades, L. W. Strathie, M. D. Day, & R. Muniappan), pp. 20–27. ARC-PPRI, Pretoria.

16 · *Conserving Africa's Mega-Diversity in the Anthropocene: The Hluhluwe-iMfolozi Park Story*

JORIS P. G. M. CROMSIGT, SALLY ARCHIBALD, AND NORMAN OWEN-SMITH

16.1 Introduction

The Hluhluwe-iMfolozi Park (HiP) is relatively small in its area in comparison with well-known protected areas elsewhere in Africa: a mere 950 km^2, compared with the 14,763 km^2 of Serengeti National Park and 19,500 km^2 of Kruger National Park. Moreover, its boundaries are completely fenced, meaning that animals cannot migrate or readily disperse beyond them. Yet within its limits it contains a full representation of the megafauna typical of the African savanna biome: African elephant, black and white rhino, hippo, giraffe, and all five large mammalian carnivores. Alongside them is a diversity of less-large grazing and browsing ungulates (16 species) and numerous smaller organisms. Vegetation formations contained within HiP are exceptionally diverse, ranging from semi-arid thorn savanna in lowlands in the south through mesic savanna and dense thickets in the wetter north and a grassland–forest mosaic on the highest hills. Contributing to this is a rainfall gradient from under 600 mm to almost 1000 mm over a distance of only 35 km, along with an underlying diversity in geological substrates and soils. Traversing the region are two major rivers, one perennial until recently, the other always seasonal in its flow. These features of HiP provided an exceptional natural laboratory for investigating processes such as the impacts of mega-grazers on grasslands, biome transitions, competition between woody plants and

Conserving Africa's Mega-Diversity in the Anthropocene, ed. Joris P. G. M. Cromsigt, Sally Archibald and Norman Owen-Smith. Published by Cambridge University Press.

grasses, the role of fire, impacts of predators and pathogens, and alien plant invasions. The scientific understanding gained has guided 'process-based management' interventions such as predator restoration, re-introduction of elephants, population management through dispersal sinks and live capture, applications of fire, and eradication of invasive plants. The park may be viewed as a successful example of how Africa's mega-diversity of large mammals can be conserved within a remnant of formerly vaster ecosystems, but how sustainable will these conservation efforts be?

In this final overview, we highlight research findings from HiP that have made major contributions to scientific understanding globally, and outline their implications for conservation practice. We conclude by looking forward from past human impacts towards a prognosis of the alternative scenarios that might eventuate in the context of the human-dominated epoch that has become labelled the Anthropocene (Crutzen and Stoermer, 2000; Corlett, 2015; Ellis, 2015). We explore how this relates to the different conservation philosophies that have shaped HiP's past and present and how they may continue to shape its future. Central in these philosophies is the question of what exactly we are trying to conserve. What traction does the idea of a 'Zululand wilderness' (Player, 1997; McCracken, 2008) still have during the 'Anthropocene' epoch (Crutzen and Stoermer, 2000), when humans and their activities both replace and modify climate as the overriding influence on ecological conditions and processes (Corlett, 2015; Ellis, 2015)?

16.2 Megaherbivore Ecology

Foundations for the megaherbivore concept emerged from a pioneering study of white rhinos conducted within HiP (Owen-Smith, 1973; Chapter 5). This led to the recognition of common features shared by rhinos with elephants, hippos, and other large herbivores attaining weights exceeding 1000 kg. These include invulnerability to predation once adult, inter-birth intervals longer than 1 year, dominance of community biomass, and capacity to transform vegetation structure (Owen-Smith, 1988, 2013). Although megaherbivores were once more common worldwide (Owen-Smith, 1987), HiP represents perhaps the only protected area in the world where an intact megaherbivore community still exists (Chapters 4 and 5). There are no other areas in Africa where black and white rhino, hippo, elephant, and giraffe co-occur at functionally relevant densities and have done so for several decades. A particularly unique feature of HiP is that it is only here that the full impacts of white rhinos on grassland structure and composition and related processes can be observed – in most other places their numbers are still recovering after re-introductions. This is relevant because the white rhino is the only extant mega-grazer with widespread impacts on grasslands, because those of hippos are restricted to the margins of rivers and lakes (Lock, 1972). Studies in HiP have shown that the white rhino plays a key, and diverse, role in the functioning of HiP's savannas (see Chapters 5, 6, and 10), increasing grassland heterogeneity, creating habitat for a suite of species from diverse taxa, restricting the spread of fires and the loss of nitrogenous material associated with burning, and potentially limiting the spread of woody plants.

The transforming impact that megaherbivores can have, not only on vegetation but also on a variety of ecosystem processes, has become widely recognized (Owen-Smith, 2013). The megaherbivore concept is increasingly being applied elsewhere, beyond the extant savannas of HiP where it originated (Mahli *et al.*, 2016). Assessments of the role of extinct Pleistocene megaherbivores in ecosystem functioning are currently booming (e.g. Barnosky *et al.*, 2015; Doughty *et al.*, 2015a,b; Bakker *et al.*, 2016; Mahli *et al.*, 2016; Smith *et al.*, 2016). These studies have linked the extinction of megaherbivores towards the end of the last ice age to continent-scale shifts in vegetation structure and composition (Zimov *et al.*, 1995; Gill, 2014; Doughty *et al.*, 2015a), large-scale changes in nutrient availability and distribution patterns (Doughty *et al.*, 2013; Doughty *et al.*, 2015b), and global-scale effects on biogeochemical cycling (Smith *et al.*, 2016) and climate warming (Doughty *et al.*, 2010; Brault *et al.*, 2013). So how do all these findings fit with the work on megaherbivore ecology from HiP? Although the white rhino has received considerable research attention in HiP, much less is known about the ecological roles of the other megaherbivore species. While elephants have been linked to significant ecosystem impacts elsewhere (Kerley *et al.*, 2008), their impact in HiP remains largely unclear, although their potential impact on vegetation is a growing concern in HiP (Chapter 14). Our understanding of the ecosystem impacts of other mega-browsers, such as giraffe and black rhino, remains poor, not only within HiP. Hence there are particularly large knowledge gaps to fill regarding the ecosystem-level impacts of several of the extant megaherbivore species, and of the megaherbivore community as whole.

Unfortunately, we may not have that much time left to learn about the role of extant megaherbivores. Across Africa, escalating poaching of rhinos and elephants is threatening to reverse the gains in conservation of these endangered species that have been achieved over past decades. Poaching of elephants is not an issue in HiP at the time of writing, but losses of both white and black rhinos have been on the rise, despite the patrolling and other steps that have been taken to restrict incursions of poachers (Chapter 11). A particular threat is that poaching limits legal sales of live rhinos via game auctions, which generate essential revenue for conservation as well as distributing the species more widely. In Kruger National Park, poaching levels are now so high that management removals, through live auction sales, have been suspended (Ferreira *et al.*, 2012). Furthermore, interest in rhino purchases by potential buyers is declining due to the huge costs involved in effectively protecting them. The fate of HiP's rhinos remains uncertain at the time of writing this book.

16.3 Redefining Biomes

Savannas have always been a contentious biome-type and the debate around how to define mixed tree–grass ecosystems has intensified in recent decades (Ratnam *et al.*, 2011). Climatic variables alone do not delineate the structure and species composition of these ecosystems very well because fire and herbivory may promote alternative vegetation states, from open grassland to closed woodland, under similar environmental conditions (Chapters 3 and 10; Staver *et al.*, 2011; Hoffmann *et al.*, 2012). Globally, similar vegetation physiognomies are often classified as different biomes (mesic savanna vs dry forest), playing havoc with global maps of savanna extent and complicating decisions around how they should be managed. Again, this global issue plays out on the small HiP stage; vegetation maps have been interpreted and reinterpreted by different scientists/managers with different agendas.

Elements of grassland, savanna, forest, and thicket are all found within the ~950 km^2 boundaries of HiP and the ratios of these different vegetation formations have clearly changed over time (Chapter 3). Many people would object to the notion that the park contains four biomes. They would argue that this undermines the concept of a biome as a global-scale vegetation unit. Chapter 8 in this book presents a different view, where biomes are seen as ecological constructs in which different processes dominate, and both ecological and abiotic controls prevent species from passing easily from one formation to the other even over the space of a few metres (cf. Crisp *et al.*, 2009). Research from HiP and elsewhere brings into question the definition of biomes as global-scale, climatically determined vegetation units. Instead, it suggests that biomes are functionally divergent units, many of which are maintained through feedbacks between vegetation and consumers (fire and/or herbivory) against the backdrop of climatic and edaphic factors (Charles-Dominique *et al.*, 2015; Moncrieff *et al.*, 2016).

Why is this debate over the biome concept important? The ease with which remotely sensed images can be produced makes it tempting to use vegetation structure alone to map vegetation units. Findings from HiP and elsewhere suggest that identifying tree cover/biomass thresholds to distinguish forests, savannas, and thickets will not work. Instead, looking at the functional attributes of the dominant species in these environments can produce maps of greater accuracy and meaning. From the perspective of HiP, this novel approach has certainly cleared up the confusion over the difference between encroached savanna grasslands dominated by savanna shrubs vs functionally distinct thickets (Chapters 2 and 8). It sets out clear guidelines for management (Chapter 10), and gives a functional basis for determining appropriate levels of change (Gillson, 2015).

This is also highly relevant at a global scale. Many of the world's grassy biomes are currently threatened by global re- and afforestation programmes, such as REDD+ and more recently the Bonn Challenge and AFR100 (www.wri.org; Veldman *et al.*, 2015a,b,c; Bond, 2016a). These programmes use global mapping exercises that model forest potential using climatic and soil data while ignoring fire and herbivory as intrinsic drivers of tree cover (Laestadius *et al.*, 2011). As a result, many of the world's savannas and grasslands are mapped as degraded land with forestation potential (see the Atlas of Forest Landscape Restoration Opportunities: www.wri.org/applications/maps/flr-atlas/#). For example, according to these maps that seem to be made to guide policy makers, almost the complete extent of HiP is classified as a degraded landscape with forest restoration opportunities (something which would decimate the biodiversity and tourism opportunities of the reserve if it were ever implemented). Global re- and afforestation programmes thus make distinctions between degraded forests and savannas highly politicized and the definition and classification of these biomes highly debated: the same patch of land can be seen as a pristine habitat or something valueless that needs to be rehabilitated, depending on the definitions used and the ecological perspective of the viewer (see the debate between De Wit *et al.*, 2016 and Bond, 2016a,b).

Finally, these novel approaches to understanding and defining vegetation formations in savannas provide an alternative framework for looking at any ecosystem. Freeing the biome concept from the constraints of being associated with regional climates requires that these ideas be re-examined in temperate and boreal ecosystems as well – potentially furthering our understanding there (see Kuijper *et al.*, 2015). At the basis of this novel thinking is the research in HiP, highlighted in Chapters 7 and 8. This emphasizes the importance of demographic bottlenecks for savanna trees at different life-history stages as a conceptual model that is both scientifically valid and easy to translate into clear management options. Key to this model is the integration of resource constraints on growth and factors that determine the disturbance regime (fire and herbivory) within which tree species complete their life histories. This allows these ideas to be generalized to systems with different resource environments and in the context of changing fire and herbivory regimes globally.

16.4 Dynamics and Population Management of Large Herbivores within Enclosed Areas

A contentious issue in wildlife management has been whether large herbivore populations within small fenced parks need to be culled to prevent overgrazing and resultant population crashes (Jewell *et al.*, 1981). This was not an issue in the early days of HiP, because back then 'game' numbers were still recovering from attempts to eradicate all hosts of tsetse flies. However, during the 1950s, following the arrival of the first park ecologist, concerns were raised about vegetation changes and the poor body condition of certain ungulates within the Hluhluwe Game Reserve, which had remained mostly protected from the shooting campaigns (Chapter 4). Given the lack of large predators at the time, culling was introduced to control the more abundant herbivores, except for the two rhino species. Animal removals expanded to such extremes that when the especially severe El Nino-related drought of 1982/3 occurred, hardly any animals died of malnutrition within HiP (Walker *et al.*, 1987). This led to the recognition that agricultural concepts of carrying capacity were inappropriate for a protected area (Caughley, 1983) and animal removals became restricted to live sales of animals for restocking other parks or commercial wildlife ranches. The low numbers of animals removed in this way meant that herbivore populations were allowed to grow and attain their own equilibrium densities within the fenced park (Chapter 4). As a result, most large herbivore species did indeed increase in numbers from the mid-1980s onwards.

The exceptions included several of the smaller antelope species, which declined to levels threatening their imminent local extinction (Chapter 4). Over time, the herbivore community in the park has changed drastically in its species composition. Bushbuck, both reedbuck species, waterbuck and blue and grey duiker were abundant at the time of the early twentieth-century shooting campaigns, but are now reaching the verge of local extinction. Eland, formerly abundant in the region, are no longer present, an attempt at re-introduction having failed. In contrast, impala, nyala, and giraffe, not historically recorded within the boundaries of HiP, have thrived.

Shifting ideas about population control went hand in hand with the restoration of large predators, starting with lions in 1965 and followed by cheetahs in 1966 and wild dogs in 1980. Initially, the presence of lions had little impact on the abundance of their prey species, and it became apparent that the lion population was suffering from the effects of inbreeding due to the small founding nucleus. To restore genetic diversity, further lions were introduced between 1999 and 2001, and after initial ups and downs lion numbers have grown to over 100 individuals at the time of writing (Chapter 12). Recent census estimates indicate sharp downturns in the abundance of the ungulate species forming the primary prey of lions, in particular wildebeest and zebra, but notably also impala and nyala (Chapter 4). It may be that wild dogs and cheetahs are also contributing to the declining abundance of the latter two herbivore species, and perhaps also to the near demise of some of the smaller antelope species. Hence the leading issue, as yet unresolved, is whether predator and prey populations can attain joint quasi-equilibrium levels within a protected area the size of HiP, or whether the persistent oscillations projected by simple models of predator–prey systems will be generated. Time will tell. Nevertheless, at no stage in the history of HiP has any species exhibited a persistent 'carrying capacity' and population fluctuations of all ungulate species during the last century have been largely a consequence of management interventions rather than environmental variables (Chapter 4).

HiP has brought about important advances in terms of the population management of megaherbivores. Park managers have applied sustainable live removal programmes for rhinos, while maintaining the functional impacts of these ecosystem engineers (Chapter 11). For white rhinos, the innovative strategy adopted was to simulate natural dispersal processes by restricting animal removals to designated dispersal sink zones (Chapter 11). For black rhinos, the aim has been to re-establish a wider metapopulation of the species by targeted translocations of animals to other areas. In the case of elephants, the management dilemma is how to restrict the vegetation impacts of such a wide-ranging species within a small protected area. The size of HiP is equivalent to just one elephant home range, meaning that all parts are vulnerable to their vegetation impacts. Live removals of elephants are currently not a feasible option because most protected areas in southern Africa now contain dense elephant populations. While lethal control of smaller ungulates has been undertaken several times in the history of HiP, this is anathema to most conservationists for highly sentient species such as the elephant. So, a leading question remains, if poaching threats are resolved, how do we deal with (over)abundant populations of both elephants and white rhinos in small protected areas in the future? Large-scale contraception of elephants is currently being tested in HiP and the effectiveness of this intervention is awaited.

16.5 Humans and the 'Zululand Wilderness'

The role and place of humans in 'nature' or 'wild, natural' areas has been a popular subject of philosophical debates. One such philosophy, the wilderness philosophy, has been very influential in HiP's conservation history, reflected by the concept of the 'Zululand Wilderness' (Player, 1997; McCracken, 2008). In fact, HiP was the first park in Africa to designate a wilderness area in 1957. Up to this day, there are no roads or other infrastructure in this area, which covers about one-third of HiP, and human activities are restricted to wilderness trails and management interventions that leave minimum evidence in the landscape (Ezemvelo KZN Wildlife, 2011).

However, this book has clearly described how the land enclosed by HiP has been subject to anthropogenic influences for many millennia (Chapter 1). At least as far back as 0.5 MYA, Stone Age hunter-gatherers were using the area for hunting, and their impacts became more significant from ~100,000 years BP through altering the fire regime (Chapter 10). During more recent millennia, iron smelting and livestock herding activities altered vegetation structure and composition and soil nutrient patterns across large parts of the landscape (Chapters 1 and 6; Feely, 1980; Hall, 1984). The area that we now know as HiP was in fact fairly densely settled by Late Iron Age people during the eighteenth and nineteenth centuries (Penner, 1970; Feely, 1980; Hall, 1984), but it is clear that up to the early to mid-1800s these people lived alongside abundant wildlife populations (see Chapter 4). Hence, past human occupation of this region largely retained landscapes with high vegetation heterogeneity and biodiversity. Humans have thus played a clear role in shaping the 'Zululand Wilderness' that we now want to conserve.

However, during more recent times, unprecedented human population densities and land-use intensification surrounding HiP are affecting ecosystems inside the park and threatening biodiversity conservation objectives. For example, alien plant invasions, mostly brought in via roads entering the park, have had dramatic effects on riverine vegetation and fire regimes (Chapter 15). Increased livestock densities outside the park have also brought new pathogens affecting wildlife inside, most notably lion, African wild dogs, and buffalo (Chapter 13). Moreover, even the 950 km^2 extent of HiP is far too small to contain the range dynamics of some (e.g. vultures, wild dogs, or elephant) and genetically viable populations of other (e.g. lion) species (Chapters 9, 12, and 14). In the light of all these pressures, conserving ecological processes and ecosystem functioning within relatively small protected areas such as HiP has required, and will continue to require, targeted interventions.

This book highlights how HiP has 'experimented' with a wide range of such management interventions. In the early days spotted hyenas (Potter, 1941) and wild dogs (Vaughan-Kirby, 1916) were poisoned within Hluhluwe GR to restrict their impacts on depressed herbivore populations (Chapter 12), but later the larger herbivores were subjected to culling when perceived overgrazing became a problem within Hluhluwe GR and later in Umfolozi GR (Chapter 4). Culling was curtailed after reassembly of the large predator assemblage (Chapter 12). Later interventions also included more direct examples of 'learning by doing', where HiP's management pioneered the concept of adaptive management. Examples include prescribed burning to reduce woody encroachment (Chapter 10), the white and black rhino offtake programmes (Chapter 11), its science-led bovine tuberculosis programme (Chapter 13), and its massive response to the *Chromolaena* problem (Chapter 15).

Increasingly, conservation management recognizes that spatial and temporal variability is part of the system that needs to be conserved (Chapter 1). In the Kruger National Park, this has resulted in a clear policy of 'thresholds of potential concern (TPCs)' – where variability in various indicators is allowed and interventions are contemplated only when thresholds of change are exceeded. A similar approach is being introduced in HiP, for example in the context of elephant management, with several indicators of concern being listed that might trigger management intervention (Chapter 14). Solid scientific understanding and responsive management are necessary, but not sufficient, for this TPC approach to work: we also need to be able to identify allowable variation and associated thresholds. The fire management chapter (Chapter 10) exemplifies this: the challenge here is not our ecological understanding or our ability to apply fire to different ends. Rather, it is our struggle to determine acceptable landscape change, and to weigh up the positive and negative impacts of interventions on different aspects of ecosystem functioning.

An arguably even larger challenge for relatively small areas such as HiP is managing relations with human communities living outside the park. Today, HiP is not only expected to (1) 'Protect a representative sample of the indigenous ecosystems, communities, ecotones and representative landscapes of the area, their indigenous biodiversity, and the ecological and evolutionary processes that generate and maintain this diversity'. It is also expected to (2) 'safeguard cultural heritage', (3) 'promote awareness of nature', (4) 'contribute to local, regional and national economies', and (5) 'enable research to improve understanding and management' (Ezemvelo KZN Wildlife, 2011). These diverse objectives are a far cry from the original intention, which was to 'prevent total destruction of game' (Chapter 1). This highlights the challenge of conservation in the Anthropocene. Not only is the park more heavily impacted by people and their activities than ever before, but it is also expected to provide more resources, education, and conservation value than before. HiP's conservation agency, Ezemvelo KZN Wildlife, runs a number of community-based conservation programmes to enable neighbouring communities to share in the revenue (through community-owned lodges and community levys), resources (access to thatching grass and seedlings), management decisions (representation on the governing board), and educational opportunities (community trails and visitor centres) that the park offers. Whether these programmes engender a sufficient sense of shared ownership and value among neighbouring communities remains to be seen.

16.6 Looking to the Future

Over a century after the original Hluhluwe and Umfolozi game reserves were proclaimed, and 65 years after management responsibility for the entire protected area was assumed by the conservation agency, the status of the biodiversity contained within HiP appears to represent an outstanding achievement. A large and thriving population of white rhinos has been established, black rhinos are more abundant than in most other places, and elephants plus all large carnivores have been restored. Almost all of the large mammals indigenous to the region are represented, albeit in variable numbers, along with a wide diversity of birds and other taxa. The park continues to attract local and international tourists.

However, key conservation challenges remain and will need resolving: (1) expanding human settlements and associated activities press strongly against the boundaries of the park (along with threats of activities such as mining); (2) water flows and water quality are declining due to increasing pressures on the larger rivers beyond the park boundary; (3) poverty and poor health are a huge problem in neighbouring communities and the park's contribution towards their remediation will increasingly be questioned; (4) the escalation in rhino poaching fuelled by the huge prices paid for their horns; (5) the ecological implications of growing elephant numbers; (6) the spread of woody plants with consequent effects on habitat conditions and game viewing; (7) the consequences of increasing numbers of the large carnivores for predator–prey dynamics; (8) the threatened local extinctions of several of the smaller ungulate species; (9) the persistence of *Chromolaena* and other alien plant invaders; (10) the persistence of bovine tuberculosis in the buffalo population; and (11) the need for ongoing augmentation of large carnivore numbers.

Among all these concerns, climate change seems quite far down the list. HiP is small, which should make it vulnerable to changes in temperature and rainfall. However, the wide range in climate and topography within the park should increase its resilience to projected changes. The potential impact of higher temperatures on the probability of extreme fire events is under investigation (Chapter 10), but more work could be done on plant and animal phenology, and the more insidious effects of elevated CO_2.

A challenge shared by all small fenced reserves is how to maintain ecosystem processes which formerly operated at larger scales than HiP encompasses. The protected area exists as a fenced fortress with wild animals within and domestic ungulates and people outside (Figure 1.6). Despite this, opportunities to extend the park area could be explored. Community-controlled wildlife reserves have been established adjoining the park, albeit small in size (Chapter 14). There is a possibility that HiP could be extended northwards via a corridor through communal land and adjoining private nature reserves to become linked with the iSimangaliso Wetland Park, which includes Mkhuze Game Reserve, on the coastal plain. This would require an underpass beneath a major highway to be constructed and used by animals, as well as dropping fences to connect with protected areas north of HiP.

This book has documented many examples of how HiP's conservation agency and its staff have responded to the challenges of conserving Africa's mega-diversity within a comparatively small area. Often this was done in close collaboration with researchers, and process-related concepts have guided management responses. Nevertheless, further challenges will need to be surmounted if this conservation success is to be maintained further. The experience gained in HiP has wider relevance for protected areas that are becoming contracted in their effective extent elsewhere in Africa. We hope that HiP will continue to be a pioneer in responding to forthcoming conservation challenges in the emerging contexts of the Anthropocene.

16.7 References

Bakker, E. S., Gill, J. L., Johnson, C. N., *et al.* (2016) Combining paleo-data and modern exclosure experiments to assess the impact of megafauna extinctions on woody vegetation. *Proceedings of the National Academy of Sciences of the USA* **113**: 847–855.

Barnosky, A. D., Lindsey, E. L., Villavicencio, N. A., *et al.* (2015) Variable impact of late-Quaternary megafaunal extinction in causing ecological state shifts in North and South America. *Proceedings of the National Academy of Sciences of the USA* **113**: 856–861.

Bond, W. J. (2016a) Ancient grasslands at risk. *Science* **351**: 120–122.

Bond, W. J. (2016b) Response to seeing the grasslands through the trees. *Science* **351**: 1036–1037.

Brault, M.-O., Mysak, L. A., Matthews, H. D., & Simmons, C. T. (2013) Assessing the impact of late Pleistocene megafaunal extinctions on global vegetation and climate. *Climate of the Past* **9**: 1761–1771.

Caughley, G. (1983) Dynamics of large mammals and their relevance to culling. In: *Management of large mammals in African conservation areas* (ed. R. N. Owen-Smith), pp. 115–126. Haum, Pretoria.

Charles-Dominique, T., Staver, A. C., Midgley, G. F., & Bond, W. J. (2015) Functional differentiation of biomes in an African savanna/forest mosaic. *South African Journal of Botany* **101**: 82–90.

Corlett, R. T. (2015) The Anthropocene concept in ecology and conservation. *Trends in Ecology & Evolution* **30**: 36–41.

Crisp, M. D., Arroyo, M. T. K., Cook, L. G., *et al.* (2009) Phylogenetic biome conservatism on a global scale. *Nature* **458**: 754–756.

Crutzen, P. J. & Stoermer, E. F. (2000) The 'Anthropocene'. *IGBP Newsletter* **41**: 17–18.

De Wit, S., Anderson, J., Kumar, C., *et al.* (2016) Seeing the grasslands through the trees. *Science* **351**: 1036.

Doughty, C. E., Wolf, A., & Field, C. B. (2010) Biophysical feedbacks between the Pleistocene megafauna extinction and climate: the first human-induced global warming? *Geophysical Research Letters* **37**: L15703.

Doughty, C. E., Wolf, A., & Malhi, Y. (2013) The legacy of the Pleistocene megafauna extinctions on nutrient availability in Amazonia. *Nature Geoscience* **6**: 761–764.

Doughty, C. E., Faurhy, S., & Svenning, J.-C. (2015a) The impact of megafauna extinctions on savanna woody cover in South America. *Ecography* **39**: 213–222.

Doughty, C. E., Roman, J., Faurby, S., *et al.* (2015b) Global nutrient transport in a world of giants. *Proceedings of the National Academy of Sciences of the USA* **113**: 868–873.

Ellis, E. C. (2015) Ecology in an Anthropocene biosphere. *Ecological Monographs* **85**: 287–332.

Ezemvelo KZN Wildlife (2011) *Integrated Management Plan: Hluhluwe-iMfolozi Park, South Africa*. Ezemvelo KZN Wildlife, Pietermaritzburg.

Feely, J. M., (1980) Did Iron Age man have a role in the history of Zululand's wilderness landscapes? *South African Journal of Science* **76**: 150–152.

Ferreira, S. M., Botha, J. M. & Emmett, M. C. (2012) Anthropogenic influences on conservation values of white rhinoceros. *PLoS ONE* **7**: e45989.

Gill, J. L. (2014) Ecological impacts of the late Quaternary megaherbivore extinctions. *New Phytologist* **201**: 1163–1169.

Gillson, L. (2015) Evidence of a tipping point in a southern African savanna? *Ecological Complexity* **21**: 78–86.

Hall, M. (1984) Prehistoric farming in the Mfolozi and Hluhluwe valleys of southeast Africa: an archaeo-botanical survey. *Journal of Archaeological Science* **11**: 223–235.

Hoffmann, W. A., Geiger, E. L., Gotsch, S. G., *et al.* (2012) Ecological thresholds at the savanna–forest boundary: how plant traits, resources and fire govern the distribution of tropical biomes. *Ecology Letters* **15**: 759–768.

Jewell, P. A., Holt, S., & Hart, D. (1981) *Problems in management of locally abundant wild mammals*. Academic Press, New York.

Kerley, G. I. H., Landman, M., Kruger, L., *et al.* (2008). Effects of elephants on ecosystems and biodiversity. In: *Elephant management: a scientific assessment for South Africa* (eds R. J. Scholes & K. G. Mennell), pp. 84–145. Wits University Press, Johannesburg.

Kuijper, D. P. J., te Beest, M., Churski, M., & Cromsigt, J. P. G. M. (2015) Bottom-up and top-down forces shaping wooded ecosystems: lessons from a cross-biome comparison. In: *Trophic ecology: bottom-up and top-down interactions across aquatic and terrestrial systems* (eds T. C. Hanley & K. J. La Pierre), pp. 107–133. Cambridge University Press, Cambridge.

Laestadius, L., Maginnis, S., Minnemeyer, S., *et al.* (2011) Mapping opportunities for forest landscape restoration. *Unasylva* **238**: 47–48.

Lock, J. M. (1972) The effect of hippopotamus grazing in grasslands. *Journal of Ecology* **60**: 445–467.

Mahli, Y., Doughty, C. E., Galetti, M., *et al.* (2016) Megafauna and ecosystem function from the Pleistocene to the Anthropocene. *Proceedings of the National Academy of Sciences of the USA* **113**: 838–846.

McCracken, D. P. (2008) *Saving the Zululand wilderness: an early struggle for nature conservation*. Jacana Media, Johannesburg.

Moncrieff, G. R., Bond, W. J., & Higgins, S. I. (2016) Revising the biome concept for understanding and predicting global change impacts. *Journal of Biogeography* **46**: 863–873.

Owen-Smith, N. (1973) The behavioural ecology of the white rhinoceros. PhD thesis, University of Wisconsin, USA.

Owen-Smith, N. (1987) Pleistocene extinctions: the pivotal role of megaherbivores. *Paleobiology* **13**: 351–362.

Owen-Smith, N. (1988) *Megaherbivores: the influence of very large body size on ecology*. Cambridge University Press, Cambridge.

Owen-Smith, N. (2013) Megaherbivores. In: *Encyclopedia of biodiversity* (ed. S. A. Levin), pp. 223–239. Academic Press, Waltham, MA.

Penner, D. (1970) Archaeological survey in Zululand game reserves. Unpublished report, Natal Parks Board, Pietermaritzburg.

Player, I. C. (1997) *Zululand wilderness: shadow and soul*. David Philip, Cape Town.

Potter, H. B. B. (1941) Report of Zululand Game Reserve and Parks Committee province of Natal – report of game conservator (Capt. Potter) for 1941. *Journal of the Society for the Preservation of the Fauna of the Empire* **43**: 35–41.

Ratnam, J., Bond, W. J., Fensham, R. J., *et al.* (2011) When is a 'forest' a savanna, and why does it matter? *Global Ecology and Biogeography* **20**: 653–660.

Smith, F. A., Doughty, C. E., Malhi, Y., Svenning, J.-C. & Terborgh, J. (2016) Megafauna in the Earth system. *Ecography* **39**: 99–108.

Staver, A. C., Archibald, S., & Levin, S. A. (2011) The global extent and determinants of savanna and forest as alternative biome states. *Science* **334**: 230–232.

Vaughan-Kirby, F. (1916) Game and game preservation in Zululand. *South African Journal of Science* **13**: 375–396.

Veldman, J. W., Buisson, E., Durigan, G., *et al.* (2015a) Toward an old-growth concept for grasslands, savannas, and woodlands. *Frontiers in Ecology and the Environment* **13**: 154–162.

Veldman, J. W., Overbeck, G. E., Negreiros, D., *et al.* (2015b) Where tree planting and forest expansion are bad for biodiversity and ecosystem services. *BioScience* **65**: 1011–1018.

Veldman, J. W., Overbeck, G., Negreiros, D., *et al.* (2015c) Tyranny of trees in global climate change mitigation. *Science* **347**: 484–485.

Walker, B. H., Emslie, R. H., Owen-Smith, R. N., & Scholes, R. J. (1987) To cull or not to cull: lessons from a southern African drought. *Journal of Applied Ecology* **24**: 381–401.

Zimov, S. A., Chuprinun, V. I., Oreshko, A., *et al.* (1995) Steppe–tundra transition: a herbivore-driven biome shift at the end of the Pleistocene. *American Naturalist* **146**: 765–794.

Index